STUDENT'S SOLUTIONS MANUAL

DAVID A. YOPP
University of Idaho

A PROBLEM SOLVING APPROACH TO MATHEMATICS FOR ELEMENTARY SCHOOL TEACHERS

ELEVENTH EDITION

Rick Billstein
University of Montana

Shlomo Libeskind
University of Oregon

Johnny W. Lott
University of Montana

PEARSON

Boston Columbus Indianapolis New York San Francisco Upper Saddle River
Amsterdam Cape Town Dubai London Madrid Milan Munich Paris Montreal Toronto
Delhi Mexico City Sao Paulo Sydney Hong Kong Seoul Singapore Taipei Tokyo

Reproduced by Pearson from electronic files supplied by the author.

Publishing as Pearson, 75 Arlington Street, Boston, MA 02116.

ISBN-13: 978-0-321-78332-5
ISBN-10: 0-321-78332-8

1 2 3 4 5 6 EBM 16 15 14 13 12

www.pearsonhighered.com

PEARSON

Table of Contents

Chapter 1

AN INTRODUCTION TO PROBLEM SOLVING

Assessment 1-1A: Mathematics and Problem solving

1. **(a)** List the numbers:

$$\begin{array}{r} 1 + 2 + \cdots + 98 + 99 \\ 99 + 98 + \cdots + 2 + 1 \\ \hline 100 + 100 + \cdots + 100 + 100 \end{array}$$

There are 99 sums of 100. Thus the total can be found by computing $\frac{99 \cdot 100}{2} = \mathbf{4950}$. (Another way of looking at this problem is to realize there are $\frac{99}{2} = 49.5$ pairs of sums, each of 100; thus $49.5 \cdot 100 = 4950$.)

(b) The number of terms in any sequence of numbers may be found by subtracting the first term from the last, dividing the result by the common difference between terms, and then adding 1 (because both ends must be accounted for). Thus $\frac{1001-1}{2} + 1 = 501$ terms.

List the numbers:

$$\begin{array}{r} 1 + 3 + \cdots + 999 + 1001 \\ 1001 + 999 + \cdots + 3 + 1 \\ \hline 1002 + 1002 + \cdots + 1002 + 1002 \end{array}$$

There are 501 sums of 1002. Thus the total can be found by computing $\frac{501 \cdot 1002}{2} = \mathbf{251{,}001}$.

3. There are $\frac{147-36}{1} + 1 = 112$ terms.

List the numbers:

$$\begin{array}{r} 36 + 37 + \cdots + 146 + 147 \\ 147 + 146 + \cdots + 37 + 36 \\ \hline 183 + 183 + \cdots + 183 + 183 \end{array}$$

There are 112 sums of 183. Thus the total can be found by computing $\frac{112 \cdot 183}{2} = \mathbf{10{,}248}$.

5. The rhyme only states that "I was going" and nothing else. So there is at least **1**.

7. Observe that $E = (1+1) + (3+1) + \cdots + (97+1) = O + 49$. Thus, E is 49 more than O.

Alternative strategy:

$$\begin{aligned} O + E &= 1 + 2 + 3 + 4 + 5 + 6 + \cdots + 97 + 98 \\ &= \frac{98(99)}{2} = 49(99) \\ E &= 2(1 + 2 + 3 + 4 + \cdots + 49) = 2\left(\frac{49(50)}{2}\right) = 49(50) \\ O &= O + E - E = 49(99) - 49(50) = 49(49). \end{aligned}$$

So O is 49 less than E.

9. Make a table.

\$20 bills	\$10 bills	\$5 bills
2	1	0
2	0	2
1	3	0
1	2	2
1	1	4
1	0	6
0	5	0
0	4	2
0	3	4
0	2	6

0	1	8
0	0	10

There are twelve rows so there are **twelve** different ways. **12**

11. Frankie and Johnny have been reading for 9 days, since 72 pages for Frankie $\div$ 8 pages per day $=$ 9 days. Thus Johnny is on 5 pages per day $\times$ 9 days $=$ **page 45**.

13. Choose the box labeled Oranges and Apples (Box B). Retrieve a fruit from Box B. Since Box B is mislabeled, Box B should be labeled as having the fruit you retrieved. For example, if you retrieved an apple, then Box B should be labeled Apples. Since Box A is mislabeled, the Oranges and Apples label should be placed on Box A. These leave only one possibility for Box C; it should be labeled Oranges. If an orange was retrieved from Box B, then Box C would be labeled Oranges and Apples and Box A should be labeled Apples.

15. Working backward: Top $-$ 6 rungs $-$ 7 rungs $+$ 5 rungs $-$ 3 rungs $=$ top $-$ 11 rungs, which is located at the middle.

From the middle rung travel up 11 to the top or down 11 to the bottom. Along with the starting rung, then, there are $11 + 11 + 1 =$ **23 rungs**.

Assessment 1-1B

1. **(a)** List the numbers:

$$\begin{array}{rcrcccrcr} 1 & + & 2 & + & \cdots & + & 48 & + & 49 \\ 49 & + & 48 & + & \cdots & + & 2 & + & 1 \\ \hline 50 & + & 50 & + & \cdots & + & 50 & + & 50 \end{array}$$

There are 49 sums of 50. Thus the total can be found by computing $\frac{49 \cdot 50}{2} =$ **1225**. (Another way of looking at this problem is to realize there are $\frac{49}{2} = 24.5$ pairs of sums, each of 50; thus $24.5 \cdot 50 = 1225$.)

(b) The number of terms in any sequence of numbers may be found by subtracting the first term from the last, dividing the result by the common difference between terms, and then adding 1 (because both ends must be accounted for). Thus $\frac{2009-1}{2} + 1 = 1005$ terms.

List the numbers:

$$\begin{array}{rcrcccrcr} 1 & + & 3 & + & \cdots & + & 2008 & + & 2009 \\ 2009 & + & 2008 & + & \cdots & + & 3 & + & 1 \\ \hline 2010 & + & 2010 & + & \cdots & + & 2010 & + & 2010 \end{array}$$

There are 1005 sums of 2010. Thus the total can be found by computing $\frac{1005 \cdot 2010}{2} =$ **1,010,025**.

3. There are $\frac{203-58}{1} + 1 = 146$ terms.

List the numbers:

$$\begin{array}{rcrcccrcr} 58 & + & 59 & + & \cdots & + & 202 & + & 203 \\ 203 & + & 202 & + & \cdots & + & 59 & + & 58 \\ \hline 261 & + & 261 & + & \cdots & + & 261 & + & 261 \end{array}$$

There are 146 sums of 261. Thus the total can be found by computing $\frac{146 \cdot 261}{2} =$ **19,053**.

5. After two socks are drawn, the two socks match or they don't. If they match, we are done. If they don't match, the next sock drawn will match one of the two socks already drawn.

7.
$$\begin{aligned} E &= (1+1) + (3+1) + (5+1) + (7+1) \\ &\quad + \cdots + (97+1) \\ &= P - 99 + 49 \\ &= P - 50 \end{aligned}$$

P is larger than E by **50**.

9. **(a)** Marc must have five pennies to make an even \$1.00. The minimum number of coins would have as many quarters as possible, or three quarters. The remaining 20¢ must consist of at least one dime and one nickel; the only possibility is one dime and two nickels. The minimum is 5 pennies, 2 nickels, 1 dime, and 3 quarters, or **11 coins**.

(b) The maximum number of coins is achieved by having as many pennies as possible. It is a requirement to have one quarter, one dime, and one nickel $= 40¢$, so there may then be 60 pennies for a total of **63 coins**.

11. Answers may vary; two solutions might be to:

(a) Put four marbles on each tray of the balance scale. Take the heavier four and weigh two on each tray. Take the heavier two and weigh one on each tray; the heavier marble will be evident on this third weighing.

(b) This alternative shows the heavier marble can be done more efficiently, two steps rather than three. Put three marbles in each tray of the balance scale.

(*i*) If the two trays are the same weight, the heavier marble is one of the remaining two. Weigh them to find the heavier.

(*ii*) If one side is heavier, take two of the three marbles and weigh them. If they are the same weight, the remaining marble is the heavier. If not, the heavier will be evident on this second weighing.

13. **(a)** There must be 1 or 3 quarters for an amount ending in 5. Then dimes can add to $1.15 plus 4 pennies to realize $1.19. Thus:

Quarters	Dimes	Pennies	Total
3	4	4	$1.19
1	9	4	$1.19

and in neither case can change for $1.00 be made.

(b) Two or zero quarters would allow an amount ending in 0. Then more combinations of dimes or pennies could add to $1.00.

15. Use a variable and a table

12 AM	5 AM	9 AM	12 PM
T	T − 15	2(T − 15)	2(T − 15) + 10

So,
$$2(T - 15) + 10 = 32$$
$$2T - 30 + 10 = 32$$
$$2T - 20 = 32$$
$$2T = 52$$
$$T = \mathbf{26\ degrees.}$$

Assessment 1-2A Explorations with Patterns

1. **(a)** Each figure in the sequence adds one box each to the top and bottom rows. The next would be:

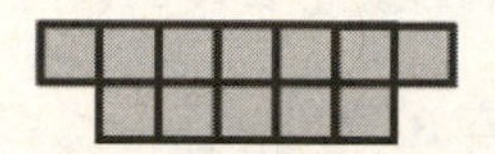

(b) Each figure in the sequence adds one upright and one inverted triangle. The next would be:

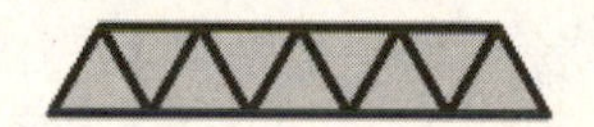

(c) Each figure in the sequence adds one box to the base and one row to the overall triangle. The next would be:

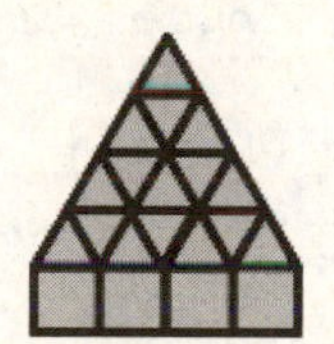

3. In these problems, let a_n represent the nth term in a sequence, a_1 represent the first term, d represent the common difference between terms in an arithmetic sequence, and r represent the common ratio between terms in a geometric sequence. In an arithmetic sequence, $a_n = a_1 + (n-1)d$; in a geometric sequence $a_n = a_1 r^{n-1}$. Thus:

(a) Arithmetic sequence: $a_1 = 1$ and $d = 2$:

(*i*) $a_{100} = 1 + (100 - 1) \cdot 2$
$= 1 + 99 \cdot 2 = \mathbf{199}.$

(*ii*) $a_n = 1 + (n - 1) \cdot 2$
$= 1 + 2n - 2 = \mathbf{2n - 1}$.

(b) Arithmetic sequence: $a_1 = 0$ and $d = 50$:

(*i*) $a_{100} = 0 + (100 - 1) \cdot 50$
$99 \cdot 50 = \mathbf{4950}$.

(*ii*) $a_n = 0 + (n - 1) \cdot 50$
$= \mathbf{50n - 50 \text{ or } 50(n - 1)}$.

(c) Geometric sequence: $a_1 = 3$ and $r = 2$:

(*i*) $a_{100} = 3 \cdot 2^{100-1} = \mathbf{3 \cdot 2^{99}}$.

(*ii*) $a_n = \mathbf{3 \cdot 2^{n-1}}$.

(d) Geometric sequence: $a_1 = 10$ and $r = 10$:

(*i*) $a_{100} = 10 \cdot 10^{100-1} = \mathbf{10^{100}}$.

(*ii*) $a_n = 10 \cdot 10^{n-1} = \mathbf{10^n}$.

(e) Arithmetic sequence: $a_1 = 9$ and $d = 4$:

(*i*) $a_{100} = 9 + (100 - 1) \cdot 4$
$= 9 + 99 \cdot 4 = \mathbf{405}$.

(*ii*) $a_n = 9 + (n - 1) \cdot 4$
$= 9 + 4n - 4 = \mathbf{4n + 5}$.

5. **(a)** Make a table.

Number of term	Term
1	$1 \cdot 1 \cdot 1 = 1$
2	$2 \cdot 2 \cdot 2 = 8$
3	$3 \cdot 3 \cdot 3 = 27$
4	$4 \cdot 4 \cdot 4 = 64$
5	$5 \cdot 5 \cdot 5 = 125$
6	$6 \cdot 6 \cdot 6 = 216$
7	$7 \cdot 7 \cdot 7 = 343$
8	$8 \cdot 8 \cdot 8 = 512$
9	$9 \cdot 9 \cdot 9 = 729$
10	$10 \cdot 10 \cdot 10 = 1000$
11	$11 \cdot 11 \cdot 11 = 1331$

The 11th term 1331 is the least 4-digit number greater than 1000.

(b) The 9th term 729 is the greatest 3-digit number in this pattern.

(c) The cell A14 corresponds to the 14th term, which is $14 \cdot 14 \cdot 14 = \mathbf{2744}$.

7. **(a)** Each cube adds four squares to the preceding figure; or 6, 10, 14, ... This is an arithmetic sequence with $a_1 = 6$ and $d = 4$. Thus $a_{10} = 6 + (10 - 1) \cdot 4 =$ **42 squares** to be painted in the 10th figure.

(b) This is an arithmetic sequence with $a_1 = 6$ and $d = 6$. The nth term is thus:
$a_n = 6 + (n - 1) \cdot 4 = \mathbf{4n + 2}$.

9. Using the general expression for the nth term of an arithmetic sequence with $a_1 = 24{,}000$ and $a_9 = 31{,}680$ yields:

$31680 = 24000 + (9 - 1)d$
$31680 = 24000 + 8d \Rightarrow d = 960$,

the amount by which Joe's income increased each year.

To find the year in which his income was \$45120:

$45120 = 24000 + (n - 1) \cdot 960$
$45120 = 23040 + 960n$
$\Rightarrow n = 23$.

Joe's income was \$45,120 in his **23rd year**.

11. **(a)** Look for the differences:

5 6 14 32 64 115 191
1 8 18 32 51 76
7 10 14 19 25
3 4 5 6
1 1 1

The third difference row is an arithmetic sequence with fixed difference of 1. Thus the 6th term in the second difference row is $25 + 7 = 32$; the 7th term in the first difference row is $76 + 32 = 108$; and the 8th term in the original sequence is $191 + 108 = 299$. Using the same reasoning, the next three terms in the original sequence will be **299, 447, 644**.

(b) Look for the differences:

0		2		6		12		20		30		42
	2		4		6		8		10		12	
		2		2		2		2		2		2

The first difference row is an arithmetic sequence with fixed difference 2. Thus the 7th term in the first difference row is $12 + 2 = 14$; the 8th term in the original sequence is $42 + 14 = 56$. Using the same reasoning, the next three terms in the original sequence are **56, 72, 90**.

13. **(a)** First term: $(1)^2 + 2 = \mathbf{3}$;
Second term: $(2)^2 + 2 = \mathbf{6}$;
Third term: $(3)^2 + 2 = \mathbf{11}$;
Fourth term: $(4)^2 + 2 = \mathbf{18}$; and
Fifth term: $(5)^2 + 2 = \mathbf{27}$.

(b) First term: $5(1) - 1 = \mathbf{4}$;
Second term: $5(2) - 1 = \mathbf{9}$;
Third term: $5(3) - 1 = \mathbf{14}$;
Fourth term: $5(4) - 1 = \mathbf{19}$; and
Fifth term: $5(5) - 1 = \mathbf{24}$.

(c) First term: $10^{(1)} - 1 = \mathbf{9}$;
Second term: $10^{(2)} - 1 = \mathbf{99}$;
Third term: $10^{(3)} - 1 = \mathbf{999}$;
Fourth term: $10^{(4)} - 1 = \mathbf{9999}$; and
Fifth term: $10^{(5)} - 1 = \mathbf{99999}$.

(d) First term: $3(1) + 2 = \mathbf{5}$;
Second term: $3(2) + 2 = \mathbf{8}$;
Third term: $3(3) + 2 = \mathbf{11}$;
Fourth term: $3(4) + 2 = \mathbf{14}$; and
Fifth term: $3(5) + 2 = \mathbf{17}$.

15. **(a)** There are 1, 5, 11, 19, 29 tiles in the five figures. Each figure adds $2n$ tiles to the preceding figure, thus a_6 the 6th term has $29 + 12 =$ **41 tiles**.

(b) $n^2 = 1, 1, 16, 25, \ldots$. Adding $(n - 1)$ to n^2 yields $1, 5, 11, 19, 29, \ldots$, which is the proper sequence. Thus the nth term has $n^2 + (n - 1)$.

(c) If $n^2 + (n - 1) = 1259$;

Then $n^2 + n - 1260 = 0$. This implies $(n - 35)(n + 36) = 0$, so $n = 35$.

There are 1259 tiles in the **35th figure**.

17. **(a)** Start with one piece of paper. Cutting it into five pieces gives us 5. Taking one of the pieces and cutting it into five pieces again gives $4 + 5 = 9$ pieces. Continuing this process gives an arithmetic sequence: **1, 5, 9, 13, ...** (Given that the scissors are small enough and sharp enough.)

(b) The number of pieces after the nth cut would be $1 + (n - 1)4 = \mathbf{4n - 3}$.

Assessment 1-2B

1. **(a)** In a clockwise direction, the shaded area moves to a new position separated from the original by one open space, then two open spaces, then by three, etc. The separation in each successive step increases by one unit; next would be:

(b) Each figure in the sequence adds one row of boxes to the base. Next would be:

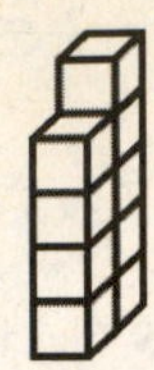

(c) Each figure in the sequence adds one box to the top and each leg of the figure. Next would be:

3. In these problems, a_n represents the nth term in a sequence, a_1 represents the first term, d represent the common difference between terms in an arithmetic sequence, and r represents the common ratio between terms in a geometric sequence.

In an arithmetic sequence, $a_n = a_1 + (n-1)d$; in a geometric sequence, $a_n = a_1 r^{n-1}$. Thus:

(a) Arithmetic sequence: $a_1 = 8$ and $d = 3$.

(*i*) $a_{100} = 8 + (100-1) \cdot 3 = \mathbf{305}$.

(*ii*) $a_n = 8 + (n-1) \cdot 3$
$= 8 + 3n - 3 = \mathbf{3n + 5}$.

(b) Geometric sequence: $a_1 = 5$ and $r = 3$.

(*i*) $a_{100} = \mathbf{5 \cdot 3^{99}}$.

(*ii*) $a_n = 5 \cdot 3^{n-1}$.

(c) Arithmetic sequence: $a_1 = 2$ and $d = 5$.

(*i*) $a_{100} = 2 + (100-1) \cdot 5 = \mathbf{497}$.

(*ii*) $a_n = 2 + (n-1) \cdot 5$
$= 2 + 5n - 5 = \mathbf{5n - 3}$.

(d) Arithmetic or Geometric. For the geometric sequence: $a_1 = 1$ and $r = 1$.

(*i*) $a_{100} = 1 \cdot (1)^{99} = \mathbf{1}$.

(*ii*) $a_n = 1 \cdot (1)^{n-1} = \mathbf{1}$.

(e) Geometric sequence: $a_1 = 2$ and $r = 5$.

(*i*) $a_{100} = \mathbf{2 \cdot (5)^{99}}$.

(*ii*) $a_n = \mathbf{2 \cdot (5)^{n-1}}$.

5. **(a)** Answers may vary:

(*i*) **The sum of the first *n* odd number is n^2**; e.g., $1 + 3 + 5 + 7 = 4^2$.

(*ii*) **Square the average of the first and last terms;** e.g.,
$1 + 3 + 5 + 7 = \left(\frac{1+7}{2}\right)^2 = 4^2$.

(b) There are $\frac{35-1}{2} + 1 = 18$ terms in this sequence.

(*i*) $1 + 3 + 5 + 7 + \cdots + 35 = 18^2$
$= \mathbf{324}$.

(*ii*) $\left(\frac{1+35}{2}\right)^2 = 18^2 = \mathbf{324}$.

7. **(a)** Looking at the third figure, there are $5 + 3 + 1 = 9$ triangles. The fourth figure would than have $7 + 5 + 3 + 1 = 16$ triangles. An alternative to simply adding 7, 5, 3, and 1 together is to note that $7 + 1 = 8$ and $5 + 3 = 8$. There are $\frac{4}{2} = 2$ of these sums, and $2 \cdot 8 = 16$. Then the 100th figure would have $100 + 99 = 199$ triangles in the base, $99 + 98 = 197$ triangles in the second row, and so on until the 100th row where there would be 1 triangle. $199 + 1 = 200$; $197 + 3 = 200$; etc. and so the sum of each pair is 200 and there are $\frac{100}{2} = 50$ of these pairs. $50 \cdot 200 = 10{,}000$, or **10,000 triangles** in the 100th figure.

(b) The number of triangles in the nth figure is $\frac{n}{2}$(number of triangles in base + 1). The number of triangles in the base is $n + (n-1)$, or $2n - 1$. $(2n-1) + 1 = 2n$. Then $\frac{n}{2}(2n) = n^2$, or $\boldsymbol{n^2}$ **triangles** in the nth figure.

9. This is an arithmetic sequence with $a_1 = 8\frac{1}{6}$ (i.e., 8 a.m. plus 10 minutes, or $\frac{10}{60}$ of an hour) and $d = \frac{5}{6}$ (or $\frac{50}{60}$ of an hour). Thus

$a_8 = 8\frac{1}{6} + (8-1) \cdot \frac{5}{6} = 14$, or **2:00 p.m.**

(14 is 2:00 p.m. on a 24-hour clock.)

10. **(a)** If the first difference of the sequence

11. **(a)** Look for the differences:

3		8		15		24		35		48
	5		7		9		11		13	

The first difference row is an arithmetic sequence with fixed difference of 2. Thus the 6th term in the first difference row is $13 + 2 = 15$; the 7th term in the original sequence is $48 + 15 = 63$. Using the same reasoning, the next three terms in the original sequence are **63, 80, 99**.

(b) Look for the differences:

1		7		18		37		67		111
	6		11		19		30		44	
		5		8		11		14		

The second difference row is an arithmetic sequence with fixed difference of 3. Thus the 5th term in the second difference row is $14 + 3 = 17$; the 6th term in the original sequence is $111 + 61 = 172$. Using the same reasoning, the next three terms in the original sequence are **172, 253, 357**.

13. **(a)**

First term:	$5(1) + 6 = \mathbf{11}$
Second term:	$5(2) + 6 = \mathbf{16}$
Third term:	$5(3) + 6 = \mathbf{21}$
Fourth term:	$5(4) + 6 = \mathbf{26}$
Fifth term:	$5(5) + 6 = \mathbf{31}$

(b)

First term:	$6(1) - 2 = \mathbf{4}$
Second term:	$6(2) - 2 = \mathbf{10}$
Third term:	$6(3) - 2 = \mathbf{16}$
Fourth term:	$6(4) - 2 = \mathbf{22}$
Fifth term:	$6(5) - 2 = \mathbf{28}$

(c)

First term:	$5 \cdot 1 + 1 = \mathbf{6}$
Second term:	$5 \cdot 2 + 1 = \mathbf{11}$
Third term:	$5 \cdot 3 + 1 = \mathbf{16}$
Fourth term:	$5 \cdot 4 + 1 = \mathbf{21}$
Fifth term:	$5 \cdot 5 + 1 = \mathbf{26}$

(d)

First term:	$1^2 - 1 = \mathbf{0}$
Second term:	$2^2 - 1 = \mathbf{3}$
Third term:	$3^2 - 1 = \mathbf{8}$
Fourth term:	$4^2 - 1 = \mathbf{15}$
Fifth term:	$5^2 - 1 = \mathbf{24}$

15. **(a)** The first figure has 2 tiles, the second has 5 tiles, the third has 8 tiles, ... This is an arithmetic sequence where the nth term is $2 + (n-1) \cdot 3$.

Thus the 6th term has $2 + (6-1) \cdot 3 =$ **17 tiles**.

(b) The nth term is $2 + (n-1) \cdot 3 = 2 + 3n - 3 = \mathbf{3n - 1}$.

(c) The question can be written as: Is there an n such that $3n - 1 = 449$. Since $3n - 1 = 449 \Rightarrow 3n = 450 \Rightarrow n = 150$, the answer is yes, the $\mathbf{150^{th}}$ **figure**.

17. Use a table of Fibonacci numbers to find the pattern, F_n is the nth Fibonacci number:

Generation	Male	Female	Number in Generation	Total
1	1	0	1	1
2	0	1	1	2
3	1	1	2	4
4	1	2	3	7
5	2	3	5	12
6	3	5	8	20
⋮	⋮	⋮	⋮	⋮
n	F_{n-2}	F_{n-1}	F_n	$F_{n+2}-1$

The sum of the first n Fibonacci numbers is $F_{n+2} - 1$. $F_{12} = 144$, so there are **143 bees** in all 10 generations.

Review Problems

11. Order the teams from 1 to 10, and consider a simpler problem of counting how many games are played if each team plays each other once. The first team plays nine teams. The second team also plays nine teams, but one of these games has already been counted. The third team also plays 9 teams, but two of these games were counted in the previous two summands. Continuing in this manner, the total is $10 + 9 + 8 + \ldots + 3 + 2 + 1 = 9(10) / 2 = 45$ games. Double this amount to obtain 90 games must be played for each team to play each other twice.

13. If the problem is interpreted to stated that at least one 12-person tent is used, then there are 10 ways. This can be seen by the table below, which illustrates the ways 2,3,5, and -6 person tents can be combined accommodate 14 people.

6-Person	5-Person	3-Person	2-Person
2	0	0	1
1	1	1	0
1	0	2	1
1	0	0	4
0	2	0	2
0	1	3	0
0	1	1	3
0	0	4	1
0	0	2	4
0	0	0	7

Assessment 1-3A: Reasoning and Logic: An Introduction

1. **(a)** **False statement**. A statement is a sentence that is either true or false, but not both.

(b) **False statement.**

(c) **Not a statement.**

(d) **True statement.**

(e) **Not a statement**. This is a paradox; if it is true, it must be false, but then it isn't true. . . .

3. **(a)** For **all** natural numbers n, $n + 8 = 11$.

(b) For **all** natural numbers n, $n^2 = 4$.

(c) There is **no** natural number x such that $x + 3 = 3 + x$.

(d) There is **no** natural number x such that $5x + 4x = 9x$.

5. **(a)** If $n = 4$, then $n < 6$ *and* $n > 3$, so the statement is **true**.

(b) If $n = 10$, then $n > 0$; *or* if $n = 1$, then $n < 5$, so the statement is **true**.

7. (a) $q \wedge r$. Both q and r are true.

(b) $r \vee \sim q$. r is true or q is not true.

(c) $\sim(q \wedge r)$. q and r are not both true.

(d) $\sim q$. q is not true.

9. (a) **False**. $p \wedge q$ is true if and only both p and q are true.

(b) **True**. Negation of p, which is false.

(c) **False**. $\sim p$ is true, so $\sim(\sim p)$ is false.

(d) **False**. The statement is false because p is false, even though $\sim q$ is true.

(e) **True.** $(\sim p \wedge q)$ is false, so the negation is true.

11.

p	q	$\sim p$	$\sim p \wedge q$
T	T	F	**F**
T	F	F	**F**
F	T	T	**T**
F	F	T	**F**

13. (a) Converse: **If $2x = 10$, then $x = 5$.**

Inverse: **If $x \neq 5$, then $2x \neq 10$.**

Contrapositive: **If $2x \neq 10$, then $x \neq 5$.**

(b) Converse: **If you do not like mathematics, then you do not like this book.**
Inverse: **If you like this book, then you like mathematics.**

Contrapositive: **If you like mathematics, then you like this book.**

(c) Converse: **If you have cavities, then you do not use Ultra Brush toothpaste.**

Inverse: **If you use Ultra Brush toothpaste, then you do not have cavities.**

Contrapositive: **If you do not have cavities, then you use Ultra Brush toothpaste.**

(d) Converse: **If your grades are high, then you are good at logic.**

Inverse**: If you are not good at logic, then your grades are not high.**

Contrapositive**: If your grades arc not high, then you are not good at logic.**

15. (a) **Valid.** All squares are quadrilaterals $\rightarrow$ all quadrilaterals are polygons $\rightarrow$ all squares are polygons.

(b) **Valid**. Some teachers arc rich $\wedge$ all teachers are intelligent $\rightarrow$ some intelligent people are rich.

(c) **Invalid.** There is no statement "sophomores, juniors, and seniors do not take mathematics."

17. (a) If a figure is a square, then it is a rectangle.

(b) If a number is an integer, then it is a rational number.

(c) If a polygon has exactly three sides, then it is a triangle.

Assessment 1-3B

1. (a) **Not a statement**. A statement is a sentence that is either true or false.

(b) **Not a statement**. A statement must be either true of false; this could be either.

(c) **True statement.**

(d) **False statement**. $2 + 3 \neq 8$.

(e) **Not a statement.**

3. **(a)** There is **no** natural number n such that $n + 0 = n$.

(b) **There exists at least one** natural number n such that $n + 1 = n + 2$.

(c) For **all** natural numbers n, $3 \cdot (n + 2) \neq 12$.

(d) For **all** natural numbers n, $n^3 \neq 8$.

5. **(a)** If $x = 10$, then $x > 5$ and $x > 2$, so the statement is **true**.

(b) x could equal 5, so the statement is **false**.

7. **(a)** $q \wedge r$.

(b) $q \wedge \sim r$.

(c) $\sim r \vee \sim q$.

(d) $\sim(q \wedge r)$.

9. **(a)** **False**. $p \vee q$ is false if both p and q are false.

(b) **True**. Negation of q, which is false.

(c) **True**. The statement is true because $\sim p$ is true.

(d) **True**. $(p \vee q)$ is false, so the negation is true.

(e) **True**. $\sim p$ and $\sim q$ are both true.

11.

p	q	$\sim q$	$p \vee \sim q$
T	T	F	**T**
T	F	T	**T**
F	T	F	**F**
F	F	T	**T**

13. **(a)** Converse: **If $x^2 = 9$, then $x = 3$.**

Inverse: **If $x \neq 3$, then $x^2 \neq 9$.**

Contrapositive: **If $x^2 \neq 9$, then $x \neq 3$.**

(b) Converse: **If classes are canceled, then it snowed.**

Inverse: **If it does not snow, then classes are not canceled.**

Contrapositive: **If classes are not canceled, then it did not snow.**

15. **(a)** **Valid**. Use modus ponens: Hypatia was a woman → all women are mortal → Hypatia was mortal.

(b) **Valid**. Use modus tollens: the conditional is true and we know the conclusion is false; i.e., today is not rainy; thus the hypothesis is false.

(c) **Not valid**. It is possible that some students do not like skiing.

17. **(a)** **If** a number is a natural number, **then** it is a real number.

(b) **If** a figure is a circle, **then** it is a closed figure.

Chapter 1 Review

1. Make a plan. Every 7 days (every week) the day will change from Sunday to Sunday. 365 days per year $\div$ 7 days per week $\approx$ 52 weeks per year $+\frac{1}{7}$ weeks per year. Thus the day of the week will change from Sunday to Sunday 52 times and then change from Sunday to Monday. July 4 will be a **Monday**.

3. The information in the rhyme is that the scholar is a "ten o'clock scholar" who "used to come at ten o'clock." The question is "what makes you come sooner." If we assume that ten o'clock means 10 AM, then the rhyme makes no sense because the scholar came later than usual. If we assume that ten o'clock means 10 PM, the question makes sense but is not answered.

5. **(a)** The successive differences are 3. Each term is 3 more than the previous term. This suggests that it is an arithmetic sequence of the form $3n + ?$. Since the first term is 5, $3(1) + ? = 5$. The nth term would be $\mathbf{3n + 2}$.

(b) Each term given is 3 times the previous term. This suggests that the sequence is geometric. The nth term will be $\mathbf{3^n}$.

(c) We are reminded of the sequence 1; 8; 27; 64; . . . , which is given by n^3. Since the terms in the original sequence are one less than the terms given by n^3, $\mathbf{n^3 - 1}$ is a possible n^{th} term.

7. **(a)** $a_1 = 2, d = 2, a_n = 200.$

So $200 = 2 + (n - 1) \cdot 2 \Rightarrow n = 100.$

Sum is $\frac{100(2+200)}{2} = \mathbf{10,100}.$

(b) $a_1 = 51,\ d = 1, a_n = 151.$

So $151 = 51 + (n - 1) \cdot 1 \Rightarrow n = 101.$

Sum is $\frac{101 \cdot (51+151)}{2} = \mathbf{10,201}.$

9. **All rows, columns, and diagonals must add to 34**; i.e., the sum of the digits in row 1. Complete rows or columns with one number missing, then two, etc. to work through the square:

16	3	2	13
5	10	**11**	**8**
9	**6**	7	12
4	**15**	14	**1**

11. If S is the cost of the shirt and T is the cost of the tie:

$S + T = 5.50 \Rightarrow$
$S = T + 5.50 \Rightarrow$
$(T + 5.50) + T = 9.50 \Rightarrow T = 2.00.$

The tie costs $2.00.

13. 1 mile $= 5280$ feet.

$5280 \div 6$ feet $= 880$ turns per mile.

880×50000 miles $=$ **44,000,000 turns.**

15. Let l be a large box, m be a medium box, and s be a small box:

$3l + (3l \times 2m \text{ each}) + [(3 \times 2)m \times 5s \text{ each}]$

$3l + 6m + 30s =$ **39 total boxes.**

17. Extend the pattern of doubling the number of ants each day. This is a geometric sequence with $a_1 = 1500,\ a_n = 100,000$, and $r = 2$.

$100,000 = 1500 \cdot 2^{n-1} \Rightarrow$

$66\frac{2}{3} = 2^{n-1}.$

Since $2^{7-1} < 66\frac{2}{3}$ and $2^{8-1} > 66\frac{2}{3}$, the ant farm will fill sometime between **the 7th and 8th day**.

19. **Yes.** Let $\ell =$ length of the longest piece,

$m =$ length of the middle-sized piece,

and

$s =$ lenth of the shortest piece.

Then $\ell = 3m$ and $s = m - 10$.

So $\ell + m + s = 90 \Rightarrow$

$3m + m + (m - 10) = 90 \Rightarrow$

$5m = 100.$

Thus $m = \mathbf{20\,cm}$;

$\ell = 3m = \mathbf{60\,cm}$; and

$s = m - 10 = \mathbf{10\ cm}.$

21. Make a table or draw a picture and write the information given to us.

Season	Person born
Winter	Al
Spring	Carl
Summer	
Fall	Betty

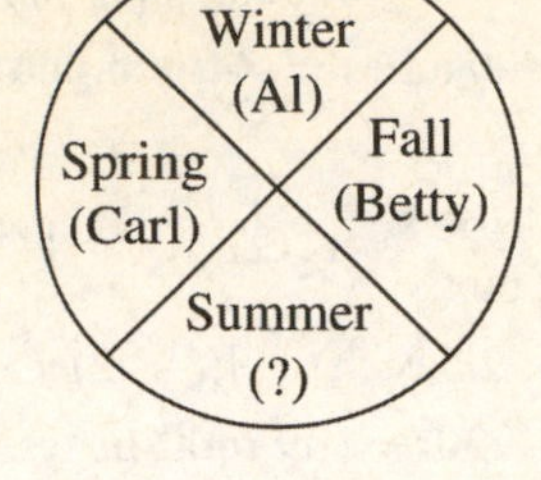

Since Dan was born in a different season, by the processor of elimination, Dan must have been born in the summer. Al-winter; Betty-fall; Carl spring; Dan-summer.

23. A possible pattern is to increase each rectangle by one row of dots and one column of dots to obtain the next term in the sequence. Make a table.

Number of the term	Row of dots	Column of dots	Term (row × column)
1	1	2	2
2	2	3	6
3	3	4	12
4	4	5	20
5	5	6	30
6	6	7	42
7	7	8	56
⋮			
100	100	101	10100
⋮			
n	n	$n+1$	$n(n+1)$

We also observe that the number of the terms corresponds to the number of rows in the arrays and that the number of columns in the array is the number of terms plus one. Thus, the next three terms are **30, 42, and 56**. The 100th term is 10100 and the nth term is $n(n+1)$.

25. **(a)** "All" means that each and every student passed the final.

(b) "Some" means that at least one, but possibly all, students passed the final.

27. **(a)** **No women smoke**, which could be stated another way as **all** women **don't** smoke. The negation of *smoke is don't smoke.*

(b) $3 + 5 \neq 8$. The negation of $=$ is $\neq$.

(c) Beethoven wrote **some** non-classical music. The negation of *only* is *some.*

29.

p	q	$\sim p$	$\sim q$	$p \to \sim q$	$q \to \sim p$
T	T	F	F	**F**	**F**
T	F	F	T	**T**	**T**
F	T	T	F	**T**	**T**
F	F	T	T	**T**	**T**

The truth values for $p \to \sim q$ and $q \to \sim p$ are the same, so the statements are **logically equivalent**.

31. **(a)** Let $p =$ all Americans love Mom and apple pie,
$q =$ Joe Czernyu is an American.

p and q are true, thus **Joe Czernyu loves Mom and apple pie.**

(b) Let $p =$ steel eventually rusts,
$q =$ the Statue of Liberty has a steel structure.

p and q are true, thus **the structure of the Statue of Liberty will eventually rust.**

(c) **Albertina passed Math 100**. She had two options; one was false so the other must be true.

33. **(a)** **True**, by modus tollens, or indirect reasoning. If it is accepted that if Bob scores at least 80 on the final he will pass the course, and it is known that Bob did not pass the course, then it must be true that he did not score at least 80.

Chapter 2

NUMERATION SYSTEMS AND SETS

Assessment 2-1A: Numeration Systems

1. **(a)** $\overline{\overline{\text{M}}}$**CDXXIV**. The double bar over the M represents $1000 \cdot 1000 \cdot 1000$ while a single bar over the M would represent only $1000 \cdot 1000$.

(b) **46,032**. The 4 in 46,032 represents 40,000 while the 4 in 4632 represents only 4000.

(c) < ▼▼. < in the first number represents 10, while in the second number it represents $10 \cdot 60$ because of the space.

(d) 𓆼∩ |. 𓆼 has a place value of 1000 while 𓍢 has a place value of only 100.

(e) ⋯ . ⋯ represents three groups of 20 plus zero l's, or 60, while ≡ represents three 5's and three l's, or 18.

3. MCMXXII is $1000 + 900 + 20 + 2$, or the year **1922**.

5. **(a)** Hindu-Arabic: **72**

Babylonian: $60 + 10 + 2$, or ▽ < ▽▽

Egyptian: $70 + 2$, or ∩∩∩∩∩∩∩||

Roman: **LXXII**

Mayan: $3 \cdot 20 + 12$, or ooo / oo / ═

(b) Hindu-Arabic: **602**

Babylonian: $10 \cdot 60 + 2$, or < ▽▽

Egyptian: $6 \cdot 100 + 2$, or 999999||

Roman: **DCII**

Mayan: $1 \cdot 360 + 12 \cdot 20 + 2$, or • / ═ / ••

(c) Hindu-Arabic: **1223**

Babylonian: $2 \cdot 10 \cdot 60 + 23$, or

<< <<▽▽▽

Egyptian: 𓆼99∩∩|||

Roman: **MCCXXIII**

Mayan: $3 \cdot 360 + 7 \cdot 20 + 3$, or ••• / •• ─ / •••

7. **(a)** $3{,}000{,}000 + 4000 + 5 =$ **3,004,005**.

(b) $20{,}000 + 1 =$ **20,001**.

9. **(a)** $(3 \cdot 25) + (2 \cdot 5) + (1 \cdot 1) =$ **86**.

(b) $(1 \cdot 8) + (0 \cdot 4) + (1 \cdot 2) + (1 \cdot 1) =$ **11**.

11. **(a)** Place values represent powers of 2; e.g.,

$1 \cdot 2^0 = 1_{two}$

$1 \cdot 2^1 + 0 \cdot 2^0 = 10_{two}$

$1 \cdot 2^2 + 0 \cdot 2^1 + 0 \cdot 2^0 = 100_{two}$, etc.

Thus the first 15 counting numbers are: $(1,10,11,100,101,110,111,1000,1001,1010,1011,1100,1101,1110,1111)_{two}$.

(b) Place values represent powers of 4; e.g.,

$1 \cdot 4^0 = 1_{four}$

$1 \cdot 4^1 + 0 \cdot 4^0 = 10_{four}$

$1 \cdot 4^2 + 0 \cdot 4^1 + 0 \cdot 4^0 = 100_{four}$, etc.

Thus the first 15 counting numbers are: $(1,2,3,10,11,12,13,20,21,22,23,30,31,32,33)_{four}$.

13. $2032_{four} = (2 \cdot 10^3 + 0 \cdot 10^2 + 3 \cdot 10^1 + 2 \cdot 10^0)_{four}$. 2032_{four} may also be expanded as $(2 \cdot 4^3 + 0 \cdot 4^2 + 3 \cdot 4^1 + 2 \cdot 4^0)_{ten} =$ **142**.

15. **(a)** $EE0_{twelve} = 11 \cdot 12^2 + 11 \cdot 12^1 + 0 \Rightarrow$

(i) $(EE0 - 1)_{twelve} =$

$11 \cdot 12^2 + 10 \cdot 12^1 + 11 = \boldsymbol{ETE_{twelve}}$.

(ii) $(EE0 + 1)_{twelve} =$

$11 \cdot 12^2 + 11 \cdot 12^1 + 1 = \boldsymbol{EE1_{twelve}}$.

(b) ***(i)*** $(100000 - 1)_{two} = 1 \cdot 2^5 + 0 \cdot 2^4 + 0 \cdot 2^3 + 0 \cdot 2^2 + 0 \cdot 2^1 + 0 - 1 = 1 \cdot 2^4 + 1 \cdot 2^3 + 1 \cdot 2^2 + 1 \cdot 2^1 + 1 =$ $\boldsymbol{11111_{two}}$ [analogous to $(100{,}000 - 1)_{ten} = 99{,}999_{ten}$].

(*ii*) $(100000 + 1)_{two} = 1 \cdot 2^5 + 0 \cdot 2^4 + 0 \cdot 2^3 + 0 \cdot 2^3 + 0 \cdot 2^1 + 0 + 1 =$ $\mathbf{100001}_{two}$.

(c) (*i*) $(555 - 1)_{six} = 5 \cdot 6^2 + 5 \cdot 6^1 + 5 - 1 = \mathbf{554}_{six}$.

(*ii*) $(555 + 1)_{six} = 5 \cdot 6^2 + 5 \cdot 6^1 + 5 + 1 = 1 \cdot 6^3 + 0 \cdot 6^2 + 0 \cdot 6^1 + 0 = 1000_{six}$ [analogous to $(999 + 1)_{ten} = 1000_{ten}$].

17. There are 3 groups of 4^3, 1 group of 4^2, 1 group of 4^1, and 2 groups of 4^0 in 214, so $214 = 3112_{four}$. Thus, where a block is $4 \times 4 \times 4$, a flat is 4×4, and a long is 1×4, **three** blocks, **one** flat, **one** long, and **two** units is the least number of base four blocks.

19. (a) (*i*) 8 pennies may be changed into 1 nickel and 3 pennies. The total is now 2 quarters, 10 nickels, and 3 pennies.

(*ii*) 10 nickels may be changed into 2 quarters. The total is now **4 quarters, 0 nickels, and 3 pennies.**

(b) 73¢, in the fewest number of coins, can be obtained with 2 quarters, 4 nickels, and 3 pennies, or $(2 \cdot 5^2 + 4 \cdot 5 + 3)$¢. This is $\mathbf{243}_{five}$.

21. (a) 10 flats = **1 block**. $1 \cdot 10^3 = \mathbf{1000}$.

(b) 20_{twelve} flats = 12 flats + 8 flats = **1 block + 8 flats**. When written in base twelve notation, it looks like $1 \cdot 12^3 + 8 \cdot 10^2 = \mathbf{1800}_{twelve}$.

23. (a) $432_{five} = 4 \cdot 5^2 + 3 \cdot 5^1 + 2 = 100 + 15 + 2 = \mathbf{117}$.

(b) $101101_{two} = 1 \cdot 2^5 + 1 \cdot 2^3 + 1 \cdot 2^2 + 1 = 32 + 8 + 4 + 1 = \mathbf{45}$.

(c) $92E_{twelve} = 9 \cdot 12^2 + 2 \cdot 12 + 11 = 1296 + 24 + 11 = \mathbf{1331}$.

25. (a) There are 8 groups of 7 days (1 week) with 2 days left over, so 58 days = **8 weeks and 2 days** (or $58 = 82_{seven}$).

(b) There is 1 group of 24 hours (1 day) with 5 hours left over, so 29 hours = **1 day and 5 hours** (or $29 = 15_{twenty\ four}$).

27. (a) $3 \cdot 5^4 + 3 \cdot 5^2 = 3 \cdot 5^4 + 0 \cdot 5^3 + 3 \cdot 5^2 + 0 \cdot 5^1 + 0 \cdot 5^0 = \mathbf{30300}_{five}$.

(b) $2 \cdot 12^5 + 8 \cdot 12^3 + 12 = 2 \cdot 12^5 + 0 \cdot 12^4 + 8 \cdot 12^3 + 0 \cdot 12^2 + 1.12^1 + 0.12^0 = \mathbf{208010}_{twelve}$.

Assessment 2-1B

1. (a) $\overline{\text{M}}$**DCXXIV**. D is greater than CD.

(b) **30,456**. The 3 in 30,456 represents 30,000 while the 3 in 3456 represents only 3000.

(c) < ▼▼. The space indicates that < is multiplied by 60; the < without the space represents just 10.

(d) 999, which represents 300; the other figure represents 211.

(e) [Mayan numeral], which represents $2 \cdot 20$. The other figure represents 13.

3. Since no numeral of lesser value is to the left of a numeral of greater value, the numerals can be combined in an additive process. MDCCLXXVI represents 1000 + 500 + 100 + 100 + 50 + 10 + 10 + 5 + 1 = **1776.**

5. (a) Hindu-Arabic: 78

Babylonian: 60 + 10 + 8, or ▼ < ▼▼▼▼▼▼▼▼

Egyptian: ∩∩∩∩∩∩∩||||||||

Roman: **LXXVIII**

Mayan: $3 \cdot 20 + 18$, or [Mayan numeral]

(b) Hindu-Arabic: **601**

Babylonian: $10 \cdot 60 + 1$, or < ▽

Egyptian: 999999 |

Roman: **DCI**

Mayan: $360 + 12 \cdot 20 + 1$, or [Mayan numeral]

(c) Hindu-Arabic: **1111**

Babylonian: $18 \cdot 60 + 31$, or
< ▽▽▽▽▽▽▽▽ <<< ▽

Egyptian: [hieroglyphs]

Roman: **MCXI**

Mayan: $3 \cdot 360 + 20 + 11$, or [Mayan numeral]

7. (a) $3000 + 500 + 60 =$ **3560**.
   (b) $9{,}000{,}000 + 90 + 9 =$ **9,000,099**.

9. **47**, or $1 \cdot 27 + 2 \cdot 9 + 2 \cdot 1 = 47$.

11. (a) Place values represent powers of 3; e.g.,
   $1 \cdot 3^0 = 1_{three}$
   $1 \cdot 3^1 + 0 \cdot 3^0 = 10_{three}$
   $1 \cdot 3^2 + 0 \cdot 3^1 + 0 \cdot 3^0 = 100_{three}$, etc.

   Thus the first 10 counting numbers are: $\mathbf{(1,2,10,11,12,20,21,22,100,101)_{three}}$.

   (b) Place values represent powers of 8; e.g.,
   $1 \cdot 8^0 = 1_{eight}$
   $1 \cdot 8^1 + 0 \cdot 8^0 = 10_{eight}$
   $1 \cdot 8^2 + 0 \cdot 8^1 + 0 \cdot 8^0 = 100_{eight}$, etc.

   Thus the first 10 counting numbers are: $\mathbf{(1,2,3,4,5,6,7,10,11,12)_{eight}}$.

13. $2022_{three} = \mathbf{(2 \cdot 10^3 + 0 \cdot 10^2 + 2 \cdot 10^1 + 2 \times 10^0)_{three}}$. 2022_{three} may alternatively be expanded as $\mathbf{(2 \cdot 3^3 + 0 \cdot 3^2 + 2 \cdot 3^1 + 2 \cdot 3^0)_{ten} = 62}$.

15. (a) (*i*) $(100 - 1)_{seven} = 1 \cdot 7^2 + 0 \cdot 7 + 0 - 1 = 6 \cdot 7 + 6 = \mathbf{66_{seven}}$.
   (*ii*) $(100 + 1)_{seven} = 1 \cdot 7^2 + 0 \cdot 7 + 0 + 1 = \mathbf{101_{seven}}$.
   (b) (*i*) $(10000 - 1)_{two} = 1 \cdot 2^4 + 0 \cdot 2^3 + 0 \cdot 2^2 + 0 \cdot 2^1 + 0 \cdot 2^0 - 1 = 1 \cdot 2^3 + 1 \cdot 2^2 + 1 \cdot 2^1 + 1 \cdot 2^0 = \mathbf{1111_{two}}$.
   (*ii*) $(10000 + 1)_{two} = 1 \cdot 2^4 + 0 \cdot 2^3 + 0 \cdot 2^2 + 0 \cdot 2^1 + 0 \cdot 2^0 + 1 = \mathbf{10001_{two}}$.
   (c) (*i*) $(101 - 1)_{two} = 1 \cdot 2^2 + 0 \cdot 2^1 + 1 \cdot 2^0 - 1 = 1 \cdot 2^2 + 0 \cdot 2^1 + 0 \cdot 2^0 = \mathbf{100_{two}}$.
   (*ii*) $(101 + 1)_{two} = 1 \cdot 2^2 + 0 \cdot 2^1 + 1 \cdot 2^0 + 1 = \mathbf{110_{two}}$.

17. There are 2 groups of 3^3, 2 groups of 3^2, 2 groups of 3^1, and 1 group of 3^0 in 79, so $79 = 2221_{three}$. Thus, where a block is $3 \times 3 \times 3$, a flat is 3×3, and a long is 1×3, **two** blocks, **two** flats, **two** longs, and **one** unit, so, **seven** is the fewest number of base three blocks.

19. $277 = 1 \cdot 12^2 + 11 \cdot 12^1 + 1 \cdot 12^0$, or **1 gross, 11 dozens, and 1 unit**.

21. (a) 10_{four} longs $= 40$ units, which can be represented by **2 flats and 2 longs**, or four pieces.
   (b) 10_{three} longs $= 30$ units, which can be represented by **1 block and 1 long**, or two pieces.

23. (a) $432_{six} = 4 \cdot 6^2 + 3 \cdot 6^1 + 2 \cdot 6^0 = \mathbf{164}$.
   (b) $11011_{two} = 1 \cdot 2^4 + 1 \cdot 2^3 + 0 \cdot 2^2 + 1 \cdot 2^1 + 1 \cdot 2^0 = \mathbf{27}$.
   (c) $E29_{twelve} = 11 \cdot 12^2 + 2 \cdot 12^1 + 9 \cdot 12^0 = \mathbf{1617}$.

25. There are $4 = 2^2$ cups per quart and $2 = 2^1$ cups per pint. Thus 1 cup, 1 pint, and 1 quart $= \mathbf{111_{two}}$.

27. If each key may be used only once, the largest four-digit number would be **9876**.

Assessment 2-2A Describing Sets

1. (a) Either a list or set-builder notation may be used: $\{\mathbf{m,a,t,h,e,i,c,s}\}$ or $\{x \mid x$ **is a letter in the word** ***mathematics***$\}$.

(b) $\{21,22,23,24,\ldots\}$ or $\{x \mid x$ **is a natural number and** $x > 20\}$ or $\{x \mid x \in N$ **and** $x > 20\}$.

3. (a) **Yes**. $\{1,2,3,4,5\} \sim \{m,n,o,p,q\}$ because both sets have the same number of elements and thus exhibit a one-to-one correspondence.

(b) **Yes**. $\{a,b,c,d,e,f,\ldots,m\} \sim \{1,2,3,\ldots,13\}$ because both sets have the same number of elements.

(c) **No**. $\{x \mid x$ is a letter in the word *mathematics*$\}$ $\nsim \{1,2,3,4,\ldots,11\}$; there are only eight unduplicated letters in the word *mathematics*.

5. (a) If x must correspond to 5, then y may correspond to any of the four remaining elements of $\{1,2,3,4,5\}$, z may correspond to any of the three remaining, etc. Then $1 \cdot 4 \cdot 3 \cdot 2 \cdot 1 =$ **24 one-to-one correspondences**.

(b) There would be $1 \cdot 1 \cdot 3 \cdot 2 \cdot 1 =$ **6 one-to-one correspondences**.

(c) The set $\{x,y,z\}$ could correspond to the set $\{1,3,5\}$ in $3 \cdot 2 \cdot 1 = 6$ ways. The set $\{u,v\}$ could correspond with the set $\{2,4\}$ in $2 \cdot 1 = 2$ ways. There would then be $6 \cdot 2 =$ **12 one-to-one correspondences**.

7. (a) Assume an arithmetic sequence with $a_1 = 101, a_n = 1100$, and $d = 1$. Thus $1100 = 101 + (n-1) \cdot 1$; solving, $n = 1000$. The cardinal number of the set is therefore **1000**.

(b) Assume an arithmetic sequence with $a_1 = 1, a_n = 1001$, and $d = 2$. Thus $1001 = 1 + (n-1) \cdot 2$; solving, $n = 501$. The cardinal number of the set is therefore **501**.

(c) Assume a geometric sequence with $a_1 = 1, a_n = 1024$, and $r = 2$. Thus $1024 = 1 \cdot 2^{n-1} \Rightarrow 2^{10} = 2^{n-1} \Rightarrow n - 1 = 10 \Rightarrow n = 11$. The cardinal number of the set is therefore **11**.

(d) If $k = 1,2,3,\ldots,100$, the cardinal number of the set $\{x \mid x = k^2, k = 1,2,3,\ldots,100\} =$ **100**, since there are 100 elements in the set.

(e) The set $\{i + j \mid i \in \{1,2,3\}$ and $j \in \{1,2,3\}\} = \{(1+1),(1+2),(1+3),(2+1),\ldots,(3+3)\}$ has only five distinguishable elements: $2,3,4,5,6$. The cardinal number is therefore **5**.

9. (a) A proper subset must have at least one less element than the set, so the maximum $\boldsymbol{n(B) = 7}$.

(b) Since $B \subset C$, then $C \neq \varnothing$; this means that C has at least one element. No matter what C is, B could be the **empty set** $\varnothing$ and satisfy $B \subset C$. In other words, the least possible number of elements in B is 0.

11. (a) A has 5 elements, thus $2^5 =$ **32 subsets**.

(b) Since A is a subset of A and A is the only subset of A that is not proper, A has $2^5 - 1 =$ **31 proper subsets**.

(c) Let $B = \{b,c,d\}$. Since $B \subset A$, the subsets of B are all of the subsets of A that do not contain a and e. There are $2^3 = 8$ of these subsets. If we join (union) a and e to each of these subsets there are still **8 subsets**.

Alternative. Start with $\{a,e\}$. For each element b, c, and d there are two options: include the element or don't include the element. So there are $2 \cdot 2 \cdot 2 = 8$ ways to create subsets of A that include a and e.

13. In roster format, $A = \{3,6,9,12,\ldots\}, B = \{6,12,18,24,\ldots\}$, and $C = \{12,24,36,\ldots\}$. Thus, $C \subset A, C \subset B$, and $B \subset A$.
Alternatively: $12n = 6(2n) = 3(4n)$.
Since $2n$ and $4n$ are natural number $\boldsymbol{C \subset A, C \subset B}$, **and** $\boldsymbol{B \subset A}$.

15. (a) $\not\subseteq$. 0 is not a set so cannot be a subset of the empty set, which has only one subset, $\varnothing$.

(b) $\subseteq$. $\{1\}$ is actually a proper subset, $\subset$, of $\{1,2\}$.

(c) $\not\subseteq$. 1024 is an element, not a subset.

(d) $\not\subseteq$. 3002 is an element, not a subset.

17. (a) Let $A = \{1,2,3,\ldots,100\}$ and $B = \{1,2,3\}$. Then $n(A) = 100$ and $n(B) = 3$. Since $B \subset A$, $\boldsymbol{n(B)} = 3 <$ **100** $= \boldsymbol{n(A)}$.

(b) $n(\emptyset) = 0$. Let $A = \{1,2,3\} \Rightarrow n(A) = 3$. $\emptyset \subset A$, which implies that there is at least one more element in A than in $\emptyset$. Thus **0 < 3**.

19. There are 9 digits that are not 0. For the tens place, we have 9 chooses. Once the tens digit is chosen, there are 9 options for the ones place to avoid repetition. So the number of ways to create two digit numbers without a 0 in the tens place and no repeated digits is $9 \cdot 9 = \mathbf{81}$.

Assessment 2-2B

1. (a) Either a list or set-builder notation may be used: $\{g, e, o, m, t, r, y\}$ or **$\{x|x$ is a letter in the word *geometry*$\}$**
   (b) **$\{8, 9, 10, 11, ...\}$** or **$\{x|x$ is a natural number and $x>7\}$** or **$\{x|x \in N$ and $x>7\}$.**

3. (a) **Yes.** $\{1,2,3,4\} \sim \{w,c,y,z\}$ because both sets have the same number of elements and thus exhibit a one-to-one correspondence.
   (b) **Yes**, because both sets have the same number of elements.
   (c) **No.** $\{x|x$ is a letter in the word *geometry*$\}$ $\nsim \{1,2,3,4,...,8\}$; there are only seven unduplicated letters in geometry.

5. (a) If b must correspond to 3, then a may correspond to any of the three remaining elements of $\{1,2,3,4\}$, and c may correspond to any of the two remaining, etc. Then $1 \cdot 3 \cdot 2 \cdot 1 =$ **6 one-to-one correspondences.**
   (b) There would be a $1 \cdot 1 \cdot 2 \cdot 1 =$ **2 one-to-one correspondences.**
   (c) The set $\{a,c\}$ could correspond to the set $\{2,4\}$ in $2 \cdot 1 = 2$ ways. The set $\{b,d\}$ could correspond with the set $\{1,3\}$ in $2 \cdot 1 = 2$ ways. There would then be $2 \cdot 2 =$ **4 one-to-one correspondences.**

7. (a) Assume an arithmetic sequence with $a_1 = 9$, $a_n = 99$, and $d = 1$. Thus $99 = 9 + (n-1) \cdot 1$; solving, $n = 91$. The cardinal number of the set is therefore **91.**
   (b) Assume an arithmetic sequence with $a_1 = 2$, $a_n = 2002$, and $d = 2$. Thus $2002 = 2 + (n-1) \cdot 2$; solving, $n = 1001$. The cardinal number of the set is therefore **1001**.
   (c) Assume an arithmetic sequence with $a_1 = 1$, $a_n = 99$, and $d = 2$. Thus $99 = 1 + (n-1) \cdot 2$; solving, $n = 50$. The cardinal number of the set is therefore **50.**
   (d) There are no natural number that have the property $x = x + 1$, so the set is empty and the cardinal number is **0**.

9. (a) The empty set is a subset of any set, so the minimum number of elements in A would be **0**.
   (b) **Yes**. Since A is not assumed to be a proper subset, A and B could be equal . Thus, both sets could be empty.

11. (a) Since $A = \{1,2,3,4,5,6,7,8,9\}$ has 9 elements, A has $2^9 = \mathbf{512}$ **subsets.**
    (b) Since A is a subset of A and not proper, A has $2^9 - 1 = \mathbf{511}$ **subsets.**

13. In roster format, $A = \{4,7,10,13,16,...\}$, $B = \{7,13,18,...\}$, and $C = \{13,25,38,...\}$. Thus, $C \subset A$, $C \subset B$, and $B \subset A$.

    Alternative: $12n + 1 = 6(2n) + 1 = 3(4n) + 1$. Since $2n$ and $4n$ are natural numbers **$C \subset A$, $C \subset B$,** and **$B \subset A$.**

15. (a) $\subseteq$. Any set is a subset of itself.
    (b) $\subseteq$. $\{2\}$ is actually a proper subset of $\{3,2,1\}$.
    (c) $\not\subseteq$. 1022 is an element, not a subset.
    (d) $\not\subseteq$. 3004 is an element, not a subset.

17. (a) $n(\emptyset) = 0$. Let $A = \{1,2\} \Rightarrow n(A) = 2$. $\emptyset \subset A$, which implies that there is at least one more element in A than in $\emptyset$. Thus **$0 < 2$.**
    (b) Let $A = \{1,2,3,\ldots 99\}$ and $B = \{1,2,3,\ldots, 100\}$. $n(A) = 99$ and $n(B) = 100$, but $A \subset B$ so $\mathbf{n(A) = 99 < 100 = n(B)}$.

19. If 0 or 1 cannot be used as the leading number but may be used in any succeeding number, there are $8 \cdot 10^6 =$ **8,000,000** possible numbers.

Review Problems

17. **1410.** $\boldsymbol{M} = 1000$, $\boldsymbol{CD} = 500 - 100$, and $\boldsymbol{X} = 10$.

19. $12^4 + 12^2 + 13 = 1 \cdot 12^4 + 0 \cdot 12^3 + 1 \cdot 12^2 + 1 \cdot 12 + 1 = \mathbf{10111}_{twelve}$.

Assessment 2-3A
Other Set Operations and Their Properties

1. $A = \{1,3,5,\ldots\}; B = \{2,4,6,\ldots\};$
$C = \{1,3,5,\ldots\}$

(a) A or C. Every element in C is either in A or C.

(b) N. Every natural number is in either A or B.

(c) $\emptyset$. There are no natural numbers in both A and B.

3. **(a)** **True**. Let $A = \{1,2\}$. $A \cup \varnothing = \{1,2\}$.

(b) **False**. Let $A = \{1,2\}$ and $B = \{2,3\}$.
$A - B = \{1\}; B - A = \{3\}$.

(c) **False**. Let $U = \{1,2\}; A = \{1\}; B = \{2\}$.
$\overline{A \cap B} = \{1,2\}; \overline{A} \cap \overline{B} = \varnothing$.

(d) **False**. Let $A = \{1,2\}; B = \{2,3,4\}$.
$(A \cup B) - A = \{1,2,3,4\} - \{1,2\} = \{3,4\} \neq B$.

(e) **False**. Let $A = \{1,2\}; B = \{2,3\}$.
$(A - B) \cup A = \{1\} \cup \{1,2\} = \{1,2\}$;
$(A - B) \cup (B - A) = \{1\} \cup \{3\} = \{1,3\}$.

5. **(a)** $A \cup B$

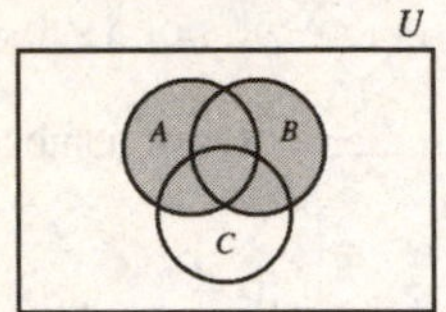

(b) $\overline{A \cap B}$

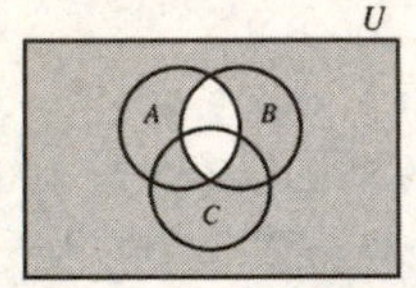

(c) $(A \cap B) \cup (A \cap C)$

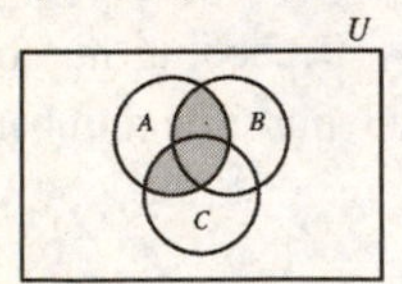

(d) $(A \cup B) \cap \overline{C}$

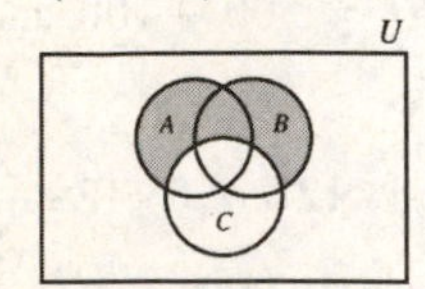

(e) $(A \cap B) \cup C$

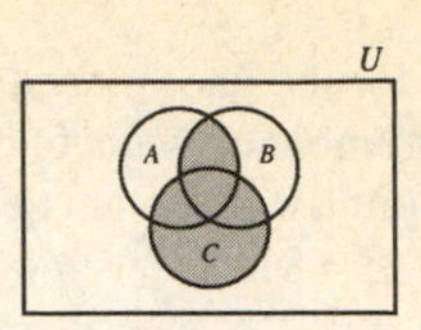

7. **(a)** If $A \cap B = \varnothing$ then A and B are disjoint sets and any element in A is not in B, so $A - B = \{x | x \in A \text{ and } x \notin B\} = A$.

(b) Since B is the empty set, there are no elements to remover from A, so $A - B = A$.

(c) If $B = U$ there are no elements in A which are not in B, so $A - B = \varnothing$.

9. Answers may vary.

(a) $\boldsymbol{B \cap \overline{A}}$ or $\boldsymbol{B - A}$; i.e., $\{x | x \in B$ but $x \notin A\}$.

(b) $\boldsymbol{\overline{A \cup B}}$ or $\boldsymbol{\overline{A} \cap \overline{B}}$; i.e., $\{x | x \notin A \text{ or } B\}$.

(c) $\boldsymbol{(A \cap B) \cap \overline{C}}$ or $\boldsymbol{(A \cap B) - C}$; i.e., $\{x | x \in A$ and B but $x \notin C\}$.

11. **(a)** **False**:

$$A \cup (B \cap C) \neq (A \cup B) \cap C$$

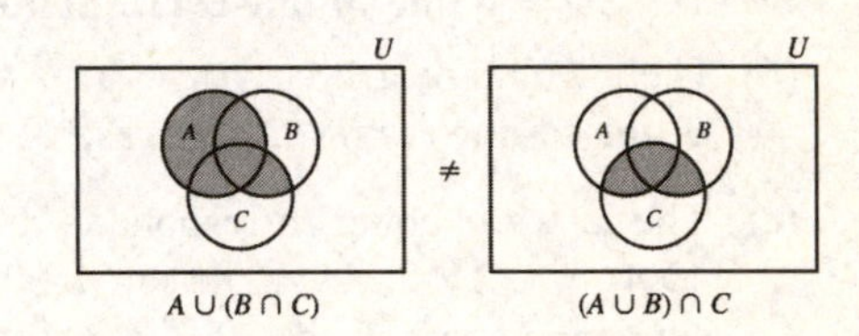

$A \cup (B \cap C)$ $(A \cup B) \cap C$

(b) **False**:

$$A - (B - C) \neq (A - B) - C$$

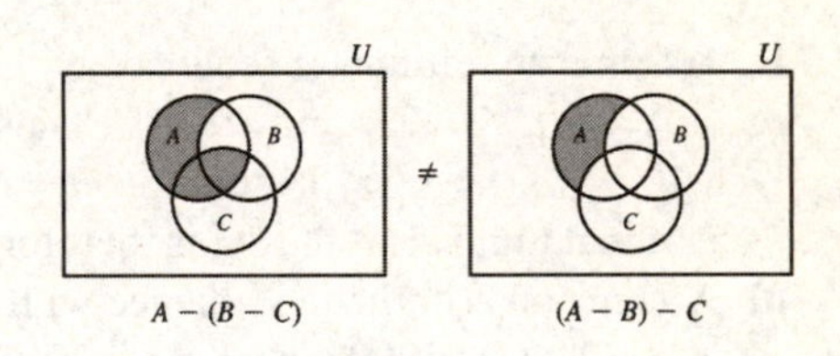

$A - (B - C)$ $(A - B) - C$

13. **(a)** ***(i)*** Greatest $n(A \cup B) = n(A) + n(B) = \mathbf{5}$ if A and B are disjoint.

(ii) Greatest $n(A \cap B) = n(B) = \mathbf{2}$ if $B \subseteq A$.

(iii) Greatest $n(B - A) = n(B) = \mathbf{2}$ if A and B are disjoint.

(iv) Greatest $n(A - B) = n(A) = \mathbf{3}$ if A and B are disjoint.

(b) (*i*) Greatest $n(A \cup B) = \boldsymbol{n + m}$ if A and B are disjoint.

(*ii*) Greatest $n(A \cap B) = \boldsymbol{m}$, if $B \subseteq A$, or $\boldsymbol{n}$, if $A \subseteq B$.

(*iii*) Greatest $n(B - A) = \boldsymbol{m}$ if A and B are disjoint.

(*iv*) Greatest $n(A - B) = \boldsymbol{n}$ if A and B are disjoint.

15. Constructing a Venn diagram will help in visualization:

(a) $B \cap S$ is the set of college **basketball players more than 200 cm tall.**

(b) $\overline{S}$ is the set of humans who are **not college students** or who are **college students less than or equal to 200 cm tall.**

(c) $B \cup S$ is the set of humans who are **college basketball players** or who are **college students taller than 200 cm.**

(d) $\overline{B \cup S}$ is the set of all humans who are **not college basketball players and who are not college students taller than 200 cm.**

(e) $\overline{B} \cap S$ is the set of all **college students taller than 200 cm who are not basketball players.**

(f) $B \cap \overline{S}$ is the set of all **college basketball players less than or equal to 200 cm tall.**

17. In the Venn diagram below:

(*i*) There were 5 members who took both biology and mathematics;

(*ii*) Of the 18 who took mathematics 5 also took biology, leaving 13 who took mathematics only;

(*iii*) 8 took neither course, so of the total of 30 members there were $30 - (5 + 13 + 8) = \boldsymbol{4}$ who took biology but not mathematics.

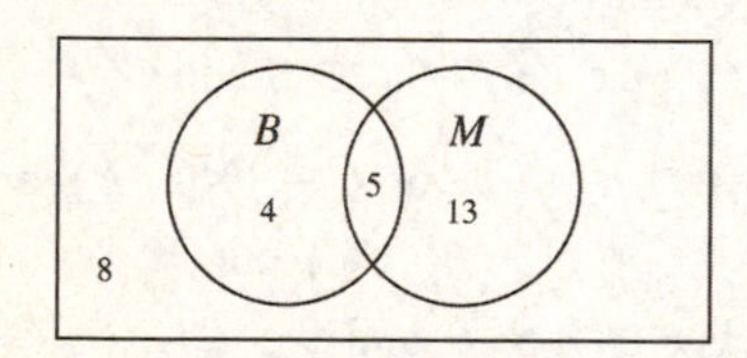

19. Generate the following Venn diagram in this order:

(*i*) The 4 who had A, B, and Rh antigens;

(*ii*) The $5 - 4 = 1$ who had A and B antigens, but who were Rh negative;

(*iii*) The $31 - 4 = 27$ who had A antigens and were Rh positive;

(*iv*) The $11 - 4 = 7$ who had B antigens and were Rh positive;

(*v*) The $40 - 27 - 4 - 1 = 8$ who had A antigens only;

(*vi*) The $18 - 7 - 4 - 1 = 6$ who had B antigens only, and;

(*vii*) The $82 - 27 - 4 - 7 = 44$ who were O positive.

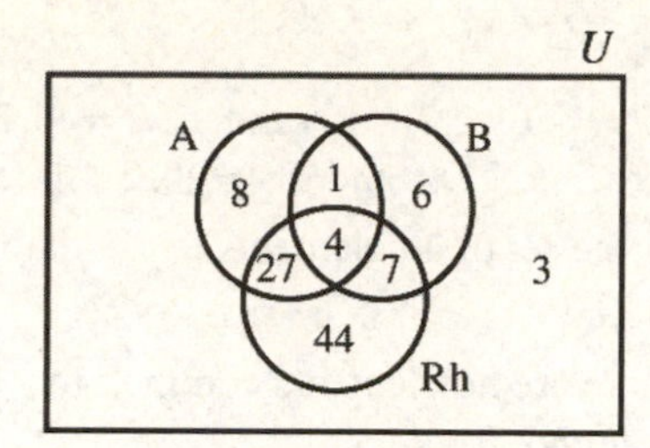

$n(A \cup B \cup \text{Rh}) = 8 + 1 + 6 + 4 + 27 + 7 + 44 = 97$. Thus the set of people who are O-negative is $100 - 97 = \boldsymbol{3}$.

21. The following Venn diagram helps in isolating the choices:

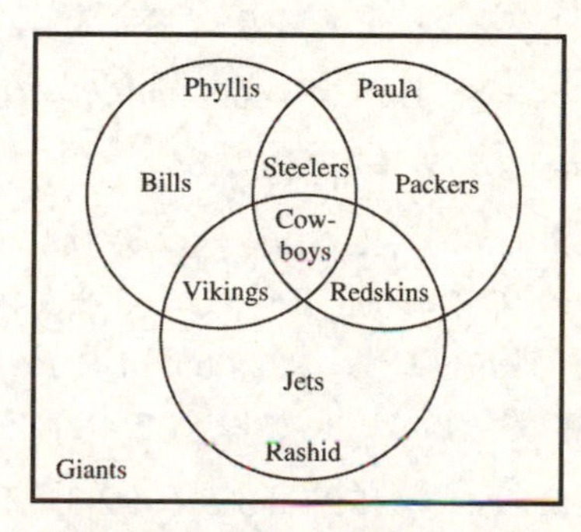

All picked the Cowboys to win their game, so their opponent cannot be among any of the other choices; the only team not picked was the Giants.

Phyllis and Paula both picked the Steelers, so their opponent cannot be among their other choices. This leaves the Jets.

Phyllis and Rashid both picked the Vikings which leaves the Packers as the only possible opponent.

Paula and Rashid both picked the Redskins which leaves the Bills as the only possible opponent.

Thus we have **Cowboys vs Giants, Vikings vs Packers, Redskins vs Bills, and Jets vs Steelers.**

23. (a) The first element of each ordered pair is a, so $\boldsymbol{C} = \{\boldsymbol{a}\}$. The second elements in the ordered pairs are, respectively, $b, c, d,$ and e, so $\boldsymbol{D = \{b, c, d, e\}}$.

(b) The first element in the first three ordered pairs is 1; in the second three is 2, so $\boldsymbol{C = \{1, 2\}}$. The second element in the ordered pairs is, respectively, 1, 2, and 3, so $\boldsymbol{D = \{1, 2, 3\}}$.

(c) The numbers 0 and 1 appear in each ordered pair, so $\boldsymbol{C = D = \{0, 1\}}$. (The order of the numbers in these sets is irrelevant.)

Assessment 2-3B

1. Given $W = \{0, 1, 2, 3, \ldots\}$ and $N = \{1, 2, 3, \ldots\}$ then $A = \{1, 3, 5, 7, \ldots\}$ and $B = \{0, 2, 4, 6, \ldots\}$:

(a) $\boldsymbol{B}$, or the set of all elements in W that are not in A.

(b) $\boldsymbol{\varnothing}$. There are no elements common to both A and B.

(c) $\boldsymbol{N}$. All elements are common to both except 0.

3. (a) False. Let $A = \{1, 2\}$ and $B = \{2\}$. Then $A - B = \{1\}$, but $A - \varnothing = \{1, 2\}$.

(b) False. Let $U = \{1, 2, 3\}, A = \{1, 2\}$, and $B = \{2, 3\}$. Then $A \cup B = \{1, 2, 3\}$ and $\overline{A \cup B} = \varnothing$. But $\overline{A} = \{3\}$ and $\overline{B} = \{1\}$, making $\overline{A} \cup \overline{B} = \{1, 3\}$.

(c) False. Let $A = \{1, 2\}, B = \{2, 3\}$, and $C = \{1, 2, 4\}$. Then $B \cup C = \{1, 2, 3, 4\}$ $\Rightarrow A \cap (B \cup C) = \{1, 2\}$ but $A \cap B$ $= \{2\} \Rightarrow (A \cap B) \cup C = \{1, 2, 4\}$.

(d) False. Let $A = \{1, 2\}$ and $B = \{2, 3\}$. Then $A - B = \{1\} \Rightarrow (A - B) \cap A = \{1\}$, but $A = \{1, 2\}$.

(e) False. Let $A = \{1, 2, 3\}, B = \{1, 2, 4\}$, and $C = \{1, 2, 3, 4\}$. Then $B \cap C = \{1, 2, 4\}$, $A - B = \{3\}$, and $A - C = \varnothing$. Thus $A - (B \cap C) = \{3\}$ but $(A - B) \cap (A - C) = \varnothing$.

5. (a) $A \cap \overline{C}$

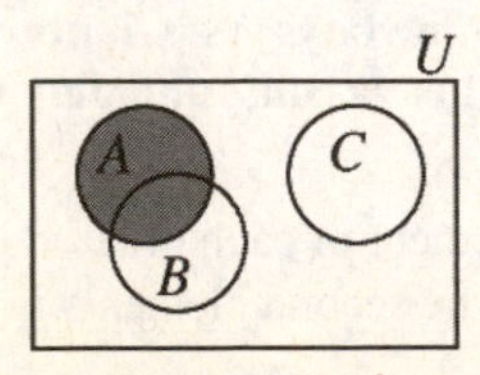

(b) $\overline{A \cup B}$

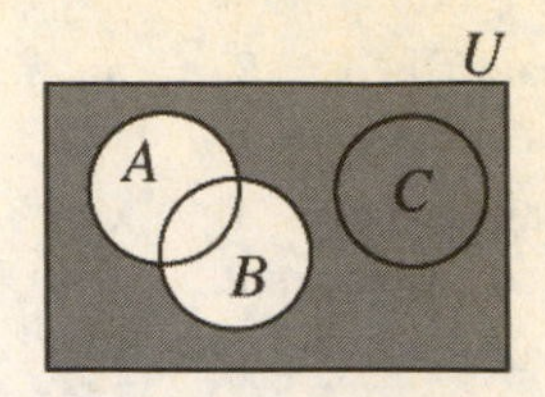

(c) $(A \cap B) \cup (B \cap C)$

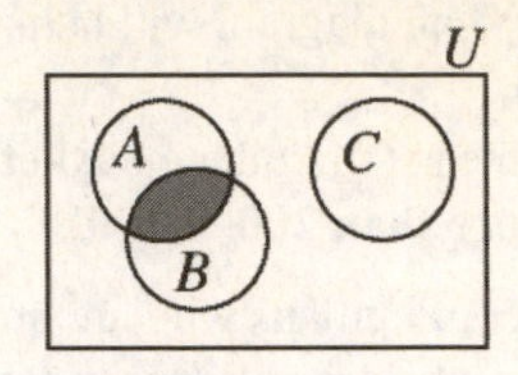

(d) $A \cup (B \cap C)$

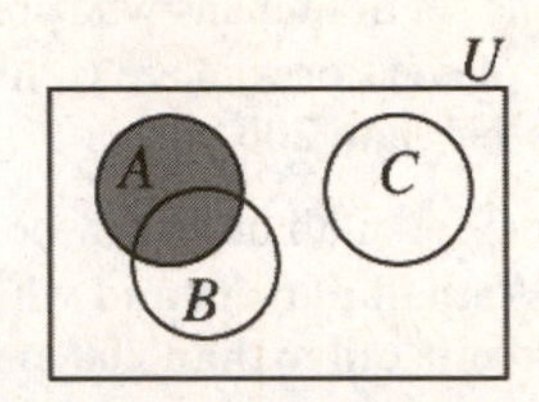

(e) $A \cup \overline{(B \cap C)}$

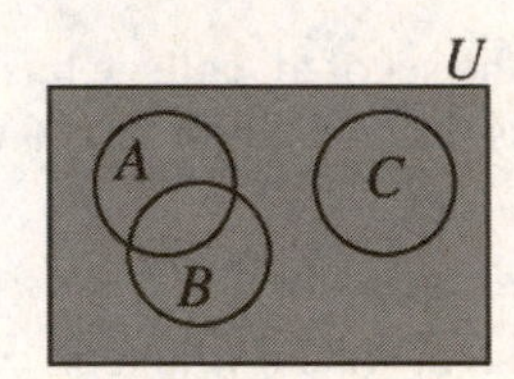

7. (a) If $A = B$ there are no elements in one set which are not also in the other, so $B - A = \varnothing$.

(b) If $B \subseteq A$ then all elements of B must also be in A, so $B - A = \{x | x \in B$ and $x \notin A\} = \varnothing$.

9. Answers may vary.

(a) $\boldsymbol{A \cap C}$; i.e., $\{x | x \in A \text{ and } C\}$.

(b) $\boldsymbol{(A \cup B) \cap C}$ or $\boldsymbol{C - \overline{(A \cup B)}}$ or $\boldsymbol{(A \cap C) \cup (B \cap C)}$; i.e., $\{x | x \in A \text{ or } B, \text{ and } x \in C\}$.

(c) In the diagram, the elements in the shaded region are in B or C but not in A. Thus, the shaded region is $\boldsymbol{(B \cup C) - A}$ **or** $\boldsymbol{(B \cup C) \cap \overline{A}}$

11. (a) False.

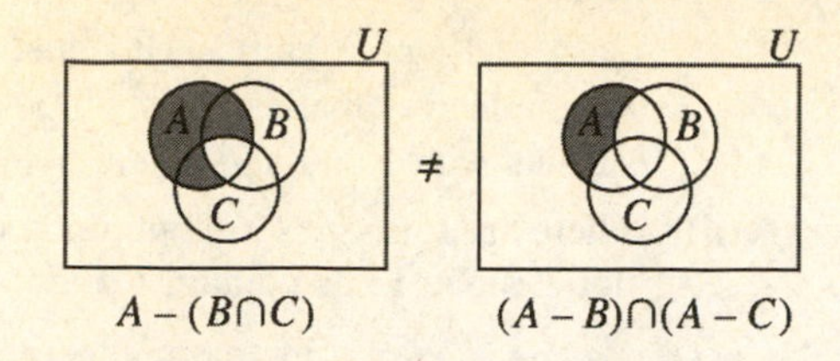

(b) **False.**

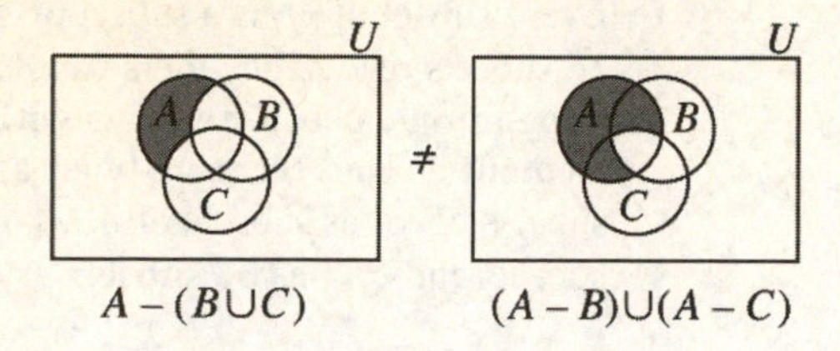

13. **(a)** Use a Venn diagram:

(i) Enter 8 as $n(A \cap B)$;

(ii) $n(B) = 12$, but 8 of these are in $A \cap B$, so there are 4 elements in B but not A;

(iii) $n(A \cup B) = 22$, but 12 are accounted for so there are 10 elements in A but not in B; so

(iv) $\mathbf{n(A) = 10 + 8 = 18}$.

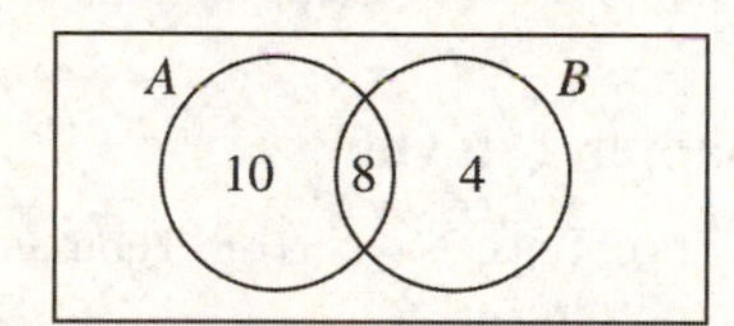

(b) Use a Venn diagram:

(i) Enter 5 as $n(A \cap B)$;

(ii) $n(A) = 8$, but 5 of these are in $A \cap B$, so there are 3 elements in A but not B;

(iii) $n(B) = 14$, but 5 of these are in $A \cap B$, so there are 9 elements in B but not in A, so;

(iv) $\mathbf{n(A \cup B) = 3 + 5 + 9 = 17}$.

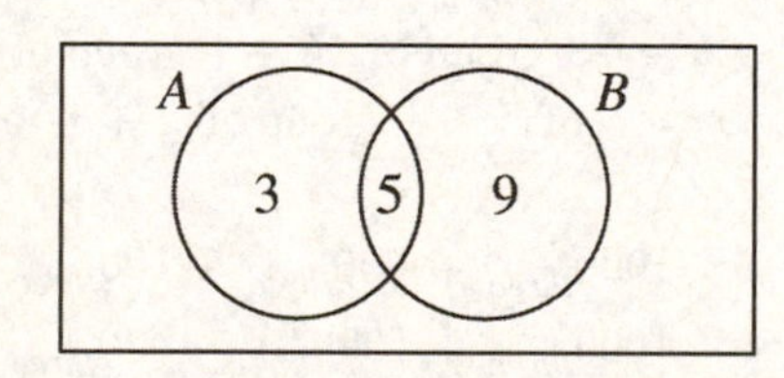

15. **(a)** The set of all Paxson 8th graders who are **members of the band but not the choir,** or $B - C$.

(b) The set of all Paxson 8th graders who are **members of both the band and the choir,** or $B \cap C$.

(c) The set of all Paxson 8th graders who are **members of the choir but not the band,** or $C - B$.

(d) The set of all Paxson 8th graders who are **neither members of the band nor of the choir,** or $\overline{B \cup C}$.

17. In the Venn diagram below:

(i) "I" is the only letter contained in the set $A \cap B \cap C$ (i.e., the only letter common to *Iowa*, *Hawaii*, and *Ohio*);

(ii) "W" and "A" are the only letters contained in the set $(A \cap B) - C$ (i.e., the letters contained in both *Iowa* and *Hawaii* other than "I");

(iii) "O" is the only letter contained in the set $(A \cap C) - B$ (i.e., the letter contained in both *Iowa* and *Ohio* other than "I");

(iv) "H" is the only letter contained in the set $(B \cap C) - A$ (i.e., the letter contained in both *Hawaii* and *Ohio* other than "I");

(v) "T", "S", "N", and "G" are the letters in *Washington* not used in *Iowa*, *Hawaii*, or *Ohio*.

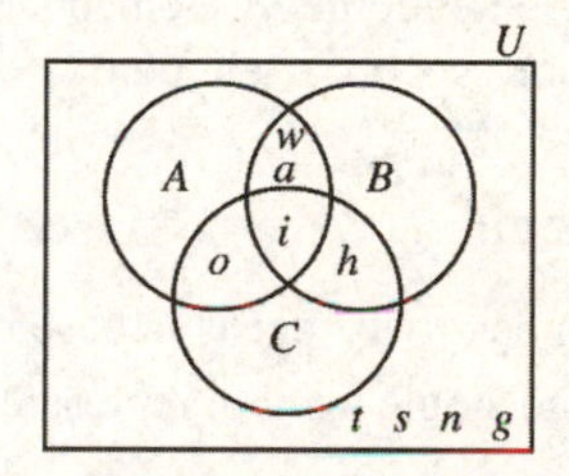

19. Generate the following Venn diagram in this order:

(i) The 50 who used all three cards;

(ii) The 60 who used only Super and Thrift cards;

(iii) The 70 who used only Gold and Thrift cards;

(iv) The 80 who used only Gold and Super cards;

(v) The $240 - (80 + 50 + 70) = 40$ who used only Gold cards;

(vi) The $290 - (80 + 50 + 60) = 100$ who used only Super cards; and

(vii) The $270 - (70 + 50 + 60) = 90$ who used only Thrift cards.

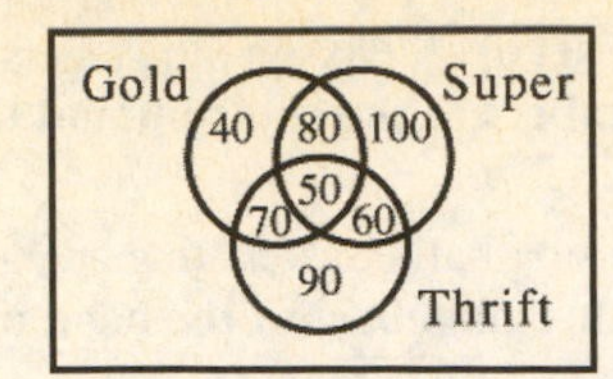

The diagram indicates only 490 cardholders accounted for, so either there is some other type credit card used by the remaining ten people or **the editor was right**.

21. (a) It is possible that no student took both algebra and biology. **0**

(b) It is possible that all **30** biology students also took algebra.

(c) If all 30 biology students also took algebra, then the numbers of students who took neither is $150 - 90 = \mathbf{60}$ students.

23. (a) Each of the five elements in A are paired with each of the four in B so there are $5 \cdot 4 = \mathbf{20}$ **elements.**

(b) Each of the m elements in A are paired with each of the n in B so there are $\boldsymbol{m \cdot n}$ **elements.**

(c) $A \times B$ has $m \cdot n$ elements, each of which are paired with the p elements in C so there are $\boldsymbol{m \cdot n \cdot p}$ **elements.**

Review Problems

15. **No**. The number "two" exists in base two as 10_{two} but there is no single symbol representing "two."

17. (a) Set-builder notation describes the elements of the set, rather than listing them; i.e., $\{\boldsymbol{x} | \boldsymbol{x} \in N \textbf{ and } \mathbf{3} < \boldsymbol{x} < \mathbf{10}\}$, where N represents the set of natural numbers, allows the set to be built.

(b) This set has only three elements; i.e., $\{\mathbf{15, 30, 45}\}$ thus the elements may easily be listed.

19. (a) These are all the subsets of $\{2,3,4\}$. There are $2^3 = \mathbf{8}$ such subsets.

(b) **8**. Every subset either contains 1 or it does not, so exactly half the $2^4 = 16$ subsets contain 1.

(c) **Twelve** subsets. There are 4 subsets of $\{3,4\}$. Each of these subsets can be appended with 1, 2, or 1 and 2 to each. By the Fundamental Counting Principle then, there are $3 \cdot 4 = 12$ possibilities. (It is also possible to simple systematically list all possible the subsets with 1 or 2 and count them.)

(d) There are **four** subsets containing neither 1 nor 2, since 12 do contain 1 or 2.

(e) **16**. B has $2^5 = 32$ subsets; half contain 5 and half do not.

(f) Every subset of A is a subset of B. List all 16 subsets of A. The others can be listed by **appending each subset of** $\boldsymbol{A}$ **with the element 5**. Thus there are twice as many subsets of B as subsets of A. A has 16 subsets and $\boldsymbol{B}$ **has 32 subsets.**

21. Answers may vary. Some possibilities:

(a) The set of basketball players on a court with the set of numbers from 1 to 10.

(b) The set of letters in the English alphabet with the set of letters in the Greek alphabet.

23. The number of combinations is equivalent to the Cartesian product of $\{SLACKS\}$, $\{SHIRTS\}$, and $\{SWEATERS\}$ so the number of elements is $n(\{SLACKS\}) \cdot n(\{SHIRTS\}) \cdot n(\{SWEATERS\}) = 4 \cdot 5 \cdot 3 = \mathbf{60}$ **combinations.**

Chapter 2 Review

1. (a) Tens. **(b)** Thousands.

(c) Hundreds.

3. (a) CM → 900; XC → 90; IX → 9, or $999 = $ **CMXCIX.**

(b) Eight groups of 10 and 6 units of one, or ∩∩∩∩∩∩∩∩||||||.

(c) Six units of 20 and three units of 1, or .

(d) 2 groups of 125 + 3 groups of 25 + 4 groups of 5 + 1, or $\mathbf{2341}_{five}$.

(e) 1 group of 16 + 1 group of 8 + 0 groups of 4 + 1 group of 2 + 1, or $\mathbf{11011}_{two}$.

5. $1000_{three} + 200_{three} + 100_{three} + 20_{three} = 1000_{three} + 1000_{three} + 20_{three} = \mathbf{2020}_{three}$.

7. (a) 123_{four}:

(b) 24_{five}:

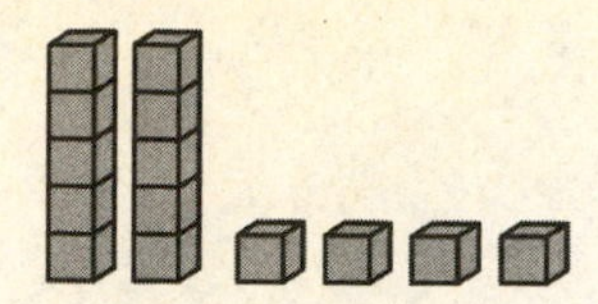

9. **(a)** $10^{10} + 23 = 1 \cdot 10^{10} + 0 \cdot 10^9 + \cdots + 2 \cdot 10 + 3$, or **10,000,000,023**.

(b) $2^{10} + 1 = 1 \cdot 2^{10} + 0 \cdot 2^9 + \cdots + 1$, or $\mathbf{10{,}000{,}000{,}001}_{two}$.

(c) $5^{10} + 1 = 1 \cdot 5^{10} + 0 \cdot 5^9 + \cdots + 1$, or $\mathbf{10{,}000{,}000{,}001}_{five}$.

(d)

$$\begin{array}{r} 1\;0\;0\;0\;0\;0\;0\;0\;0\;0\;0 \\ -\qquad\qquad\qquad\qquad\quad 1 \\ \hline 9\;9\;9\;9\;9\;9\;9\;9\;9\;9 \end{array}$$

(e) 10 must be borrowed from each 0 digit in the subtrahend, starting with the rightmost, just as with base ten.

$$\begin{array}{rl} 1\;0\;0\;0\;0\;0\;0\;0\;0\;0\;0 & _{two} \\ -\qquad\qquad\qquad\qquad\quad 1 & \\ \hline 1\;1\;1\;1\;1\;1\;1\;1\;1\;1 & _{two} \end{array}$$

(f)

$$\begin{array}{rl} 1\;0\;0\;0\;0\;0 & _{twelve} \\ -\qquad\qquad 1 & \\ \hline E\;E\;E\;E\;E & _{twelve} \end{array}$$

11. **(a)** The Egyptian system had seven symbols. It was a tally system and a grouping system, and it used the additive property. It did not have a symbol for zero which was not important inasmuch as they did not use place value.

(b) The Babylonian system used only two symbols. It was a place value system (base 60) and was additive within the positions. It lacked a symbol for zero until around 300 B.C.

(c) The Roman system used seven symbols. It was additive, subtractive, and multiplicative. It did not have a symbol for zero.

(d) The Hindu-Arabic system uses ten symbols. It uses place value involving base ten and has a symbol for zero.

13. Place value is determined by powers of each base.

(a) $2^{10} + 2^3 = \mathbf{10000001000}_{two}$.

(b) $11 \cdot 12^5 + 10 \cdot 12^3 + 20 = 11 \cdot 12^5 + 10 \cdot 12^3 + 1 \cdot 12 + 8 = \boldsymbol{E0T018}_{twelve}$.

15. A set with n elements has 2^n subsets. $2^4 = 16$. This includes A. There are $2^4 - 1 =$ **15 proper subsets**.

17. **(a)** $A \cup B = \{r,a,v,e\} \cup \{a,r,e\} = \{r,a,v,e\} = A$.

(b) $C \cap D = \{l,i,n,e\} \cap \{s,a,l,e\} = \mathbf{\{l,e\}}$.

(c) $\overline{D} = \overline{\{s,a,l,e\}} = \mathbf{\{u,n,i,v,r\}}$.

(d) $A \cap \overline{D} = \{r,a,v,e\} \cap \{u,n,i,v,r\} = \mathbf{\{r,v\}}$.

(e) $\overline{B \cup C} = \overline{\{a,r,e\} \cup \{l,i,n,e\}}$
$= \overline{\{a,e,i,l,n,r\}}$
$= \mathbf{\{s,u,v\}}$.

(f) $B \cup C = \{a,e,i,l,n,r\} \Rightarrow \{B \cup C\} \cap D$
$= \{a,e,i,l,n,r\} \cap \{s,a,l,e\}$
$= \mathbf{\{a,l,e\}}$.

(g) $\overline{A} \cup B = \{u,n,i,s,l\} \cup \{a,r,e\}$
$= \{u,n,i,e,r,s,a,l\}$.
$C \cap \overline{D} = \{l,i,n,e\} \cap \{u,n,i,v,r\} = \{i,n\}. \Rightarrow (\overline{A} \cup B) \cap (C \cap \overline{D}) = \mathbf{\{i,n\}}$.

(h) $(C \cap D) \cap A = (\{l,i,n,e\} \cap \{s,a,l,e\}) \cap \{r,a,v,e\} = \{l,e\} \cap \{r,a,v,e\} = \mathbf{\{e\}}$.

(i) $n(B - A) = \mathbf{0}$.

(j) $n(\overline{C}) = n\{u,v,r,s,a\} = \mathbf{5}$.

(k) $n(C \times D) = 4 \cdot 4 = \mathbf{16}$. Each of the four elements in C can be paired with each of the four in D.

19. Since all 7 letters are distinct, consider seven "slots" in which to put the letters. There are 7 letters which could go in the first slot, then 6 left which could go in the second slot, and so on. Assuming any ordered arrangement of seven distinct letters is a word, the number of possible words is then $7 \cdot 6 \cdot 5 \cdot 4 \cdot 3 \cdot 2 \cdot 1 = \mathbf{5040}$.

21. $A \cap (B \cup C) \neq (A \cap B) \cup C$

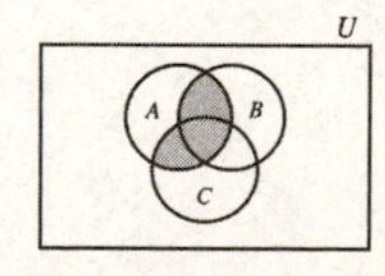

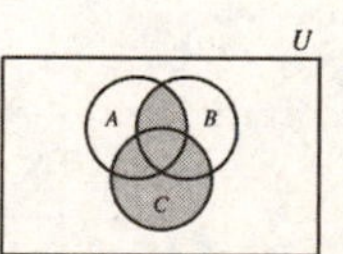

23. Answers may vary. Possibilities are:

(a) The shaded areas show $\boldsymbol{B \cup (A \cap C)}$

(b) The shaded areas show $\boldsymbol{B - C}$ or $\boldsymbol{B \cap \overline{C}}$

25. **(a)** **True**. Venn diagrams show that $A - B$, $B - A$, and $A \cap B$ are all disjoint sets, so $n(A - B) + n(B - A) + n(A \cap B) = n(A \cup B)$.

(b) **True**. Venn diagrams show that $A - B$ and B are disjoint sets, so $n(A - B) + n(B) = n(A \cup B)$. Likewise, Venn diagrams show that $(B - A)$ and A are disjoint, so $n(B - A) + n(A) = n(A \cup B)$.

27. $n(\text{Crew}) + n(\text{Swimming}) + n(\text{Soccer}) = 57$. The 2 lettering in all three sports are counted three times, so subtract 2 twice, giving 53. $n(\text{Awards}) = 46$, so $53 - 46 = 7$ were counted twice; i.e., **7 lettered in exactly two sports**.

29. $A \times B \times C$ is the set of ordered triples (a,b,c), where $a \in A$, $b \in B$, and $c \in C$. There are 3 possibilities for a, the first entry, 4 possibilities for the second, and 2 for the third. So $n(A \times B \times C) =$ **24**.

31. **(a)** Let $A = \{1,2,3,...,13\}$ and $B = \{1,2,3\}$. $B \subset A$, so B has fewer elements than A. Then $n(B) < n(A)$ and thus $3 < 13$.

(b) Let $A = \{1,2,3,...,12\}$ and $B = \{1,2,3,...,9\}$. $B \subset A$, so A has more elements than B. Then $n(A) > n(B)$ and thus $12 > 9$.

CHAPTER 3

WHOLE NUMBERS AND THEIR OPERATIONS

Assessment 3-1A: Addition and Subtraction of Whole Numbers

1. (a) **True**. $n(A) = 3; n(b) = 2; n(a) + n(B) = 5$.
$A \cup B = \{a,b,c,d,e\}; n(A \cup B) = 5$.
$\boldsymbol{n(A) + n(B) = n(A \cup B)}$ because the sets are disjoint.

(b) **False**. $n(A) = 3; n(B) = 2; n(A) + n(B) = 5$.
$A \cup B = \{a, b, c\}; n(A \cup B) = 3$.
$\boldsymbol{n(A) + n(B) \neq n(A \cup B)}$ because the sets are not disjoint.

(c) **True**. $n(A) = 3; n(B) = 0; n(A) + n(B) = 3$.
$A \cup B = \{a,b,c\}; n(A \cup B) = 3$.
$\boldsymbol{n(A) + n(B) = n(A \cup B)}$ because the sets are disjoint.

3. If A and B are not disjoint:
Let $A = \{1, 2\}, B = \{2, 3\}$, so $A \cup B = \{1, 2, 3\}$.
Then $n(A) = 2, n(B) = 2, n(A \cup B) = 3$.
But $n(A) + n(B) = 4 \neq 3 = n(A \cup B)$.

5. (a) **Closed**. $0 + 0 = 0$, and $0 \in \{0\}$.

(b) **Closed**. Assuming the arithmetic sequence $0, 3, 6, \ldots$, any element of T added to any other element in T is whole and divisible by 3.

(c) **Closed**. N is the set of natural numbers and any natural number added to any other natural number is an element of N.

(d) **Not Closed**. $3 + 7 \notin \{3,5,7\}$.

(e) **Closed**. $\{W\}$ is the set of whole numbers and any whole number greater than 10 added to any other whole number greater than 10 is an element of W.

(f) **Not Closed.** $1 + 1 \notin \{0,1\}$.

7. (a) 1, commutative property of addition.

(b) 7, commutative property of addition.

(c) 0, additive identity.

(d) 7, associative and commutative properties of addition.

9. **No**. If $k = 0$ (a whole number) and $a = b = 0$, then the $a = b + k \Rightarrow a < b$, which would imply that $0 < 0$, a contradiction.

11. (a) Each term is found by adding 5 to the previous term. Thus the next three are $28 + 5 = \mathbf{33}$, $33 + 5 = \mathbf{38}$, $38 + 5 = \mathbf{43}$.

(b) Each term is found by subtracting 7 from the previous term. Thus the next three are $63 - 7 = \mathbf{56}$, $56 - 7 = \mathbf{49}$, $49 - 7 = \mathbf{42}$.

13. Since the way the domino is positioned doesn't matter, i.e., is the same domino as , each number put on the right side gets paired with each of the 9 choices for the right

Number printed on left	Number of choices for right
0	9
1	8
2	7
3	6
4	5
5	4
6	3
7	2
8	1

Sum the right column and we have $\frac{10(9)}{2} = 45$ dominos.

15. (a) Add 7 to each side: $\mathbf{9 = 7 + x}$.

(b) Add 6 to each side: $\mathbf{x = 6 + 3}$.

(c) Add x to each side: $\mathbf{9 = x + 2}$.

17. (a) By the associative property of addition, $x + (y + z) = (x + y) + z$. By the commutative property of addition, $(x + y) + z = z + (x + y)$.

(b) (There is a possible typo in the text.)
By the commutative property of addition,
$x + (y + z) = (y + z) + x$.

By the associative property of addition,
$(y + z) + x = y + (z + x)$.

By the commutative property of addition,
$y + (z + x) = y + (x + z)$.

19. (a) $3 + (4 + 7) = (3 + x) + 7$
$(3 + 4) + 7 = (3 + x) + 7$
$3 + 4 = 3 + x$
$\mathbf{4 = x}$.

(b) $8 + 0 = x \Rightarrow \mathbf{x = 8}$.

(c) $5 + 8 = 8 + x$
$5 + 8 = x + 8$
$\mathbf{5 = x}$.

(d) $x + 8 = 12 + 5$
$x + 8 = 17$
$x + 8 - 8 = 17 - 8$
$\mathbf{x = 9}$.

(e) $x + 8 = 5 + (x + 3)$
$x + 8 = x + 8$
$\mathbf{x = 0}$.

(f) $x - 2 = 9$
$\mathbf{x = 11}$.

(h) $x - 3 = x + 1$
$-3 = 1$
There are no solutions
in set of natural numbers.

(f) $0 + x = x + 0$
$\mathbf{x = x}$.
All natural numbers
are solutions.

Assessment 3-1B

1. (a) True. $n(A) = 2; n(B) = 2; n(A) + n(B) = 4$.
$A \cup B = \{a,b,d,e\}; n(A \cup B) = 4$.
$\mathbf{n(A) + n(B) = n(A \cup B)}$ because the sets are disjoint.

(b) False. $n(A) = 3; n(B) = 3; n(A) + n(B) = 6$.
$A \cup B = \{a,b,c,d\}; n(A \cup B) = 4$.
$\mathbf{n(A) + n(B) \neq n(A \cup B)}$ because the sets are not disjoint.

(c) True. $n(A) = 1; n(B) = 0; n(A) + n(B) = 1$.
$A \cup B = \{a\}; n(A \cup B) = 1$.
$\mathbf{n(A) + n(B) = n(A \cup B)}$ because the sets are disjoint.

3. If $A \cap B \neq \emptyset$, the elements in the intersection would be counted twice in the operation $n(A) + n(B)$. In general, $n(A \cup B) = n(A) + n(B) - n(A \cap B)$. This method subtracts the number of elements that are counted twice. Thus, $n(A) + n(B) = n(A \cup B)$ if and only if $A \cap B = \varnothing$.

5. (a) Not Closed. $1 + 1 = 2$, and $2 \notin \{0,1\}$.

(b) Closed. Assuming the arithmetic sequence $0, 4, 8, \ldots$, any element of T added to any other element in T is whole and divisible by 4.

(c) Closed. Any element of E added to any other element of E is > 5 and is thus $\in E$.

(d) Closed. W is the set of whole numbers and any whole number greater than 100 added to any other whole number greater than 100 is $\in W$.

7. (a) $3 + 4 = \mathbf{4} + 3$; **commutative property of addition.**

(b) $5 + (4 + 3) = (4 + 3) + \mathbf{5}$; **commutative property of addition**.

(c) $8 + \mathbf{0} = 8$; **identity property of addition**.

(d) $3 + (4 + 5) = (3 + \mathbf{4}) + 5$; **associative property of addition**.

(e) $3 + 4$ is a unique **whole** number; **closure property of addition**.

9. In this chapter we are only considering whole numbers a and b. So k must be a **whole** number.

11. (a) Each term is found by adding 7 to the previous term. Thus the next three are $33 + 7 = \mathbf{40}$, $40 + 7 = \mathbf{47}$, $47 + 7 = \mathbf{54}$.

(b) Each term is found by subtracting 4 from the previous term. Thus the next three are $47 - 4 = \mathbf{43}, 43 - 4 = \mathbf{39}, 39 - 4 = \mathbf{35}$.

13. (a) The number 0 can be paired with all digits from 0 to 6 (7 pairs), the number 1 can additionally be paired with all digits from 1 to 6 (6 pairs), then 2 can be paired with all digits from 2 to 6, . . ., 6 is paired with 6. The total of unique pairings is thus $7 + 6 + 5 + 4 + 3 + 2 + 1 = \mathbf{28}$.

(b) If the domino is horizontal and the sum of the dots in the right square plus the sum of the dots in the left square is known, and the domino is then turned 180°, the sum is the same.

15. (a) Add 3 to each side: $\mathbf{9 = x + 3}$.

(b) Add 5 to each side: $\mathbf{x = 8 + 5}$.

(c) Add x to each side: $\mathbf{11 = 2 + x}$.

17. (a) $a + (b + c) = (a + b) + c$: Associative property of addition.

$(a + b) + c = c + (a + b)$: Commutative property of addition.

(b) $a + (b + c) = a + (c + b)$: Commutative property of addition.

$a + (c + b) = (c + b) + a$: Commutative property of addition.

19. (a) $12 - x = x + 6$

$12 - x + x = x + x + 6$

$12 = 2x + 6$

$12 - 6 = 2x + 6 - 6$

$6 = 2x \Rightarrow \frac{6}{2} = \frac{2}{2}x \Rightarrow \mathbf{x = 3}$.

(b) $9 - x - 6 = 1$

$3 - x = 1$

$3 - x + x = 1 + x$

$3 = 1 + x$

$3 - 1 = 1 - 1 + x \Rightarrow \mathbf{x = 2}$.

(c) $3 + x = x + 3$

$x + 3 = x + 3 \Rightarrow$

All whole numbers.

(d) $11 - x = 0$

$11 - x + x = 0 + x$

$11 = x \Rightarrow \mathbf{x = 11}$

(e) $14 - x = 7 - x$

$14 - x + x = 7 - x + x$

$14 = 7 \Rightarrow$ **No solution.**

(f) $x - 3 = 17$

$x - 3 + 3 = 17 + 3$

$\mathbf{x = 20}$

(g) $x + 3 = x - 1$

$x + 3 - x = x - 1 - y$

$3 = -1 \Rightarrow$ **No solution.**

(h) $0 + x = x - 0$

$x = x \Rightarrow$ **All whole numbers.**

Assessment 3-2A
Algorithms for Whole Number Addition and Subtraction

1. (a) In the units column, $1 + \underline{1} = 2$.
In the tens column, $\underline{8} + 2 = 10$ (regroup).
In the hundreds column, $1 + \underline{9} + 4 = 14$.

$$\begin{array}{rrrr} & \underline{\mathbf{9}} & \underline{\mathbf{8}} & 1 \\ + & 4 & 2 & \underline{1} \\ \hline \underline{\mathbf{1}} & 4 & 0 & 2 \end{array}$$

(b) In the units column, $5 + 6 + 8 = 19$ (regroup).
In the tens column, $1 + 2 + \underline{9} + 4 = 16$ (regroup).
In the hundreds column, $1 + 0 + 1 + 1 = \underline{3}$.
In the thousands column, $\underline{2} + 1 + 3 = 6$.

$$\begin{array}{rrrrr} & \underline{\mathbf{2}} & 0 & 2 & 5 \\ & 1 & 1 & \underline{\mathbf{9}} & 6 \\ + & 3 & 1 & 4 & 8 \\ \hline & 6 & \underline{\mathbf{3}} & 6 & \underline{\mathbf{9}} \end{array}$$

3. (a) Answers may vary: for the unique greatest sum, the larger numbers must be in the hundreds column:

$$\begin{array}{rrrr} & \boxed{7} & \boxed{6} & \boxed{2} \\ + & \boxed{8} & \boxed{5} & \boxed{3} \\ \hline 1 & 6 & 1 & 5 \end{array}$$

(b) Answers may vary; for the unique least sum the smaller numbers must be in the hundreds column:

$$\begin{array}{rrrr} & \boxed{2} & \boxed{6} & \boxed{7} \\ + & \boxed{3} & \boxed{5} & \boxed{8} \\ \hline & 6 & 2 & 5 \end{array}$$

5. (a)
$$\begin{array}{r} 93 \\ -\ 37 \\ \hline \end{array} \Rightarrow \begin{array}{r} 93 + 3 \\ -\ 37 + 3 \\ \hline \end{array} \Rightarrow \begin{array}{r} 96 \\ -\ 40 \\ \hline 56 \end{array}$$

(b)
$$\begin{array}{r} 321 \\ -\ 38 \\ \hline \end{array} \Rightarrow \begin{array}{r} 321 + 2 \\ -\ 38 + 2 \\ \hline \end{array} \Rightarrow \begin{array}{r} 323 \\ -\ 40 \\ \hline \end{array}$$

$$\Rightarrow \quad \begin{array}{r} 323 + 60 \\ -\quad 40 + 60 \\ \hline \end{array} \Rightarrow \quad \begin{array}{r} 383 \\ -\ 100 \\ \hline 283 \end{array}$$

7. Answers may vary; some possibilities are:

(a) $8 + 5 = 13$ and 13 was written down with no regrouping. $2 + 7 = 9$ and 9 was simply placed in front of the 13.

(b) $8 + 5 = 13$, but instead of writing 3 and regrouping with the 1, the 1 was written and the 3 was regrouped.

(c) Only the difference in the units $(9 - 5 = 4)$, tens $(5 - 0 = 5)$, and the hundreds $(3 - 2 = 1)$ was recorded, without taking into account the signs of the numbers.

(d) Three hundreds was regrouped as 2 hundreds and 10 tens, but 10 tens was not regrouped as $9 \cdot 10 + 15$ in order to obtain $15 - 9 = 6$ in the ones place.

9. Step 1 → Expanded form;

Step 2 → Commutative and associative properties of addition;

Step 3 → Distributive property of multiplication over addition;

Step 4 → Closure property of addition

Step 5 → Expanded form condensed.

11. (a)

$$\begin{array}{rcccc} & 4 & 3 & 5 & 8 \\ + & 3 & 8 & 6 & 4 \\ \hline & 0/7 & 1/1 & 1/1 & 1/2 \\ & 8 & 2 & 2 & 2 \end{array}$$

(b)

$$\begin{array}{rcccc} & 4 & 9 & 2 & 3 \\ + & 9 & 8 & 9 & 7 \\ \hline & 1/3 & 1/7 & 1/1 & 1/0 \\ 1 & 4 & 8 & 2 & 0 \end{array}$$

13. For an example of how to use the table, move down the rows in the + column to 3 and then across that row to the column headed by 6 . This will give the sum of $3 + 6$, or 11_{eight}.

+	0	1	2	3	4	5	6	7
0	0	1	2	3	4	5	6	7
1	1	2	3	4	5	6	7	10
2	2	3	4	5	6	7	10	11
3	3	4	5	6	7	10	11	12
4	4	5	6	7	10	11	12	13
5	5	6	7	10	11	12	13	14
6	6	7	10	11	12	13	14	15
7	7	10	11	12	13	14	15	16

(a)

$$\begin{array}{rcccl} & \overset{4}{\not 5} & \overset{16}{\not 7} & \overset{13}{\not 3} & _{eight} \\ - & & 7 & 7 & _{eight} \\ \hline & 4 & 7 & 4 & _{eight}. \end{array}$$

We used the table to complete the subtraction. For example $13_{eight} - 7_{eight}$ can be found by observing that $4_{eight} + 7_{eight} = 13_{eight}$ in the table.

(b)

$$\begin{array}{rcccl} & \overset{6}{\not 7} & \overset{15}{\not 6} & \overset{15}{\not 5} & _{eight} \\ - & & 7 & 6 & _{eight} \\ \hline & 6 & 6 & 7 & _{eight} \end{array}$$

15. (a) The sum of each row, column, and diagonal is **34**. For example, the sum of the entries in row 1 is $1 + 15 + 14 + 4 =$ **34**.

(b) $6 + 7 + 10 + 11 =$ **34**.

(c) $1 + 4 + 16 + 13 =$ **34**.

(d) **Yes**. Adding five to each number in the square will increase the sum of any four numbers in the square by 20.

(e) **Yes**. Subtracting 1 from each number in the square will decrease the sum of any four numbers in the square by 4.

17. There is **no numeral 5** in base five; $2_{five} + 3_{five} = 10_{five}$; $22_{five} + 33_{five} = 110_{five}$.

19. The information in (*a*) and (*b*) complete the first column.

Teams	1
Hawks	14
Elks	18 = 14 + 4

The information in (*c*) and (*d*) complete the second column.

Teams	1	2
Hawks	14	31 = 23 + 8
Elks	18	23 = 18 + 5

The information in (*g*) and (*h*) complete the third column.

Teams	1	2	3
Hawks	14	31	36 = 2(18)
Elks	18	23	45 = 14 + 31

The information in (*f*) tells us that $14 + 22 + 36 + (4^{th}$ quarter score$) = 120$.

So the 4th quarter score for the Hawks is 39. The information in (*g*) tells us that the Elks scored $48 + 6$ points in the 4th quarter.

	Quarter				
Teams	1	2	3	4	Final
Hawks	14	31	36	39	120
Elks	18	23	45	45	131

21. **(a)** $93 + 39 = 132; 132 + 231 = 363$, which is a palindrome.

(b) $588 + 885 = 1473; 1473 + 3741 = 5214;$ $5214 + 4125 = 9339$, which is a palindrome.

(c) $2003 + 3002 = 5005$, which is a palindrome.

Assessment 3-2B

1. **(a)** In the units column, $1\underline{3} - 9 = 4$ (trade from the tens column).

In the tens column, $7 - 5 = 2$, (1 has been traded from $\underline{8}$).

In the hundreds column, $3 - 1 = \underline{2}$.

$$\begin{array}{r} 3\ \underline{\mathbf{8}}\ \underline{\mathbf{3}} \\ -\ 1\ 5\ 9 \\ \hline \underline{2}\ 2\ 4 \end{array}$$

(b) In the units column, 10 has been traded from the tens column.

In the tens column, $8 - 0 = 8$ but 1 has been traded from $\underline{9}$.

In the hundreds column, $12 - 3 = 9$ (trade 1000 from the thousands column).

In the thousands column, $12 - 8 = 4$, (1 has been traded from $\underline{3}$).

$$\begin{array}{r} 1\ \underline{\mathbf{3}}\ \underline{\mathbf{2}}\ \underline{\mathbf{9}}\ 6 \\ -\ 8\ 3\ 0\ 9 \\ \hline 4\ 9\ 8\ 7 \end{array}$$

3. If whole numbers are used:

(a)

$$\begin{array}{r} \boxed{8}\ \boxed{7}\ \boxed{6} \\ -\ \boxed{2}\ \boxed{3}\ \boxed{5} \\ \hline 6\ \ 4\ \ 1 \end{array}$$

(b)

$$\begin{array}{r} \boxed{6}\ \boxed{2}\ \boxed{3} \\ -\ \boxed{5}\ \boxed{8}\ \boxed{7} \\ \hline 3\ \ 6 \end{array}$$

5. **(a)**

$$\begin{array}{r} 86 \\ -\ 38 \\ \hline \end{array} \Rightarrow \begin{array}{r} 86 + 2 \\ -\ 38 + 2 \\ \hline \end{array} \Rightarrow \begin{array}{r} 88 \\ -\ 40 \\ \hline \mathbf{48} \end{array}$$

(b)

$$\begin{array}{r} 582 \\ -\ 44 \\ \hline \end{array} \Rightarrow \begin{array}{r} 582 + 6 \\ -\ 44 + 6 \\ \hline \end{array} \Rightarrow \begin{array}{r} 588 \\ -\ 50 \\ \hline \mathbf{538} \end{array}$$

7. Answers may vary, for example:

(a) A tens digit was not regrouped as 1 ten when the sum of the units digits was more than 9.

(b) Partial sums are not in the correct place value position.

(c) The units minuend is subtracted from the units subtrahend.

(d) One tens value should have been traded from the 5 in the minuend's ten position.

9. Step 1 → Expanded form;

Step 2 → Commutative and associative properties of addition;

Step 3 → Distributive property of multiplication over addition;

Step 4 → Closure property of addition, one digit addition facts;

Step 5 → Expanded form condensed.

11. (a)

$$\begin{array}{ccccc} & 2 & 3 & 4 & 5 \\ + & 8 & 8 & 8 & 8 \\ \hline & \boxed{1/0} & \boxed{1/1} & \boxed{1/2} & \boxed{1/3} \\ 1 & 1 & 2 & 3 & 3 \end{array}$$

(b)

$$\begin{array}{ccccc} & 8 & 7 & 1 & 3 \\ + & 4 & 2 & 1 & 4 \\ \hline & \boxed{1/2} & \boxed{0/9} & \boxed{0/2} & \boxed{0/7} \\ 1 & 2 & 9 & 2 & 7 \end{array}$$

13. For an example of use of the table, moving down the rows in the + column to 3 and then across the row to the column headed by 4 will give the sum of $3 + 4$, or 11_{six}.

+	0	1	2	3	4	5
0	0	1	2	3	4	5
1	1	2	3	4	5	10
2	2	3	4	5	10	11
3	3	4	5	10	11	12
4	4	5	10	11	12	13
5	5	10	11	12	13	14

$$\begin{array}{cccc} & 1 & 12 & 1 \\ & \not{2} & \not{3} & \not{1}_{six} \\ - & 1 & 4 & 4_{six} \\ \hline & & 4 & 3_{six} \end{array}$$

Check:

$$\begin{array}{cccc} & 1 & 1 & \\ & 1 & 4 & 4_{six} \\ + & & 4 & 3_{six} \\ \hline & 2 & 3 & 1_{six} \end{array}$$

$$\begin{array}{cccc} & 2 & 13 & 12 \\ & \not{3} & \not{4} & \not{2}_{six} \\ - & 1 & 4 & 4_{six} \\ \hline & 1 & 5 & 4_{six} \end{array}$$

Check:

$$\begin{array}{cccc} & 1 & 1 & \\ & 1 & 5 & 4_{six} \\ + & 1 & 4 & 4_{six} \\ \hline & 3 & 4 & 2_{six} \end{array}$$

15. (a) Each row, column, and diagonal sums to **34**.

(b) The sum is **34**.

(c) The sum is **34**.

(d) **Yes**. All rows, columns, and diagonals still contain four numbers; adding 11 to each adds 44 to all sums and they are still equal (to 78).

(e) **Yes**. This subtracts 44 from all sums and they are still equal (to $^{-}10$).

17. There is **no numeral 6 in base six**. $23_{six} + 43_{six} = 110_{six}$.

19.

- The Hawks scored 15 points in the first quarter, so 15 goes in that block.
- The Hawks were behind by 5 points at the end of the first quarter, so 20 goes to the Elks in the first.
- The Elks scored 5 more points in the second quarter than in the first, so 25 goes in that block.
- The Hawks scored 7 more points than the Elks in the second quarter, so 32 goes to the Hawks in the second.
- The Hawks scored twice as many points in the third quarter as the Elks did in the first, so 40 goes in that block.
- The Hawks scored 120 points in the game, so $120 - (15 + 32 + 40) = 33$ goes to the Hawks in the fourth quarter.
- The Elks outscored the Hawks by 6 points in the fourth quarter, so 39 goes in that block.
- The Elks scored as many points in the third quarter as the Hawks did in the first two; $15 + 32 = 47$ goes in that block.
- The Elks scored a total of $20 + 25 + 47 + 39 = 131$ in the game. Thus:

TEAMS	**QUARTERS** 1	2	3	4	**FINAL SCORE**
Hawks	**15**	**32**	**40**	**33**	**120**
Elks	**20**	**25**	**47**	**39**	**131**

Review Problems

15. For example, $(2 + 3) + 4 = 2 + (3 + 4)$.

Assessment 3-3A Multiplication and Division of Whole Numbers

1. (a) Use the repeated addition model: $3 + 3 + 3 + 3 + 3 = 15$; there are five threes, so $3 \cdot \boxed{5} = 15$.

(b) $18 - 6 = 6 - 6 + 3 \cdot \square$ (note the order of operations specifying that $6 + 3$ is not permitted) $\Rightarrow 12 = 3 \cdot \square$. Now use the repeated addition model: $3 + 3 + 3 + 3 = 12$; there are four threes, so $18 = 6 + 3 \cdot \boxed{4}$.

(c) The distributive property of multiplication over addition, where n is **any whole number,** specifies that $a(b + c) = ab + ac$. Thus $\boxed{n} \cdot (5 + 6) = \boxed{n} \cdot \mathbf{5} + \boxed{n} \cdot \mathbf{6}$.

3. (a) **No**. $2 + 3 = 5$.

(b) **Yes**. There will be no numbers in the set that will multiply to give a product of 5.

5. (a) $\mathbf{(5 + 6) \cdot 3} = 33$. Without parentheses, the result would be 23.

(b) **No parentheses** are needed; the order of operations specifies that addition and subtraction are performed in order from left to right.

(c) **No parentheses** are needed; the order of operations specifies that division is performed before addition or subtraction.

(d) $\mathbf{(9 + 6) \div 3} = 5$. Without parentheses, the result would be 11.

7. (a) $18 \div 3 = \boxed{6}$ (since $3 \cdot 6 = 18$).

(b) $\boxed{0} \div 76 = 0$ (since $0 \cdot 76 = 0$).

(c) $28 \div \boxed{4} = 7$ (since $7 \cdot 4 = 28$).

9. (a) Illustrated is 4 groups of 2 *x*s: $\mathbf{4 \cdot 2}$.

(b) Illustrated is a 4 by 2 array or a 2 by 4 array: $\mathbf{2 \cdot 4}$ **or** $\mathbf{4 \cdot 2}$.

11. (a) **Closure property of multiplication of whole numbers;** i.e., for any whole numbers a and b, $a \cdot b$ is a *unique* whole number.

(b) **Zero multiplication property of whole numbers;** i.e., $a \cdot 0 = 0$.

(c) **Identity property of multiplication of whole numbers;** i.e., $a \cdot 1 = a$.

13. (a) $9(10 - 2) = 9 \cdot 10 - 9 \cdot 2 = 90 - 18 = \mathbf{72}$.

(b) $20(8 - 3) = 20 \cdot 8 - 20 \cdot 3 = 160 - 60 = \mathbf{100}$.

15. The question is really to show that the area of the large square minus the area of the small square is the same as the area of the four rectangles.

The area of the large square is $(a + b)^2$. The area of the small square is $(a - b)^2$. The difference between the two areas is the four rectangles, each with area ab; the total area of the four is $4ab$.

Therefore $(a + b)^2 - (a - b)^2 = 4ab$.

17. (a) ***(i)*** $(ab)c = c(ab)$ by the commutative property of multiplication of whole numbers.

(ii) $c(ab) = (ca)b$ by the associative property of multiplication of whole numbers.

(b) ***(i)*** $(a + b)c = c(a + b)$ by the commutative property of multiplication.

(ii) $c(a + b) = c(b + a)$ by the commutative property of addition.

19. (a) $40 \div 8 = 5 \Rightarrow \mathbf{40 = 8 \cdot 5}$.

(b) $326 \div 2 = x \Rightarrow \mathbf{326 = 2 \cdot x}$.

21. Answers may vary, but for example:

(a) There is no associative property in division; e.g., $(8 \div 4) \div 2 \neq 8 \div (4 \div 2)$.

(b) There is no distributive property in division; e.g., $8 \div (2 + 2) \neq (8 \div 2) + (8 \div 2)$.

23. (a) $5x + 2 = 22 \Rightarrow 5x + 2 - 2 = 22 - 2 \Rightarrow 5x = 20 \Rightarrow \frac{5}{5}x = \frac{20}{5} \Rightarrow \mathbf{x = 4}$.

(b) $3x + 7 = x + 13 \Rightarrow 3x - x + 7 = x - x + 13 \Rightarrow 2x + 7 = 13 \Rightarrow 2x + 7 - 7 = 13 - 7 \Rightarrow 2x = 6 \Rightarrow \frac{2}{2}x = \frac{6}{2} \Rightarrow \mathbf{x = 3}$.

(c) $3(x + 4) = 18 \Rightarrow 3x + 12 = 18 \Rightarrow 3x + 12 - 12 = 18 - 12 \Rightarrow 3x = 6 \Rightarrow \frac{3}{3}x = \frac{6}{3} \Rightarrow \mathbf{x = 2}$.

(d) $(x-5) \div 10 = 9 \Rightarrow \left(\frac{x-5}{10}\right)10 = 9(10) \Rightarrow$
$x - 5 = 90 \Rightarrow x - 5 + 5 = 90 + 5 \Rightarrow$
$x = \mathbf{95}$.

25. (a) $28 \div 5 = 5.6 \rightarrow 5 \times 5 = 25 \rightarrow 28 - 25 = 3$ (the remainder is **3**).

(b) $32 \div 10 = 3.2 \rightarrow 10 \times 3 = 30 \rightarrow$ $32 - 30 = 2$ (the remainder is **2**).

(c) $29 \div 3 = 9.\overline{6} \rightarrow 3 \times 9 = 27 \rightarrow 29 - 27 =$ 2 (the remainder is **2**).

(d) $41 \div 7 = 5.\overline{857142} \rightarrow 7 \times 5 = 35 \rightarrow$ $41 - 35 = 6$ (the remainder is **6**).

(e) $49382 \div 14 = 3527.2\overline{85714} \rightarrow 14 \times 3527 =$ $49378 \rightarrow 49382 - 49378 = 4$ (the remainder is **4**).

27. There were 10 teams with 12 on each team. So, there were $10(12) = 120$ people. Divide them into teams of 8 people and there are $120 \div 8 =$ **15 teams**.

29. These numbers will be multiples of 4 plus 1, $\{4m + 1 | m \in W\}$ or $\{1,5,9,13,\ldots\}$.

Assessment 3-3B

1. (a) Use the repeated addition model: $8 + 8 + 8 = 24$; there are three eights, so $8 \cdot \boxed{3} = 24$.

(b) $28 - 4 = 4 - 4 + 6 \cdot \square$ (note the order of operations indicates that multiplication is performed before addition, so that $4 + 6$ is not permitted) $\Rightarrow 24 = 6 \cdot \square$. Now use the repeated addition model: $6 + 6 + 6 + 6 = 24$; there are four sixes, so $28 = 4 + 6 \cdot \boxed{4}$.

(c) The distributive property of multiplication over addition, where n is any whole number, specifies that $a(b + c) = ab + ac$. Thus $\boxed{n} \cdot \mathbf{(8 + 6)} = \boxed{n} \cdot \mathbf{8} + \boxed{n} \cdot \mathbf{6}$.

3. (a) **No**, $1 + 1 = 2$.

(b) **Yes**, the product of two whole numbers greater than 1 is greater than one and still a whole number.

5. (a) $\mathbf{(4 + 3) \cdot 2 = 14}$. Without parentheses the result would be $4 + 6 = 10$.

(b) $\mathbf{9 \div 3 + 1 = 4}$. Parentheses are unnecessary.

(c) $\mathbf{(5 + 4 + 9) \div 3 = 6}$. Without parentheses the result would be $5 + 4 + 3 = 12$.

(d) $\mathbf{3 + 6 - 2 \div 1 = 7}$. Parentheses are unnecessary.

7. (a) $27 \div 9 = \boxed{3}$ (since $3 \cdot 9 = 27$).

(b) $\boxed{52} \div 52 = 1$ (since $1 \cdot 52 = 52$).

(c) $13 \div \boxed{1} = 13$ (since $13 \cdot 1 = 13$).

9. (a) Fred has four bags of marbles, and each bag contains three marbles. How many marbles does Fred have?

(b) Mary has three piggy banks, each with seven quarters. How many quarters does Mary have?

11. (a) **Commutative property of multiplication of whole numbers**; i.e., $a \cdot b = b \cdot a$.

(b) **Distributive property of multiplication over addition for whole numbers**; i.e., $9 \cdot 6 = 9 \cdot 5 + 9 \cdot 1 = 45 + 9 = 54$.

13. (a) $15(10 - 2) = 15 \cdot 10 - 15 \cdot 2$
$= 150 - 30 = \mathbf{120}$.

(b) $30(9 - 2) = 30 \cdot 9 - 30 \cdot 2$
$= 270 - 60 = \mathbf{210}$.

15. The first two figures on the left illustrates a^2, viewed as the area of the large square, minus b^2, viewed as the area of the small square on the bottom right. The figure on the right illustrates how the pieces can be rearranged to form a rectangle with height $a + b$ and length $a - b$.

17. (a) **(*i*)** $(ab)c = (ba)c$ by the commutative property of multiplication of whole numbers.

(*ii*) $(ba)c = b(ac)$ by the associative property of multiplication of whole numbers.

(b) **(*i*)** $a(b + c) = ab + ac$ by the distributive property of multiplication over addition.

(*ii*) $ab + ac = ac + ab$ by the commutative property of addition.

19. (a) $48 \div x = 16 \Rightarrow \mathbf{48 = x \cdot 16}$.

(b) $x \div 5 = 17 \Rightarrow \mathbf{x = 5 \cdot 17}$.

21. Answers may vary, but for example:

(a) There is no commutative property in division; e.g., $(8 \div 4) \neq (4 \div 8)$.

(b) There is no commutative property in subtraction; e.g., $8 - 4 \neq 4 - 8$.

23. (a) $5x + 8 = 28 \Rightarrow 5x + 8 - 8 = 28 - 8 \Rightarrow 5x = 20 \Rightarrow \frac{5}{5}x = \frac{20}{5} \Rightarrow \mathbf{x = 4}$.

(b) $5x + 6 = x + 14 \Rightarrow 5x - x + 6 = x - x + 14 \Rightarrow 4x + 6 = 14 \Rightarrow 4x + 6 - 6 = 14 - 6 \Rightarrow 4x = 8 \Rightarrow \frac{4}{4}x = \frac{8}{4} \Rightarrow \mathbf{x = 2}$.

(c) $5(x + 3) = 35 \Rightarrow 5x + 15 = 35 \Rightarrow 5x + 15 - 15 = 35 - 15 \Rightarrow 5x = 20 \Rightarrow \frac{5}{5}x = \frac{20}{5} \Rightarrow \mathbf{x = 4}$.

(d) $(x - 6) \div 3 = 1 \Rightarrow x - 6 = 3 \Rightarrow \mathbf{x = 9}$.

25. (a) $28 - 8(3) = \mathbf{4}$.

(b) $42 - 10(4) = \mathbf{2}$.

(c) $29 - 13(2) = \mathbf{3}$.

(d) $45 - 7(6) = \mathbf{3}$.

(e) $59382 - 14(4241) = \mathbf{8}$.

27. 8 teams $\times$ 9 players per team $=$ 72 players $\Rightarrow$ 72 players $\div$ 6 players per team $=$ **12 teams**.

29. Given any whole numbers a and $b (b \neq 0)$, there exist unique whole numbers q (quotient) and r (remainder) such that $a = bq + r \Rightarrow 5n + 3 \in W$. Those numbers are **3, 8, 13, ...**.

Review Problems

15. No. For example, $5 - 2 \neq 2 - 5$.

Assessment 3-4A Algorithms for Whole-Number Multiplication and Division

1. (a) Start with multiplication of 4_6 by 3. The _ must be 2, since 1 must be regrouped from $3 \cdot 6$ and only $3 \cdot 2 + 1 = 7$. Then $3 \cdot 4 = 12$ for 2 in the _ of the first partial product. Similar reasoning gives:

$$\begin{array}{rrrrrr} & & & 4 & \underline{2} & 6 \\ & & \times & 7 & 8 & 3 \\ \hline & & 1 & \underline{2} & 7 & 8 \\ & 3 & 4 & 0 & 8 & \\ \underline{2} & 9 & 8 & 2 & & \\ \hline 3 & 3 & 3 & 5 & \underline{5} & 8 \end{array}$$

(b) The _ in the multiplier must be 4, since only $4 \cdot 7$ gives 8 in the units place of the second partial product. Similar reasoning gives:

$$\begin{array}{rrrrrr} & & & 3 & 2 & 7 \\ & & \times & 9 & \underline{4} & 1 \\ \hline & & & 3 & 2 & 7 \\ & 1 & \underline{3} & 0 & 8 & \\ \underline{2} & 9 & \underline{4} & 3 & & \\ \hline 3 & 0 & \underline{7} & \underline{7} & 0 & 7 \end{array}$$

3. (a) On average, the U.S. consumes $310{,}000{,}000 \times 160 = 49{,}600{,}000{,}000$ lbs of sugar per year. $49{,}600{,}000{,}000 \text{ lbs} \times \frac{1 \text{ metric ton}}{2205 \text{ lbs}} \approx$ **22,494,331 metric tons/year**.

(b) $160 \text{ lbs/year} \times \frac{120 \text{ teaspoons}}{1 \text{ lb}} =$ **19,200 teaspoons/year**.

(c) $4 \text{ lbs/year} \times \frac{120 \text{ teaspoons}}{1 \text{ lb}} =$ **480 teaspoons/year**.

(d) The average American consumer $160 \frac{\text{lbs}}{\text{year}} \times \frac{454 \text{ grams}}{1 \text{ lb}} \times \frac{1 \text{ year}}{365 \text{ days}} \approx$ 199 grams/day. This is **5** times the recommended amount.

5. (a) $\mathbf{2^{100}}$ is greater. $2^{80} + 2^{80} = 2^{80}(1 + 1) = 2^{80} \cdot 2 = 2^{81} < 2^{100}$.

(b) $\mathbf{2^{102}}$ is greatest. $2^{102} = 2^2 \cdot 2^{100} > 3 \cdot 2^{100} > 2^{101} = 2 \cdot 2^{100}$.

7. (a)

$$\begin{array}{rrrrrl} & 1 & 1 & 0 & & _{two} \\ \times & & 1 & 1 & & _{two} \\ \hline & 1 & 1 & 0 & & _{two} \\ 1 & 1 & 0 & 0 & & _{two} \\ \hline \mathbf{1} & \mathbf{0} & \mathbf{0} & \mathbf{1} & \mathbf{0} & _{two} \end{array}$$

(b) Annexation can be interpreted as append. In base two, two is 10_{two}. To illustrate the property consider a three digit number in base two: $abc_{two} \cdot 10_{two} = (a \cdot 10^2_{two} + b \cdot 10_{two} +$

$c \cdot 1_{two}) \cdot 10_{two} = a \cdot 10^3_{two} + b \cdot 10^2_{two} + c \cdot 10_{two} + 0 \cdot 1_{two} = abc0_{two}$. So multiplying abc_{two} by 10_{two} "annexed" the numeral abc_{two} with a 0 in the "ones" place.

(c) In base two, 4 can be expressed as $10_{two} \cdot 10_{two}$. Thus, given what we learned in (b), multiplying by 4 in base two "annexes" a base two numeral by 00 by annexing the original number by 0 twice.

(d) $110_{two} \cdot 11_{two} = 110_{two}(10_{two} + 1_{two})$
$= 110_{two} \cdot 10_{two} + 110_{two} \cdot 1_{two}$
$= 1100_{two} + 110_{two}$
$= \mathbf{10010_{two}}$.

9. (a) 3 hrs skiing × 444 calories/hr = **1332** calories.

(b) Jane: 2 hrs × 462 calories/hr = 924 calories.

Carolyn: 3 hrs × 198 calories/hr = 594 calories.

Thus Jane burned 924 − 594 = **330** more.

(c) Lyle: 3 hrs × 708 calories/hr = 2124 calories.

Maurice: 5 hrs × 444 calories/hr = 2220 calories.

Thus Maurice burned 2220 − 2124 = **96** more.

11. Assuming the price is \$30 per \$1000 per year, there are \$50,000 ÷ \$1,000 = 50 installments of \$30. Each quarter, the cost is 50 · \$30 ÷ 4 = **\$375**.

13. (a) The division will be the least digit value and the dividend will be arranged so that the greatest is placed for the greatest place value. **3)754**

(b) The thinking in (a) is reversed. **7)345**

15. (a) Answers may vary. One such is 36 · 84 = 3024 and 63 · 48 = 3024.

(b) Let the digits be a, b, c and d. Then $(10a + b) \cdot (10c + d) = (10b + a) \cdot (10d + c) \Rightarrow 100ac + 10bc + 10ad + bd = 100bd + 10bc + 10ad + ac \Rightarrow 99ac = 99bd \Rightarrow ac = bd$.

So if $a \cdot c = b \cdot d$ then the products will always be the same when the digits are reversed (e.g., in part (a) above, $a \cdot c = b \cdot d \Rightarrow 3 \cdot 8 = 6 \cdot 4$).

17. 36 apples × 50 boxes = 1800 apples, or 600 3-apple bags. She had 18 ÷ 3 = 6 bags left over, so she sold 600 − 6 = 594 bags at \$1 per bag. Thus her profit was \$594 − \$452 = **\$142**.

19. 56 · 10

$= (5 \cdot 10 + 6) \cdot 10$	expanded form.
$= (5 \cdot 10) \cdot 10 + 6 \cdot 10$	distributive property.
$= 5(10 \cdot 10) + 6 \cdot 10$	associative property.
$= 5 \cdot 10^2 + 6 \cdot 10$	definition of a^n.
$= 5 \cdot 10^2 + 6 \cdot 10 + 0 \cdot 1$	additive identity.
$= \mathbf{560}$	place value.

21. (a) **Base nine**. In the units column 3 + 8 = 12 with the two brought down and the 1 regrouped, so $(3 + 8)_{nine} = 12_{nine}$.

(b) **Base six**. 2 · 3 produced 0 in the units column, so $(2 \cdot 3)_{six} = 10_{six}$.

23. (a) The greatest product requires the largest multiplicands which can be formed using the four numbers: 8 × 763 > 7 × 863 because 8 × 700 = 7 × 800 but 8 × 63 > 7 × 63. Thus:

```
  [7] [6] [3]
 ×        [8]
 ------------
 6  1  0  4
```

(b) The least product requires the smallest multiplicands which can be formed using the four numbers; 3 × 678 < 6 × 378 because 3 × 600 = 6 × 300 but 3 × 78 < 6 × 78. Thus:

```
  [7] [6] [8]
 ×        [3]
 ------------
 2  0  3  4
```

Assessment 3-4B

1. Start by multiplication of 4_4 by 7. The _ must be 8 since the 2 must be regrouped from 7 · 4 and only 7 · 8 + 2 yields an 8 in the tens column. Now that the factors are 484 and 327, compute the partial products in the usual way to solve for the other missing terms.

$$\begin{array}{r} 4\ \underline{8}\ 4 \\ \times\ 3\ 2\ 7 \\ \hline 3\ \underline{3}\ 8\ 8 \\ 9\ 6\ 8\ \ \\ \underline{1}\ 4\ 5\ 2\ \ \ \ \\ \hline 1\ 5\ 8\ 2\ \underline{6}\ 8 \end{array}$$

3. **(a)** Answers may vary. Assume the daily averages for taking a shower, toilet flushing, washing of hands and face, drinking, and brushing teeth, and assume three dish washes and two cookings (have cold cereal for breakfast). Then $75 + 22 + 7 + 1 + 1 + (3 \cdot 30) + (2 \cdot 18) =$ 232 liters.

(b) The hypothetical person in (a) uses slightly more water than average.

(c) About 310,000,000 people × 200 liters per day = **about 62,000,000,000 liters** per day.

5. **(a)** $2^{20} + 2^{20} = 2^{20}(1 + 1) = 2^{20} \cdot 2 = 2^{21}$, so **they are equal**.

(b) $9 \cdot 3^{30} = 3^2 \cdot 3^{30} = 3^{32}$, so $\mathbf{3^{31} < 9 \cdot 3^{30} < 3^{33}}$.

7. **(a)**

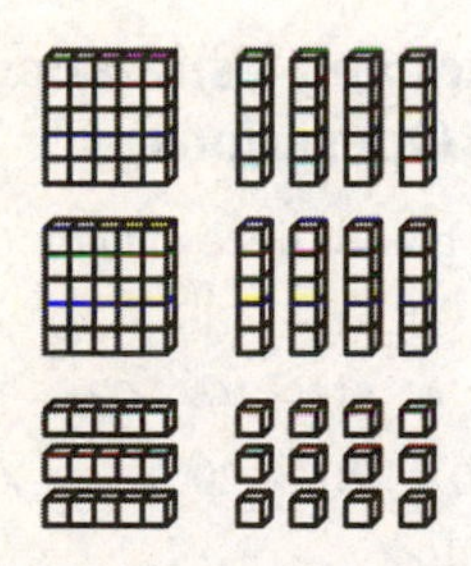

There are $2 \cdot 1$ blocks, $(2 \cdot 4 + 3 \cdot 1)$ longs, and $4 \cdot 3$ units $=$ 2 blocks $+$ (2 blocks + 1 long) + (2 longs + 2 units) = 4 blocks + 3 longs + 2 units $= \mathbf{432_{five}}$.

(b) Use a three-digit number in base five, abc_{five}.
$abc_{five} \cdot 10_{five} = (a \cdot 10^2 + b \cdot 10^1 + c \cdot 10^0)_{five} \cdot 10_{five} = (a \cdot 10^3 + b \cdot 10^2 + c \cdot 10^1)_{five} = \mathbf{abc0_{five}}$.

(c) Multiplying by 100_{five} is the same as multiplying twice by 10_{five} (i.e., $10_{five} \cdot 10_{five} = 100_{five}$). Since a zero is annexed each time a number is multiplied by 10, multiplication by 100 will result in annexation of two zeros (this is true for multiplying by 100 in any base, not just base five).

(d) $14_{five} \cdot 23_{five} = (10 + 4)_{five} \cdot 23_{five} = (10 \cdot 23 + 4 \cdot 23)_{five} = (230 + 202)_{five} = \mathbf{432_{five}}$.
Observe that in base five notation
$4_{five} \cdot 23_{five} = [4 \cdot (2 \cdot 10 + 3)]_{five}$
$= (130 + 22)_{five} = 202_{five}$.

9. **(a)** 444 calories per hour · 4 hours = **1776 calories**.

(b) Jane burned $462 \cdot 3 = 1386$ calories and Carolyn burned $198 \cdot 4 = 792$ calories. **Jane** burned more calories.

(c) Lyle burned $708 \times 4 = 2832$ calories and Maurice burned $444 \times 5 = 2220$ calories. **Lyle** burned more calories.

11. \$24/thousand × 30 thousands = \$720 annual premiums. \$720 ÷ 12 months = **\$60**/month in installment payments.

13. **(a)** The greatest quotient follows from dividing the smallest number possible into the largest number possible $(876 \div 3 = 292)$:

$\boxed{3}\ \overline{)\ \boxed{8}\ \boxed{7}\ \boxed{6}}$

(b) The least quotient follows from dividing the largest number possible into the smallest number possible $(367 \div 8 = 45$ remainder 7):

$\boxed{8}\ \overline{)\ \boxed{3}\ \boxed{6}\ \boxed{7}}$

15. Because sum of two numbers less than 10,000 is less than 20,000, ***M* must be 1**. To produce an exchange (carry), *S* must be 8 or 9. Since $S + M$ is 9 or 10, *O* must be 0 or 1. But *M* is 1, so ***O* is 0**. At this point we have

$$\begin{array}{r} S\ E\ N\ D \\ +\ \ 1\ 0\ R\ E \\ \hline 1\ 0\ N\ E\ Y \end{array}$$

and we know that S is 8 or 9.

If there is an exchange from column three to four, then $E = 9$ (since $E + 0 = E$) and the $N = 0$. But $O = 0$. So there is no exchange from column three to four and thus, $\mathbf{S = 9}$.

If there is no exchange from column two to column three, then $E = N$, which is not possible. There is a carry and thus $E = N + 1$.

We have

$$\begin{array}{rcccc} & 9 & E & E+1 & D \\ + & 1 & 0 & R & E \\ \hline 1 & 0 & E+1 & E & Y \end{array}$$

If there were no exchange from column 1, then R is 9. But S is 9. So $\mathbf{R = 8}$.

Similar reasoning results in $\mathbf{E = 5, N = 6, D = 7, \& Y = 2}$

$$\begin{array}{rrrrr} & & 1 & 1 & \\ & 9 & 5 & 6 & 7 \\ & 1 & 0 & 8 & 5 \\ \hline 1 & 0 & 6 & 5 & 2 \end{array}$$

17. $5340 − (12 months × $95 per month) = $4200 saved in the final two years. $4200 ÷ 24 months = **$175** per month.

19. $35 \cdot 100 = (3 \cdot 10 + 5) \cdot 100$ expanded form

$= (3 \cdot 10 + 5) \cdot 10^2$ definition of a^n

$= (3 \cdot 10) \cdot 10^2 + 5 \cdot 10^2$ distributive property

$= 3(10 \cdot 10^2) + 5 \cdot 10^2$ associative property of multiplication

$= 3 \cdot 10^3 + 5 \cdot 10^2 + 0 \cdot 10 + 0 \cdot 1$

$= 3500$ place value

21. (a) **Base four**. In the units column it was necessary to regroup to subtract; since $6 - 3 = 3$, a 4 must have been regrouped. Then $(4 + 2)_{ten} = 12_{four}$ and $(12 - 3)_{four} = 3_{four}$.

(b) **Any base $\geq$ two** would yield this quotient.

23. (a) The greatest product requires the largest multiplicands which can be formed using the five numbers; $83 \times 762 > 73 \times 862$ because $80 \times 700 = 70 \times 800$ but $83 \times 62 > 73 \times 62$:

$$\begin{array}{rrrrr} & & \boxed{7} & \boxed{6} & \boxed{2} \\ & & \times & \boxed{8} & \boxed{3} \\ \hline 6 & 3 & 2 & 4 & 6 \end{array}$$

(b) The least product requires the smallest multiplicands which can be formed using the five numbers; $26 \times 378 < 36 \times 278$ because $20 \times 300 = 30 \times 200$ but $26 \times 78 < 36 \times 78$:

$$\begin{array}{rrrr} & \boxed{3} & \boxed{7} & \boxed{8} \\ & \times & \boxed{2} & \boxed{6} \\ \hline 9 & 8 & 2 & 8 \\ & & 2 & 3 \\ & & 2 & 3 \\ \hline & & & 0 \end{array}$$

Review Problems

15. (a) $ax + bx + 2x = \mathbf{(a + b + 2)x}$.

(b) $3(a + b) + x(a + b) = \mathbf{(3 + x)(a + b)}$.

17. (a) $36 \div 4 = 9 \Rightarrow \mathbf{36 = 4 \cdot 9}$.

(b) $112 \div 2 = x \Rightarrow \mathbf{112 = 2x}$.

(c) $48 \div x = 6 \Rightarrow \mathbf{48 = x6}$.

(d) $x \div 7 = 17 \Rightarrow \mathbf{x = 7 \cdot 17}$.

Assessment 3-5 A
Mental Mathematics and Estimation for whole-Number Operations

1. Answers may vary; possibilities include:

(a) $180 + 97 - 23 + 20 - 140 + 26$

$= 180 + 97 + 20 + 26 - 23 - 140$

$= (180 + 90 + 20 + 20 + 7 + 6) - 23 - 140$

$= (310 + 13) - 23 - 140$

$= 323 - 23 - 140 = 300 - 140 = \mathbf{160}$.

(b) $87 - 42 + 70 - 38 + 43$

$= 87 + 70 + 43 - 42 - 38$

$= (80 + 70 + 40 + 7 + 3) - 40 - 30 - 2 - 8$

$= (190 + 10) - 70 - 10$

$= 200 - 80 = \mathbf{120}$.

3. (a) Answers may vary; possibilities include:
Compatible numbers

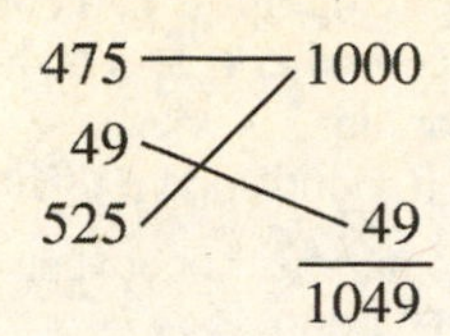

(b) Answers may vary; possibilities include:
Breaking up and bridging

$$375 - 76 = 375 - 75 - 1$$
$$= 300 - 1 = 299.$$

5. (a) $85 - 49 \Rightarrow (85 + 1) - (49 + 1) = 86 - 50 = \mathbf{36}$.

(b) $87 + 33 \Rightarrow (87 + 3) + (33 - 3) = 90 + 30 = \mathbf{120}$.

(c) $143 - 97 = (143 + 3) - (97 + 3) = 146 - 100 = \mathbf{46}$.

(d) $58 + 39 \Rightarrow (58 + 2) + (39 - 2) = 60 + 37 = \mathbf{97}$.

7. 8 hours $\times$ 62 mph $= (8 \cdot 60) + (8 \cdot 2) = 480 + 16 =$ **496 miles** (i.e., the front-end multiplying method).

9. (a) $5\underline{2}80$: The number is between 5200 and 5300;
The midpoint is 5250;
The number is greater than the midpoint;
So it rounds up to **5300**.

(b) $\underline{1}15{,}234$: The number is between 100,000 and 200,000;
The midpoint is 150,000;
The number is less than the midpoint;
So it rounds down to **100,000**.

(c) $1\underline{1}5{,}234$: The number is between 110,000 and 120,000;
The midpoint is 115,000;
The number is greater than the midpoint;
So it rounds up to **120,000**.

(d) $23\underline{2}5$: The number is between 2320 and 2330;
The midpoint is 2325;
When the number is at the midpoint it is conventional to round up (to **2330**).

11. (a) $2215 + 3023 + 5987 + 975$:

(*i*) $2 + 3 + 5 + 0 = 10$ (add front-end digits);

(*ii*) 10,000 (place value);

(*iii*) $215 + 23 + 987 + 975 \approx 200 + 0 + 1000 + 1000 = 2200$ (adjust);

(*iv*) $10{,}000 + 2200 = \mathbf{12{,}200}$ (adjusted estimate).

(b) $234 + 478 + 987 + 319 + 469$:

(*i*) $2 + 4 + 9 + 3 + 4 = 22$ (add front-end digits)

(*ii*) 2200 (place value);

(*iii*) $34 + 78 + 87 + 19 + 69 \approx 30 + 80 + 90 + 20 + 70 = 290$ (adjust);

(*iv*) $2200 + 290 = \mathbf{2490}$ (adjusted estimate).

13. (a) The range is $20 \cdot 30 = 600$ to $30 \cdot 40 = 1200$. Then $600 < (22 \cdot 38) < 1200$.

(b) The range is $100 + 600 = 700$ to $200 + 700 = 900$. Then $700 < (145 + 678) < 900$.

(c) The range is $200 + 0 = 200$ to $300 + 100 = 400$. Then $200 < (278 + 36) < 400$.

15. Estimates may vary. The range would be from $30 \cdot 20 = 600$ to $40 \cdot 30 = 1200$. Rounding, $38 \approx 35$ or 40 and $23 \approx 20$ or 25, thus about $35 \cdot 20 = 700$ seats, or $40 \cdot 25 = 1000$ seats. 700 would be low (rounded down) and 1000 would be high (rounded up).

17. (a) **False**. Sweden would be estimated to be about 44,000 square miles larger than Finland.

(b) **False**. Twice Norway's size would be about 250,000 square miles.

(c) **False**. France would be estimated to be about 85,000 square miles larger than Norway.

(d) **True**. About 195,000 square miles to about 174,000 square miles.

19. Note that in parts (b) through (d), a larger divisor produces a lower quotient when the dividend stays the same and a larger dividend produces a higher quotient when the divisor stays the same.

(a) **High**. $299 \cdot 300 < 300 \cdot 300$.

(b) **Low**. $6001 \div 299 > 6000 \div 300$.

(c) **Low**. $6000 \div 299 > 6000 \div 300$.

(d) **Low**. $10 \cdot 99 = 990 < 999 \Rightarrow 999 \div 99 > 10$.

Assessment 3-5B

1. Answers may vary; possibilities include:

(a) $160 + 92 - 32 + 40 - 18$
$= 160 + 92 + 40 - 32 - 18$
$= (160 + 90 + 40 + 2) - 32 - 18$
$= (290 + 2) - 32 - 18$
$= 292 - 32 - 18 = 292 - 50 = \mathbf{242}$.

(b) $36 + 97 - 80 + 44$
$= 36 + 97 + 44 - 80$
$= (30 + 90 + 40 + 6 + 7 + 4) - 80$
$= (160 + 17) - 80$
$= 177 - 80 = \mathbf{97}$.

3. Step 1: **Associative** property of addition.
Step 2: **Commutative** property of addition.
Step 3: **Associative** property of addition.
Step 4: **Closure property of whole-number addition.**

5. (a) $75 - 38 \Rightarrow (75 + 2) - (38 + 2) = 77 - 40 = \mathbf{37}$.

(b) $57 + 35 \Rightarrow (57 + 3) + (35 - 3) = 60 + 32 = \mathbf{92}$.

(c) $137 - 29 \Rightarrow (137 + 1) - (29 + 1) = 138 - 30 = \mathbf{108}$.

(d) $78 + 49 \Rightarrow (78 + 2) + (49 - 2) = 80 + 47 = \mathbf{127}$.

7. Methods may vary.

(a) $81 - 46 = (81 - 1) - (46 - 1) = 80 - 45 = \mathbf{35}$ (trading off).

(b) $98 - 19 \Rightarrow 19 + 1 = 20$; $20 + 70 = 90$; $90 + 8 = 98$. Then $1 + 70 + 8 = \mathbf{79}$ (cashier's algorithm).

(c) $9700 - 600 \Rightarrow 97 - 6 = 91 \Rightarrow \mathbf{9100}$ (drop the zeros).

9. (a) $3\underline{5}87$: The number is between 3500 and 3600;
The midpoint is 3550;
The number is greater than the midpoint;
So it rounds up to **3600**.

(b) $\underline{1}48{,}213$: The number is between 100,000 and 200,000;
The midpoint is 150,000;
The number is less than the midpoint;
So it rounds down to **100,000**.

(c) $2\underline{3}{,}785$: The number is between 23,000 and 24,000;
The midpoint is 23,500;
The number is greater than the midpoint;
So it rounds up to **24,000**.

(d) $23\underline{5}7$: The number is between 2350 and 2360;
The midpoint is 2355;
The number is greater than the midpoint;
So it rounds up to **2360**.

11. Estimates may vary.

(a) $2345 + 5250 + 4210 + 910$:

(*i*) $2 + 5 + 4 + 0 = 11$ (add front-end digits);

(*ii*) 11,000 (place value);

(*iii*) $345 + 250 + 210 + 910 \approx 350 + 250 + 200 + 900 = 1700$ (adjust);

(*iv*) $11{,}000 + 1700 = \mathbf{12{,}700}$ (adjusted estimate).

(b) $345 + 518 + 655 + 270$:

(*i*) $3 + 5 + 6 + 2 = 16$ (add front-end digits);

(*ii*) 1600 (place value);

(*iii*) $50 + 20 + 60 + 70 = 200$ (adjust);

(*iv*) $1600 + 200 = 1800$ (adjusted estimate).

13. Estimates may vary:

(a) The range is $30 \cdot 40 = 1200$ to $40 \cdot 50 = 2000$. Then $1200 < (32 \cdot 47) < 2000$.

(b) The range is $100 + 700 = 800$ to $200 + 800 = 1000$. Then $800 < (123 + 780) < 1000$.

(c) The range is $400 + 200 = 600$ to $500 + 300 = 800$. Then $600 < (482 + 246) < 800$.

15. About **21,000** calories. 3540 calories per pound is about 3500 calories, and $6 \cdot 3500 = 3 \cdot (2 \cdot 3500) = 3 \cdot (7000) = 21{,}000$ calories. (Actual is 21,240.)

17. (a) **Yes**. Rounding, Josh plans to write checks for about \$40, \$30, \$60, and \$250, or about \$380. Since the check amounts were rounded up, he will have enough.

(b) **Yes**. (assuming a non-negative beginning balance). $\$981 + \$1140$ is greater than $\$900 + \$1100 = \$2000$.

(c) **Alberto**. He received 10 more votes than Juan from the first district, but only 1 less from the second.

(d) **The second**. Each dimension of the second parcel is greater than the corresponding dimension of the first.

19. Note that in parts (a) through (d), a larger divisor produces a lower quotient when the dividend stays the same and a larger dividend produces a higher quotient when the divisor stays the same.

(a) **High**. $398 \cdot 500 < 400 \cdot 500$.

(b) **Low**.
$8001 \div 398 > 8001 \div 400 > 8000 \div 400$.

(c) **Low**. $10{,}000 \div 999 > 10{,}000 \div 1000$.

(d) **High**. $1999 \div 201 < 2000 \div 200$.

Review Problems

13. (a) Repeated subtraction:

$$\begin{array}{rrl}
18\overline{)623} & & \\
\underline{540} & 30 & \text{eighteens} \\
83 & & \\
\underline{72} & \underline{4} & \text{eighteens} \\
11 & 34 & \text{remainder } 11
\end{array}$$

Standard Algorithm:

$$\begin{array}{rl}
34 & \text{remainder } 11 \\
18\overline{)623} & \\
\underline{540} & \\
83 & \\
\underline{72} & \\
11 &
\end{array}$$

(b) Repeated subtraction:

$$\begin{array}{rrl}
21\overline{)493} & & \\
\underline{420} & 20 & \text{twenty ones} \\
73 & & \\
\underline{63} & \underline{3} & \text{twenty ones} \\
10 & 23 & \text{remainder } 10
\end{array}$$

Standard algorithm:

$$\begin{array}{rl}
23 & \text{remainder } 10 \\
21\overline{)493} & \\
\underline{420} & \\
73 & \\
\underline{63} & \\
10 &
\end{array}$$

(c) Repeated subtraction:

$$\begin{array}{rrl}
97\overline{)1000} & & \\
\underline{970} & \underline{10} & \text{ninety sevens} \\
30 & 10 & \text{remainder } 30
\end{array}$$

Standard algorithm:

$$\begin{array}{rl}
10 & \text{remainder } 30 \\
97\overline{)1000} & \\
\underline{970} & \\
30 &
\end{array}$$

Chapter 3 Review

1. (a) **Distributive** property of multiplication over addition.

(b) **Commutative** property of addition.

(c) **Identity** property of multiplication for whole numbers.

(d) **Distributive** property of multiplication over addition.

(e) **Commutative** property of multiplication.

(f) **Associative** property of multiplication.

3. (a) $4 \cdot \boxed{10 \le \textit{whole number} \le 15} - 37 < 27$.

(b) $398 = \boxed{10} \cdot 37 + 28$.

(c) $\boxed{n} \cdot (3 + 4) = \boxed{n} \cdot 3 + \boxed{n} \cdot 4$, where $n \in W$ is any whole number.

(d) $42 - \boxed{\textit{Any whole number} \le 26} \ge 16$.

5. 60 people $\times$ 8 ounces $= 480$ ounces required.
$480 \div 12$ ounces per can $=$ **40** twelve-ounce cans.

7. Work backward from 93 using inverse operations:
Subtract 89, giving 4;
Add 20, giving 24;
Divide by 12, giving 2; then
Multiply by 13, giving **26** as the original number.

9. 30 hours per week × \$5 per hour + 8 hours overtime × \$8 per overtime hour = **\$214.**

11. (a) Let n be the original number. Then

$$\frac{2[2(n+17)-4]+20}{4} - 20 =$$

$$\frac{4(n+17)-8+20}{4} - 20 =$$

$$\frac{4n+80}{4} - 20 = n + 20 - 20 = \mathbf{n}.$$

(b) Answers may vary. For example, if n is the original number:
$4(n+18) - 7 = 4n + 65.$

Then two more steps might be:
$4n + 65 - 65$ (subtract 65);
$\frac{4n}{4}$ (divide by 4).

(c) Answers may vary; use the techniques of parts (a) and (b).

13. Traditional:

$$\begin{array}{r} 6\;1\;3 \\ \times\quad 9\;8 \\ \hline 4\;9\;0\;4 \\ 5\;5\;1\;7 \\ \hline 6\;0\;0\;7\;4 \end{array}$$

Lattice:

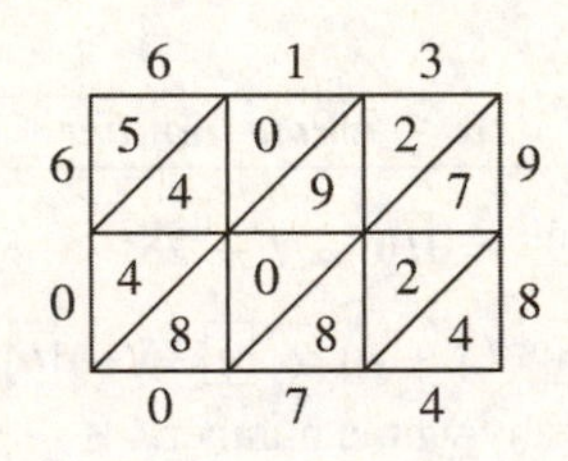

15. (a) If $4803 \div 912 = 5$ remainder 243
Then $912 \cdot 5 + 243 = 4803.$

(b) If $1011 \div 11 = 91$ remainder 10 Then $11 \cdot 91 + 10 = 1011.$

(c) If $(3312 \div 23)_{five} = (120 \text{ remainder } 2)_{five}$
Then $(23 \cdot 120)_{five} + 2_{five} = 3312_{five}.$

(d) If $(1011 \div 11)_{two} = (11 \text{ remainder } 10)_{two}$
Then $(11 \cdot 11)_{two} + 10_{two} = 1011_{two}.$

17. (\$320 × 6 mos) + (\$410 × 6 mos) = **\$4380.**

19. There are 8 groups of 3 in 24 (or two dozen) apples. 8 × 69¢ per group = 552¢ on sale; 32¢ each × 24 apples = 768¢ regular price. 768¢ − 552¢ = **216¢ saved** (or \$2.16).

21. (a)

$$\begin{array}{rl} {\scriptstyle 1\;\;1} & \\ 1\;2\;3 & _{five} \\ +\;\;3\;4 & _{five} \\ \hline \mathbf{2\;1\;2} & _{five} \end{array}$$

(b)

$$\begin{array}{rl} {\scriptstyle 10\;\;0\;\;10} & \\ \not{1}\;0\;\not{1}\;0 & _{two} \\ -\;1\;0\;1 & _{two} \\ \hline \mathbf{1\;0\;1} & _{two} \end{array}$$

(c)

$$\begin{array}{rl} 2\;3 & _{five} \\ \times\;3\;4 & _{five} \\ \hline 2\;0\;2 & \\ 1\;2\;4 & \\ \hline \mathbf{1\;4\;4\;2} & _{five} \end{array}$$

(d)

$$\begin{array}{rl} 1\;0\;0\;1 & _{two} \\ \times\;1\;0\;1 & _{two} \\ \hline 1\;0\;0\;1 & \\ 0\;0\;0\;0 & \\ 1\;0\;0\;1 & \\ \hline \mathbf{1\;0\;1\;1\;0\;1} & _{two} \end{array}$$

23.

$$\begin{array}{rl} 4_{five}\overline{)\,434_{five}} & \\ -\,400_{five} & (100 \cdot 4)_{five} \\ \hline 34_{five} & \\ -\,31_{five} & (4 \cdot 4)_{five} \\ \hline 3_{five} & \end{array}$$

Thus, $\mathbf{434_{five} = 104_{five} \cdot 4_{five} + 3_{five}}$

25. Methods may vary; for example:

(a) $63 \cdot 7 = (7 \cdot 60) + (7 \cdot 3) = 420 + 21 =$ **441** (front-end multiplying).

(b) $85 - 49 = (85 + 1) - (49 + 1) = 86 - 50 =$ **36** (trading off).

(c) $(18 \cdot 5) \cdot 2 = 18 \cdot (5 \cdot 2) = 18 \cdot 10 =$ **180** (using compatible numbers).

(d) $2436 \div 6 = (2400 \div 6) + (36 \div 6) =$ $400 + 6 =$ **406** (breaking up the dividend).

27. The addends cluster around 2400, so one would estimate the sum to be $4 \cdot 2400 = \mathbf{9600}$.

29. (a)

$$\begin{array}{rrrrrrrr} & 3x^3 & + & 4x^2 & + & 7x & + & 8 \\ + & & & 5x^2 & + & 2x & + & 1 \\ \hline & 3x^3 & + & 9x^2 & + & 9x & + & 9 \end{array}$$

(b) Answers may vary. For example:

$$\begin{array}{rrrrrrrr} & 3\cdot 10^3 & + & 5\cdot 10^2 & + & 7\cdot 10 & + & 8 \\ - & & & 4\cdot 10^2 & + & 2\cdot 10 & + & 1 \\ \hline & 3\cdot 10^3 & + & 1\cdot 10^2 & + & 5\cdot 10 & + & 7 \end{array}$$

Is equivalent (when $x = 10$) to:

$$\begin{array}{rrrrrrrr} & 3x^3 & + & 5x^2 & + & 7x & + & 8 \\ - & & & 4x^2 & + & 2x & + & 1 \\ \hline & 3x^3 & + & x^2 & + & 5x & + & 7 \end{array}$$

(c) Answers may vary. For example:

$25 \cdot 10^2 = (2 \cdot 10 + 5) \cdot 10^2 = 2 \cdot 10^3 +$

$5 \cdot 10^2 + 0 \cdot 10 + 0 = 2500.$

Is equivalent to:

$(2x + 5)x^2 = 2x^3 + 5x^2.$

Chapter 4

NUMBER THEORY

Assessment 4.1A Divisibility

1. (a) **True**. $30 = 5 \cdot 6$.
   (b) **True**. $30 \div 6 = 5$.
   (c) **True**. $2|30$ and $3|30$, so $6|30$.
   (d) **True**. $30 \div 6 = 5$.
   (e) **True**. $6 \cdot 5 = 30$.
   (f) **False**. 6 is a factor of 30, not a multiple of 30. Moreover, since $6 < 30$, 30 times another natural number cannot equal 6.

3. Use these tests for each number:
   (*i*) $2|n$ if the units digit is divisible by 2.
   (*ii*) $3|n$ if the sum of the digits is divisible by 3.
   (*iii*) $4|n$ if the last two digits are divisible by 4.
   (*iv*) $5|n$ if the units digit is 0 or 5.
   (*v*) $6|n$ if $2|n$ and $3|n$.
   (*vi*) $8|n$ if the last three digits are divisible by 8.
   (*vii*) $9|n$ if the sum of the digits is divisible by 9.
   (*viii*) $10|n$ if the units digit is 0.
   (*ix*) $11|n$ if the sum of the digits in places that are even powers of 10 minus the sum of the digits in the places that are odd powers of 10 is divisible by 11.

		2	3	4	5	6	8	9	10	11
(a)	746,988	Y	Y	Y	N	Y	N	N	N	Y
(b)	81,342	Y	Y	N	N	Y	N	Y	N	N
(c)	15,810	Y	Y	N	Y	Y	N	N	Y	N

5. (a) **True** by Theorem 4-1. For any integers a and d, if $d|a$ and n is any integer, then $d|na$.
   (b) **True** by Theorem 4-2(b). If $d|a$ and $d \nmid b$, then $d \nmid (a + b)$.
   (c) **None**. In fact, $4|1300$.
   (d) **True** by Theorem 4-2(b). Consider $a + b$ a single integer.
   (e) **True** by Theorem 4-1.

7. (a) $7|210$ because $\mathbf{210 = 7 \cdot 30}$ (definition of division).
   (b) $19|(1900 + 38)|$ because $\mathbf{19|1900}$ and $\mathbf{19|38}$. [Theorem 4-2(a)].
   (c) $6|(2^3 \cdot 3^2 \cdot 17^4)$ because $\mathbf{2^3 \cdot 3^2 \cdot 17^4} = \mathbf{(2 \cdot 3) \cdot 2^2 \cdot 3 \cdot 17^4}$ (definition of division).
   (d) $7 \nmid (4200 + 22)$ because $\mathbf{7|4200}$ but $\mathbf{7 \nmid 22}$ [Theorem 4-2(b)].

9. (a) $3|74\underline{7}$. $3|(7 + 4 + 7 = 18)$. The _ could be filled by 1, 4, or 7, but 7 is the greatest.
   (b) $9|83\underline{7}45$. $9|(8 + 3 + 7 + 4 + 5 = 27)$. Only 7 makes them viable.
   (c) $11|6\underline{6}55$. $11|(6 + 5) - (6 + 5)$. 6_55 in this case must result in the sum of the digits in places that are even powers of 10 being equal to the sum of the digits in places that are odd powers of 10.

11. (a) **True**. Break into parts of compatible number: $390026 = 390000 + 26 = 13(30000) + 13(2) = 13(30002)$.
    (b) **True**. By Theorem 4-2(b), $13 \mid 260000$ and $13 \nmid 33$. Thus $13 \nmid (260000 + 33)$.
    (c) **True**. $93^{11} = 93 \cdot 93^{10} = 31 \cdot 3 \cdot 93^{10} = 31(3 \cdot 93^{10})$. Therefore, since $3 \cdot 93^{10} \in w$, there exist a $g \in w$ such that $93^{11} = 31(g)$.
    (d) **True**. By Theorem 4-2(b), $23|690000$ and $23 \nmid 68$, so $23 \nmid (690000 + 23)$.

13. **19¢.** $209 = 11 \cdot 19$ (both prime numbers). For whole cent pricing 1¢, 11¢, 19¢, and 209¢ are possibilities, but the problem statement is for

pencils (plural) and the pencils must cost more than 12¢.

15. The term "casting out nines" implies that any 9 or sum of digits equaling 9 in n may be "cast out." The remaining digit is the remainder when n is divided by 9. Then the remainder when n is divided by 9 is the same as the remainder when the sum of the digits of n is divided by 9 ("casting out" as needed).

(a) *(i)* $12{,}343 + 4546 + 56 = 16{,}945$.

(ii) $12343 = 9 \cdot 1371 + 4$

$4546 = 9 \cdot 505 + 1$

$56 = 9 \cdot 6 + 2$

and $4 + 1 + 2 = 7$.

(iii) $16945 = 9 \cdot 1882 + 7$.

(b) *(i)* $987 + 456 + 8765 = 10{,}208$.

(ii) $987 = 9 \cdot 109 + 6$

$456 = 9 \cdot 50 + 6$

$8765 = 9 \cdot 973 + 8$

and $6 + 6 + 8 = 20; 20 = 9 \cdot 2 + 2$.

(iii) $10208 = 9 \cdot 1134 + 2$.

(c) *(i)* $10{,}034 + 3004 + 400 + 20 = 13{,}458$.

(ii) $100349 \cdot 1114 + 8$

$3004 = 9 \cdot 333 + 7$

$400 = 9 \cdot 44 + 4$

$20 = 9 \cdot 2 + 2$ and

$8 + 7 + 4 + 2 = 21; 21 = 9 \cdot 2 + 3$.

(iii) $13458 = 9 \cdot 1495 + 3$.

(d) $1003 - 46 = 957.1003$ has a remainder of 4 when divided by 9; 46 has a remainder of 1 when divided by 9; $4 - 1 = 3$ has a remainder of 3 when divided by 9. $9 + 5 + 7 = 21$ with a remainder of 3 when divided by 9.

(e) $345 \cdot 56 = 19{,}320$. 345 has a remainder of 3 when divided by 9; 56 has a remainder of 2 when divided by 9; $3 \cdot 2 = 6$ has a remainder of 6 when divided by 9. $1 + 9 + 3 + 2 + 0 = 15$ has a remainder of 6 when divided by 9.

(f) Answers may vary. The division may not have an integer quotient, in which case the test fails.

17. **(a)** **1, 3, and 7 divide *n*.**

(i) 1 divides every number.

(ii) $21|n \Rightarrow n = 21 \cdot d, d \in \mathbb{N} \Rightarrow n = (3 \cdot 7)d = 3(7d) = 7(3d)$ which implies that n is divisible by 3 and 7.

19. **(a)** **False**. $2|4$ but $2 \nmid 3$ and $2 \nmid 1$.

(b) **False**. $2|8$ but $2 \nmid 5$ and $2 \nmid 3$.

(c) **False**. $12|72$ but $12 \nmid 8$ and $12 \nmid 9$.

(d) **True**. If $ab|c$ then c must be a multiple of $a \cdot b$.

(e) **False.** If $a = 5$ and $b = -5$ then $a|b$ and $b|a$ but $a \neq b$.

Assessment 4-1B

1. **(a)** **False**. 5 is a factor of 20.

(b) **True**. $30 \div 10 = 3$.

(c) **True**. $32 \div 8 = 4$.

(d) **True**. Any integer is divisible by 1 because of the multiplication identity property.

(e) **True**. $5 \cdot 6 = 30$.

(f) **False**. No number multiplied by 20 yields 6.

3. Use these tests for each number:

(i) $2|n$ if the units digit is divisible by 2.

(ii) $3|n$ if the sum of the digit is divisible by 3.

(iii) $4|n$ if the last two digits are divisible by 4.

(iv) $5|n$ if the units digit is 0 or 5.

(v) $6|n$ if $2|n$ and $3|n$.

(vi) $8|n$ if the last three digits are divisible by 8.

(vii) $9|n$ if the sum of the digits is divisible by 9.

(viii) $10|n$ if the units digit is 0.

(ix) $11|n$ if the sum of the digits in places that are even powers of 10 minus the sum of the digits in the places that are odd powers of 10 is divisible by 11.

	2	3	4	5	6	8	9	10	11
(a) 4,201,012	Y	N	Y	N	N	N	N	N	N
(b) 1001	N	N	N	N	N	N	N	N	Y
(c) 10,001	N	N	N	N	N	N	N	N	N

5. (a) $26|(13^4 \cdot 100)$ because $13^4 \cdot 100 = 13 \cdot 2 \cdot 13^3 \cdot 50 = \mathbf{26(13^3 \cdot 50)}$ (Theorem 4-1).

(b) $13 \nmid (2^4 \cdot 5^3 \cdot 26 + 1)$ because $\mathbf{13|(2^4 \cdot 5^3 \cdot 26)}$ but $\mathbf{13 \nmid 1}$ [Theorem 4-2 (b)].

(c) $2^4 \nmid (2 \cdot 4 \cdot 6 \cdot 8 \cdot 17^{10} + 1)$ because $2^4|(1 \cdot 2 \cdot 3 \cdot 4) \cdot 2^4 \cdot 17^{10}$ but $\mathbf{2^4 \nmid 1}$ [Theorem 4-2(b)].

(d) $2^4|(10^4 + 6^4)$ because $\mathbf{2^4|10^4} = \mathbf{2^4 \cdot 5^4}$ and $\mathbf{2^4|6^4} = \mathbf{2^4 \cdot 3^4}$ [Theorem 4-2 (a)].

7. (a) $280 = 28 \cdot 10$. Since $7|28$, $7|28 \cdot 10$.

(b) $3800 = 38 \cdot 100 = 2 \cdot 19 \cdot 100$. Since $19|3800$ and $19|19$, $19|(3800 + 19)$.

(c) $2^4 \cdot 3^5 \cdot 5 = 2^4 \cdot 3^4 \cdot 3 \cdot 5 = 2^4 \cdot 3^4 \cdot 15$. Since $2^4 \cdot 3^4$ is a whole number, $15|2^4 \cdot 3^5 \cdot 5$.

(d) $37 = 1(19) + 18$. Thus $19 \nmid 37$. Since $19 \nmid 37$ and $19|3800$, $19 \nmid (3800 + 37)$.

9. (a) **16|*n* if 16|(last four digits of *n*)**. This continues the pattern of divisibility by 2, 4, and 8.

(b) **25|*n* if 25|(last two digits of *n*)**; i.e., if *n* ends in 00, 25, 50, or 75.

11. (a) **True**. $7|280000$ and $7|21$ so $7|280021$ [Theorems 4-1 and 4-2(a)]

(b) **True**. $19|3{,}800{,}000$ but $19 \nmid 18$ so $19 \nmid 3{,}800{,}018$ [Theorems 4-1 and 4-2(b)].

(c) **True**. $46^{10} = 2^{10} \cdot 23^{10} = 23(2^{10} \cdot 23^9)$ (Theorem 4-1).

(d) **True**. $23|460{,}000$ and $23 \nmid 45$, so $23 \nmid 460{,}045$ [Theorem 4-1and 4-2(a).]

(e) **True**. $23|46$. Thus, $23|460{,}000$. Therefore, $23|(460{,}000 + 46)$.

13. **85,041** The number must be divisible by 9 and 11; if it is written as $85ab1$:

$9|(8 + 5 + a + b + 1) \Rightarrow 9|(14 + a + b)$ and

$11|[(8 + a + 1) - (5 + b)] \Rightarrow 11|(4 + a - b)$

So $a + b = 4$ or 13 and $a - b = {}^-4$ or 7. Solving, and using trial and error $\Rightarrow a = 0$ and $b = 4$.

15. (a) The sum of the digits in $99 + 28$ is $9 + 9 + 2 + 8$, which leaves remainder 1 upon division by 9. However, 227 leaves the same remainder as $2 + 2 + 7$, i.e., 2 upon division by 9.

(b) 11199 – 21 leaves remainder 3 – 3 or 0 upon division by 9 (using the casting out technique). However, 11168 has remainder 8, when divided by 9.

(c) $99 \cdot 26$ divided by 9 has remainder 0, while 2575 divided by 9 has remainder 1.

17. If $16|n$, then $n = 16k$, where *k* is a whole number. Thus, $n = 2^4 \cdot k$. This shows that **1, 2, 2^2, and 2^3** also divide *n*.

19. (a) **True**. $d|a$ means $a = dk$ for some whole number *k*. Let *x* be any whole number $ax = d(kx)$. Thus, $d|ax$. Similarly, $d|by$. Thus, $d|(ax + by)$.

(b) **False**. Let $d = 2, a = 3$ and $b = 5$. $d \nmid a$ and $d \nmid b$, but $d|(a + b)$, i.e., $2|8$.

(c) **False**. $50|10^2$, but $50 \nmid 10$.

(d) **False**. $50 \nmid 10$, but $50|10^2$.

Assessment 4-2A
Prime and Composite Numbers

1. The three smallest primes are 2, 3, and 5, so the least number divisible by three different primes is $2 \cdot 3 \cdot 5 = \mathbf{30}$.

3. (a) $2|504$; the factor tree starts with $2 \cdot 252 = 504$.
$2|252$; the next level is $2 \cdot 126 = 252$.
$2|126$; the third level is $2 \cdot 63 = 126$.
$3|63$; the fourth level is $3 \cdot 21 = 63$.
$3|21$; the final level is $3.7 = 21$, since 3 and 7 are prime.

Thus the factor tree is:

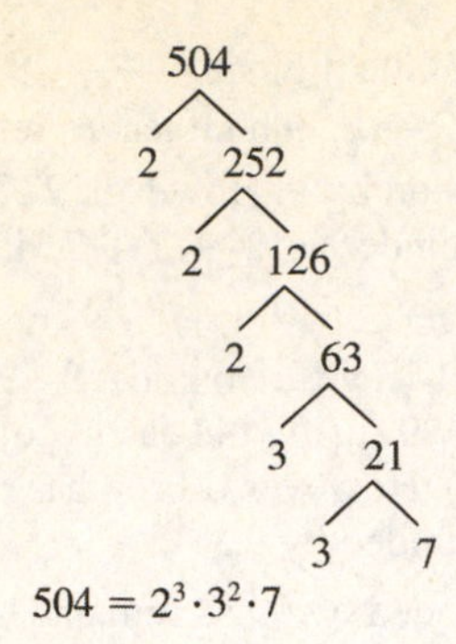

$504 = 2^3 \cdot 3^2 \cdot 7$

(b)

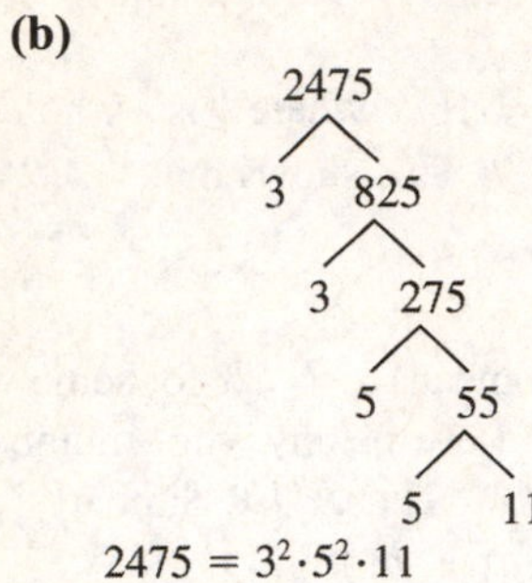

$2475 = 3^2 \cdot 5^2 \cdot 11$

(c)

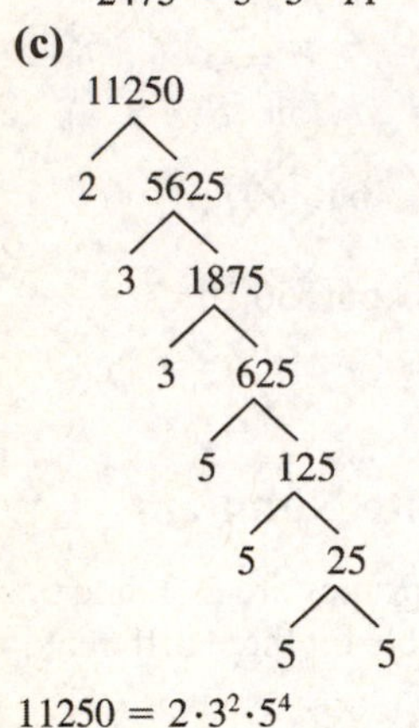

$11250 = 2 \cdot 3^2 \cdot 5^4$

5. **73**. $73^2 < 5669$, but $79^2 > 5669$ (73 is prime and 79 is the next largest prime).

7. **(a)** **$1 \cdot 48$; $2 \cdot 24$; $3 \cdot 16$; or $4 \cdot 12$**, all pairs of divisors of 48.

(b) **One**. 47 is prime, so the only possibility would be 1×47.

9. **(a)** The Fundamental Theorem of Arithmetic tells us that each composite number, n, can be written as a product of primes in one and only one way. Since $2|n$ and $3|n$, we know that 2 and 3 must both occur in ns prime decomposition. Therefore, $6|n$.

(b) **True.** If $a \mid n$ and b|n then there exist natural numbers k and l such that $n = ak$ and $n = bl$. Thus, $n^2 = ak \cdot bl = ab \cdot kl$.

11. **No**. $2 \cdot 3 \cdot 5 \cdot 7 \cdot 11 \cdot 13 + 1$ is not the product of prime numbers and so it can not be the prime factorization of 30,031. In fact, $30{,}031 = 59 \cdot 509$.

13. **(a)** This is an arithmetic sequence that can be expressed by $\{3n + 4 \mid n \in W\}$. Since there are infinitely many whole numbers of the form $4k$, the sequence has infinitely many whole numbers of the form $3(4k) + 4 = 4(3k + 1)$, which are composite since they are divisible by 4. Alternative: Every other number is divisible by 2.

(b) Notice that 111 is divisible by 3 since the sum of its digits is divisible by 3. Notice every third term will have this property.

15. **Yes**. $7^5 \cdot 11^3 = 7(7^4 \cdot 11^3)$.

17. **(*i*)** $2^3 \cdot 3^2 \cdot 25^3$ is not a prime factorization because 25 is not prime.

(*ii*) $25^3 = (5^2)^3 = 5^6$, so **prime factorization** is $2^3 \cdot 3^2 \cdot 5^6$.

19. **(a)** **1, 2, 3, 4, 6, 9, 12, 18, or 36** rows. The prime factorization of 36 is $2^2 \cdot 3^2$, so there are $(2 + 1) \cdot (2 + 1) = 9$ divisors. The divisors of 2^2 are $2^0 = 1, 2^1 = 2$, and $2^2 = 4$ while the divisors of 3^2 are $3^0 = 1, 3^1 = 3$, and $3^2 = 9$. The divisors of 36 are all possible combinations of these two lists of divisors.

(b) **1, 2, 4, 7, 14, or 28** rows. The prime factorization 28 is $2^2 \cdot 7$, so there are $(2 + 1) \cdot (1 + 1) = 6$ divisors.

(c) **1 or 17** rows. 17 is prime, so there are only two divisors.

(d) **1, 2, 3, 4, 6, 8, 9, 12, 16, 18, 24, 36, 48, 72, or 144** rows. The prime factorization of 144 is $2^4 \cdot 3^2$, so there are $(4 + 1) \cdot (2 + 1) = 15$ divisors.

Assessment 4-2B

1. $2 \cdot 3 \cdot 5 \cdot 7 =$ **210**.

3. **(a)** 2|304; the factor tree starts with $2 \cdot 152 = 304$.

2|152; the next level is $2 \cdot 76 = 152$.

2|76; the next level is $2 \cdot 38 = 76$.

2|76; the final level is $2 \cdot 19 = 38$, since 19 is prime.

Thus the factor tree is:

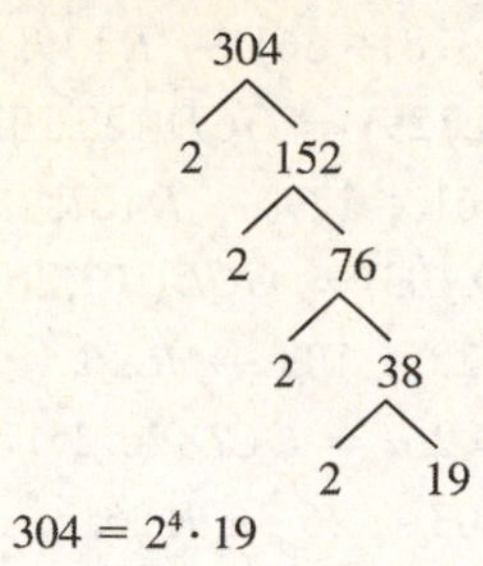

$304 = 2^4 \cdot 19$

(b)

1570
2 785
5 157

$1570 = 2 \cdot 5 \cdot 157$

(c)

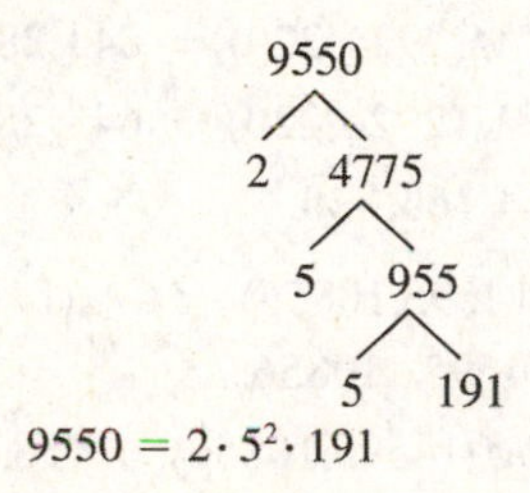

$9550 = 2 \cdot 5^2 \cdot 191$

5. **19**. $19^2 < 503$, but $23^2 > 503$. (19 is prime and 23 is the next largest prime.)

7. **(a)** **3, 5, 15, or 29** members. Prime factorization of $435 = 3 \cdot 5 \cdot 29$, so committees could have these numbers of members or products thereof.

(b) **145** three-member committees ($435 \div 3$), **87** five-member committees, **29** fifteen-member committees, or **15** twenty-nine member committees.

9. **(a)** Since $4|n$, 2^2 must be in the prime factorization of n. Since $9|n$, 3^2 must be in the prime factorization of n. Because 2^2 and 3^2 have no common factors, $2^2 \cdot 3^2 = 36$ must divide n.

(b) Part (a) has the key idea: that a and b have no common factors. This notion is after expressed as $GCD(a,b) = 1$.

11. Every number would have its "usual" factorization, $1 \cdot (p_1 \cdot p_2 \cdot \cdots \cdot p_n)$, along with infinitely many other such factorizations (because $1^n = 1$, where n may be any natural number).

13. The numbers 9, 21, 33, 45,... which are an arithmetic sequence with difference 12, belong to the original sequence and are divisible by 3.

15. **No**. No integer times $3^2 \cdot 2^4 = 144$ will yield $3^3 \cdot 2^2 = 108$.

17. **(*i*)** $2^2 \cdot 5^3 \cdot 9^2$ is not a prime factorization because 9 is not prime.

(*ii*) The prime factorization is $2^2 \cdot 5^3 \cdot (3^2)^2 = \mathbf{2^2 \cdot 5^3 \cdot 3^4}$.

19. For all of (a)–(d), the property "each row is to have the same number of trees" tells us that the number at rows is a divisor of the number of trees.

(a) $15 = 3 \cdot 5$. So there can be **1, 3, 5, or 15** rows

(b) $20 = 2^2 \cdot 5$. There can be **1, 2, 4, 5, 10, or 20** rows.

(c) 19 is prime. There can be **1 or 19** rows.

(d) $100 = 2^2 \cdot 5^2$. There can be **1, 2, 4, 5, 10, 20, 25, 50, or 100** rows.

Review Problems

19. Use the standard divisibility tests for all except 7; for 7 just perform the division:

(a) Divisors are **2, 3, and 6**.

(b) Divisors are **2, 3, 5, 6, 9, and 10**.

21. Only among **8** people; each would receive $422 (8|3376 but $7 \nmid 3376$).

Assessment 4-3A Greatest Comman Divisor and Least Comman Multiple

1. **(a)** **(*i*)** $D_{18} = \{1,2,3,6,9,18\}$;

$D_{10} = \{1,2,5,10\};$

$\Rightarrow GCD(18,10) = \mathbf{2}.$

(*ii*) $M_{18} = \{18,36,54,72,90,...\};$

$M_{10} = \{10,20,30,...,90,...\};$

$\Rightarrow LCM(18,10) = \mathbf{90}.$

(b) (*i*) $D_{24} = \{1,2,3,4,6,8,12,24\};$

$D_{36} = \{1,2,3,4,6,9,12,18,36\};$

$\Rightarrow GCD(24,36) = \mathbf{12}.$

(*ii*) $M_{24} = \{24,48,72,96,...\};$

$M_{36} = \{36,72,108,...\};$

$\Rightarrow LCM(24,36) = \mathbf{72}.$

(c) (*i*) $D_8 = \{1,2,4,8\};$

$D_{24} = \{1,2,3,4,6,8,12,24\};$

$D_{52} = \{1,2,4,13,26,52\};$

$\Rightarrow GCD(8,24,52) = \mathbf{4}.$

(*ii*) $M_8 = \{8,16,24,...,312,...\};$

$M_{24} = \{24,48,72,...,312,...\};$

$M_{52} = \{52,104,156,...,312,...\};$

$\Rightarrow LCM(8,24,52) = \mathbf{312}.$

(d) (*i*) $D_7 = \{1,7\};$

$D_9 = \{1,3,9\};$

$\Rightarrow GCD(7,9) = \mathbf{1}.$

(*ii*) $M_7 = \{7,14,21,...,63,...\};$

$M_9 = \{9,18,27,...,63,...\};$

$\Rightarrow LCM(7,9) = \mathbf{63}.$

3. $\rightarrow R$ symbolizes the reminder left after the indicated divisions.

(a) $GCD(2924,220) = GCD(220,64)$ since $2924 \div 220 \rightarrow R\ 64;$

$GCD(220,64) = GCD(64,28)$ since $220 \div 64 \rightarrow R\ 28;$

$GCD(64,28) = GCD(28,8)$ since $64 \div 28 \rightarrow R\ 8;$

$GCD(28,8) = GCD(8,4)$ since $28 \div 8 \rightarrow R\ 4;$

$GCD(8,4) = GCD(4,0)$ since $8 \div 4 \rightarrow R\ 0;$

$\Rightarrow GCD(2924,220) = \mathbf{4}.$

(b) $GCD(14595,10856) = GCD(10856,3739)$ since $14595 \div 10856 \rightarrow R\ 3739;$

$GCD(10856,3739) = GCD(3739,3378)$ since $10856 \div 3739 \rightarrow R\ 3378;$

$GCD(3739,3378) = GCD(3378,361)$ since $3739 \div 3378 \rightarrow R\ 361;$

$GCD(3378,361) = GCD(361,129)$ since $3378 \div 361 \rightarrow R\ 129;$

$GCD(361,129) = GCD(129,103)$ since $361 \div 129 \rightarrow R\ 103;$

$GCD(129,103) = GCD(103,26)$ since $129 \div 103 \rightarrow R\ 26;$

$GCD(103,26) = GCD(26,25)$ since $103 \div 26 \rightarrow R\ 25;$

$GCD(26,25) = GCD(25,1)$ since $26 \div 25 \rightarrow R\ 1;$

$GCD(25,1) = GCD(1,0)$ since $25 \div 1 \rightarrow R\ 0;$

$\Rightarrow GCD(14595,10856) = \mathbf{1}.$

5. **(a)** $GCD(2924,220) \cdot LCM(2924,220) = 2924 \cdot 220;$

$4 \cdot LCM(2924,220) = 643,280;$

$\Rightarrow LCM(2924,220) = 643,280 \div 4 = \mathbf{160,820}.$

(b) $GCD(14595,10856) \cdot LCM(14595,10856) = 14595 \cdot 10856;$

$1 \cdot LCM(14595,10856) = 158,443,320;$

$\Rightarrow LCM(14595,10856) = \mathbf{158,443,320}.$

7. (*i*) The 2 rods can be used to build both the 6 rod and the 10 rod $\Rightarrow GCD(6,10) = \mathbf{2}.$

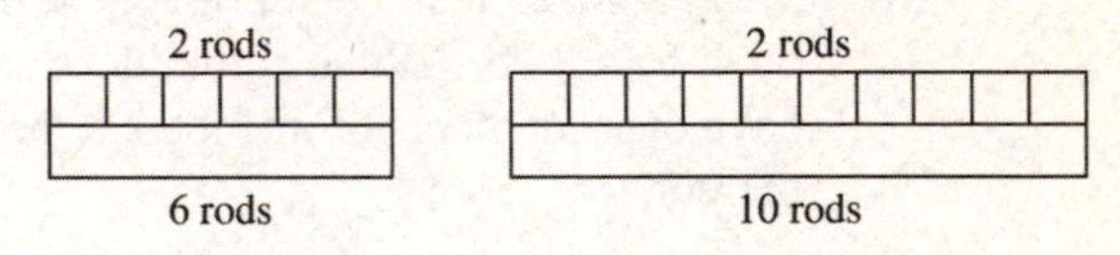

(*ii*) Five 6 rods (length 30) are the same length as three 10 rods (length 30) $\Rightarrow LCM(6,10) = \mathbf{30}.$

9. $LCM(24, 45) = 360$; i.e., 360¢ worth of cookies was sold; 360¢ ÷ 24¢ per cookie = **15 cookies**.

11. The question is really, "What is $LCM(12, 18, 16)$?" since that is when the times will coincide $LCM(12, 18, 16) = \mathbf{144}$ **minutes.**

13. **(a)** **True**. If both a and b are even, then $GCD(a, b) \geq 2$.

(b) **True**. $GCD(a, b) = 2$ implies that $2|a$ and $2|b$.

(c) **False**. GCD could be any larger multiple of 2; e.g., $GCD(8, 20) = 4$.

15. 2 is the only prime factor of 4 and $2 \nmid 97, 219, 988, 751$, so 1 is their only common division; i.e., they are relatively prime.

17. One revolution of gear 2, rotates 28 teeth. One revolution of gear 1, rotates 48 teeth. For each of the gears, the arrows are back in their original positions every multiple of 28 teeth for gear 2 and 48 teeth for gear 1. Thus, the arrows realign when $LCM(48, 28) = 336$ teeth have rotated. So gear 2 must make $336 \div 28 = \mathbf{12}$ **revolutions**.

19. This will be the set of all numbers between 1 and 49 that are relatively prime to 49. Since $49 = 7^2$, the set will contain all numbers between 1 and 49 **except** $1 \cdot 7, 2 \cdot 7, 3 \cdot 7, 4 \cdot 7, 5 \cdot 7, 6 \cdot 7$, and $7 \cdot 7$.

21. **(a)** $12x^4y^3 + 18x^3y^4 = \mathbf{6x^3y^3(2x + 3y)}$; i.e., $GCD(12x^4y^3, 18x^3y^4) = 6x^3y^3$.

(b) $12x^3y^2z^2 + 18x^2y^4z^3 + 24x^4y^3z^4 = \mathbf{6x^2y^2z^2(2x + 3y^2z + 4x^2yz^2)}$.

Assessment 4-3B

1. **(a)** **(*i*)** $D_{12} = \{1, 2, 3, 4, 6, 12\}$

$D_{18} = \{1, 2, 3, 6, 9, 18\}$

$\Rightarrow GCD(12,18) = \mathbf{6}$.

(*ii*) $M_{12} = \{12, 24, 36, 48, 60, 72, 84,...\}$

$M_{18} = \{18, 36, 54, 72, 90, 108,...\}$

$\Rightarrow LCM(12,18) = \mathbf{36}$.

(b) **(*i*)** $D_{18} = \{1, 2, 3, 6, 9, 18\}$

$D_{36} = \{1, 2, 3, 4, 6, 9, 12, 18, 36\}$

$\Rightarrow GCD(18,36) = \mathbf{18}$.

(*ii*) $M_{18} = \{18, 36, 54, 72, 90, 108, \ldots\}$

$M_{36} = \{36, 72, 108, 144, 180,...\}$

$\Rightarrow LCM(18, 36) = \mathbf{36}$.

(c) **(*i*)** $D_{12} = \{1, 2, 3, 4, 6, 12\}$

$D_{18} = \{1, 2, 3, 6, 9, 18,\}$

$D_{24} = \{1, 2, 3, 4, 6, 8, 12, 24\}$

$\Rightarrow GCD(12,18, 24) = \mathbf{6}$.

(*ii*) $M_{12} = \{12, 24, 36, 48, 60, 72, 84,...\}$

$M_{18} = \{18, 36, 54, 72, 90, 108,...\}$

$m_{24} = \{24, 48, 72, 96, 120, 144,...\}$

$\Rightarrow LCM(12,18, 24) = \mathbf{72}$.

(d) **(*i*)** $D_6 = \{1, 2, 3, 6\}$

$D_{11} = \{1, 11\}$

$\Rightarrow GCD(6,11) = \mathbf{1}$.

(*ii*) $M_6 = \{6, 12, 18, 24, 30,....., 66, 72,...\}$

$M_{11} = \{11, 22, 33, 44, 55, 66, 77,...\}$

$\Rightarrow LCM(6, 11) = \mathbf{66}$.

3. $\rightarrow R$ symbolizes the reminder left after the indicated divisions.

(a) $GCD(14560, 8250) = GCD(8250, 6310)$ since $14560 \div 8250 \rightarrow R\,6310$

$GCD(8250, 6310) = GCD(6310, 1940)$ since $8250 \div 6310 \rightarrow R\,1940$

$GCD(6310, 1940) = GCD(1940, 490)$ since $6310 \div 1940 \rightarrow R\,490$

$GCD(1940, 490) = GCD(490, 470)$ since $1940 \div 490 \rightarrow R\,470$

$GCD(490, 470) = GCD(470, 20)$ since $490 \div 470 \rightarrow R\,20$

$GCD(470, 20) = GCD(20, 10)$ since $470 \div 20 \rightarrow R\,10$

$GCD(20, 10) = GCD(10, 0)$ since $20 \div 10 \rightarrow R\,0$

$\Rightarrow GCD(14560, 8250) = \mathbf{10}$.

(b) $GCD(8424, 2520) = GCD(2520, 864)$ since $8424 \div 2520 \rightarrow R\,864$

$GCD(2520,864) = GCD(864,792)$ since
$2520 \div 864 \rightarrow R\,792$
$GCD(864,792) = GCD(792,72)$ since
$864 \div 792 \rightarrow R\,72$
$GCD(792,72) = GCD(72,0)$ since
$792 \div 72 \rightarrow R\,0$
$\Rightarrow GCD(8424,2520) = \mathbf{72}.$

5. **(a)** $GCD(14560,8250) \cdot LCM(14560,8250) = 14560 \cdot 8250$
$10 \cdot LCM(14560,8250) = 120{,}120{,}000$
$\Rightarrow LCM(14560,8250) = \mathbf{12{,}012{,}000}.$

(b) $GCD(8424,2520) \cdot LCM(8424,2520) = 8424 \cdot 2520$
$72 \cdot LCM(8424,2520) = 21{,}228{,}480$
$\Rightarrow LCM(8424,2520) = \mathbf{294{,}840}.$

7. ***(i)*** $GCD(4, 10) = \mathbf{2}$:

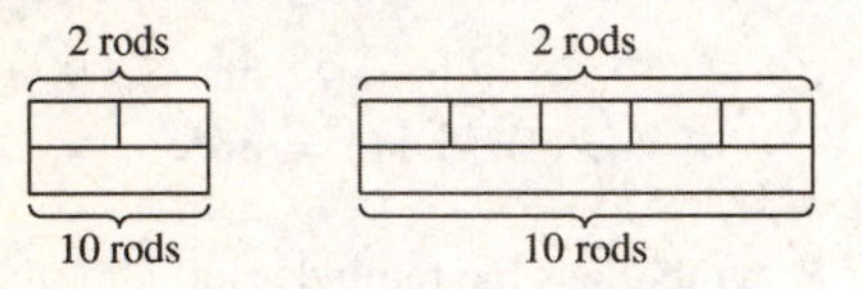

(ii) $LCM(4, 10) = \mathbf{20}$:

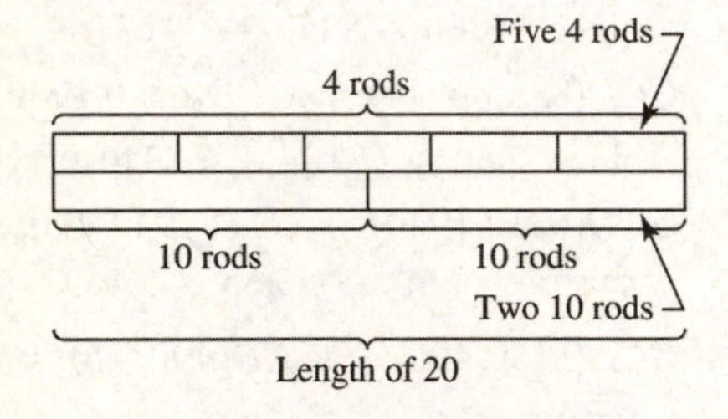

9. The question is really "What is $LCM(6, 8)$?" since that is when the off nights will coincide: $LCM(6, 8) =$ **24 nights**.

11. Divisors common to 18, 24, and 36 are 2, 3, and 6. Thus the classes may be divided into groups of **2**, **3**, or **6 students**.

13. **(a)** **False**. For $a \neq b$, $LCM(a,b) > GCD(a,b)$.

(b) **True**. Theorem 4-18.

(c) **True**. If $GCD(a, b) >$ a, it could not divide a.

(d) **True**. If $LCM(a, b) <$ a, it could not be a multiple.

15. 11 is its own only prime factor and $11 \nmid 181{,}345{,}913$ $[11 \nmid (1 + 1 + 4 + 9 + 3) - (8 + 3 + 5 + 1)]$, so 1 is their only common divisor; i.e., they are relatively prime.

17. $LCM(40, 24, 60) = 120$. **Gear 1** must make $120 \div 40 =$ **3 revolutions**; **gear 2** must make $120 \div 24 =$ **5 revolutions**; and **gear 3** must make $120 \div 60 =$ **2 revolutions** to line up the arrows.

19. $GCD(25, x) = 1$ means that x can have no factors in common with 25, so x cannot have a factor of 5; the solution set is thus $\{1, 2, 3, 4, 6, 7, 8, 9, 11, 12, 13, 14, 16, 17, 18, 19, 21, 22, 23, 24\}$.

21. **(a)** $18x^4y^3 + 12x^2y^4 = \mathbf{6x^2y^3(3x^2 + 2y)}$.

(b) $18x^3y^2z^3 + 48x^2y^3z^2 + 12x^4y^3z^4 = \mathbf{6x^2y^2z^2(3xz + 8y + 2x^2yz^2)}$.

Review Problems

17. $1{,}000{,}000 = 10^6 = 2^6 \cdot 5^6$. If a 2 is paired with a 5 in any factor a 0 results; no 0's as digits implies 2^6 and 5^6 must be separate factors. Therefore $x = 2^6 = \mathbf{64}$; $y = 5^6 = \mathbf{15{,}625}$.

19. **No.** $3|(3 + 1 + 1 + 1)$ so $3|3111$; it is thus not prime.

21. The question is really "What is $LCM(2, 3, 4, \ldots, 11)$?". Their factors are $\{2, 3, 2^2, 5, 2 \cdot 3, 7, 2^3, 3^2, 2 \cdot 5, 11\}$, thus $LCM(2, 3, 4, \ldots, 11) = 2^3 \cdot 3^2 \cdot 5 \cdot 7 \cdot 11 = \mathbf{27{,}720}$.

Chapter 4 Review

1. **(a)** **False**. $8 \nmid 4$ since $4 = 8x$ has no integer solution for x.

(b) **False**. $0 \nmid 4$ since $4 = 0 \cdot x$ has no integer solution for x.

(c) **True**. 0 divided by any non-zero integer is 0.

(d) **False**. E.g., 12. The test works only when the two numbers have no common factors.

(e) **False**. E.g., 9.

3. (a) $m = 83{,}160$.

$2|m$ because $2|0$.

$3|m$ because $3|(8 + 3 + 1 + 6 + 0)$.

$4|m$ because $4|60$.

$5|m$ because $5|0$.

$6|m$ because $2|m$ and $3|m$.

$8|m$ because $8|160$.

$9|m$ because $9|(8 + 3 + 1 + 6 + 0)$.

$11|m$ because $11|(0 + 1 + 8) - (6 + 3)$

(b) $m = 83{,}193$.

$2\nmid m$ because $2\nmid 3$.

$3|m$ because $3|(8 + 3 + 1 + 9 + 3)$.

$4\nmid m$ because $4\nmid 93$.

$5\nmid m$ because $5\nmid 3$.

$6\nmid m$ because $3|m$ but $2\nmid m$.

$8\nmid m$ because $8\nmid 193$.

$9\nmid m$ because $9|(8 + 3 + 1 + 9 + 3)$.

$11|m$ because $11|(3 + 1 + 8) - (9 + 3)$.

5. (a) write the number as $87a4$.

$6|87a4$ if $2|87a4$ and $3|87a4$;
$2|4$ so $2|87a4$, thus
$6|87a4$ if
$3|87a4 \Rightarrow 3|(8 + 7 + a + 4 = 19 + a)$;

$3|(21, 24, \text{or } 27)$ when $a = 2, 5, \text{or } 8$;
so **6|(87<u>2</u>4, 87<u>5</u>4, or 87<u>8</u>4)**.

(b) Write the number as $4a856$.

$24|4a856$ if $3|4a856$ and $8|4a856$;
$8|856$ so $8|4a856$, thus
$24|4a856$ if
$3|(4 + a + 8 + 5 + 6) = 23 + a$;

$3|(24, 27, \text{or } 30)$ when $a = 1, 4, \text{or } 7$;
so **24|(4<u>1</u>856, 4<u>4</u>856, or 4<u>7</u>856)**.

(c) write the number as $87ab4$.

$29|87{,}000$ so $29|ab4$.

The only integer to return 4 in the units position when multiplied by 9 is 6. So the possibilities are $6 \cdot 29, 16 \cdot 29$, etc. until the product of $n \cdot 29 > 999$.

$6 \cdot 29 = 174; 16 \cdot 29 =$
$464; 26 \cdot 29 = 754;$
$36 \cdot 29 > 999.$

Thus $ab = 17, 46, \text{or } 75$.

So **29|(87<u>17</u>4, 87<u>46</u>4, or 87<u>75</u>4)**.

7. (a) **Composite**. $11|143$.

(b) **Prime**. Primes through 13 do not divide 223; $17^2 > 223$.

9. **No**. *LCM* and *GCD* are the same if the number are equal.

11. The number is not divisible by 2, 3, 5, 7, 11, and 13 because each of these primes divides one product in the sum $2 \cdot 3 \cdot 5 \cdot 7 + 11 \cdot 13$ but not the other. The student checked that $17\nmid 353$ and because $19^2 > 353$ no other primes needed to be tested.

13. (a) $LCM(2^3 \cdot 5^2 \cdot 7^3, 2 \cdot 5^3 \cdot 7^2 \cdot 13,$
$2^4 \cdot 5 \cdot 7^4 \cdot 29) =$
$\mathbf{2^4 \cdot 5^3 \cdot 7^4 \cdot 13 \cdot 29}$.

(b) $GCD(279, 278) = GCD(278, 1) = 1$
Therefore no common factors in 279 and 278.
Thus $LCM(278, 279) \cdot 1 = 278 \cdot 279$
$= \mathbf{77{,}562}$.

15. $144 = (2^2 \cdot 3)^2 = 2^4 \cdot 3^2$. There are thus $(4 + 1) \cdot (2 + 1) = 15$ divisors: **1, 2, 3, 4, 6, 8, 9, 12, 16, 18, 24, 36, 48, 72 and 144**.

17. Product (all positive integers ≤ 10) $=$
$1 \cdot 2 \cdot 3 \cdot 4 \cdot 5 \cdot 6 \cdot 7 \cdot 8 \cdot 9 \cdot 10 =$
$1 \cdot 2 \cdot 3 \cdot 2^2 \cdot 5 \cdot (2 \cdot 3) \cdot 7 \cdot 2^3 \cdot 3^2 \cdot (2 \cdot 5)$.

Thus *LCM* (all positive integers ≤ 10) $=$
$1 \cdot 2^3 \cdot 3^2 \cdot 5 \cdot 7 = \mathbf{2520}$.

19. LCM $(45,30) = 90$ minutes. 8:00 A.M. + 90 minutes = **9:30 A.M.**

21. There are 15 children. The candy will come out evenly for $LCM(15,12) = 60$; 60 candies $\div$ 12 candies per package = **5 packages**.

23. $9869 = 71 \cdot 139$ (both prime) implies **71 lattes** at \$1.39 each. She could not sold 139 lattes at \$0.71 each, since she never sells for less than \$1.

25. By the division algorithm, every prime number > 3 can be written as $12q + r$ where $r = 1, 5, 7, \text{or } 11$.

For other values of r, $12q + r$ is composite because 12 and r will share a common factor, making it not prime.

27. First show that among any three consecutive odd whole numbers there is always one divisible by 3.

Suppose that the first whole number in the triplet is not divisible by 3. By the Division Algorithm that whole number can be written in the form $3n + 1$ or $3n + 2$ for some whole number n.

Then the three consecutive odd whole numbers are $(3n + 1, 3n + 3, 3n + 5)$ or $(3n + 2, 3n + 4, 3n + 6)$.

In the first triplet $3|3n + 3$; in the second $3|3n + 6$. If the first whole number is greater than 3 and not divisible by 3 then the second or third must be divisible by 3 and so cannot be prime.

Chapter 5

INTEGERS

Assessment 5-1A: Integers and the Operations of Addition and Subtraction

1. For every integer a, there exists a unique integer ^{-}a, such that $a + {}^{-}a = 0 = {}^{-}a + a$. ^{-}a is called the additive inverse of a.

   (a) The unique integer $^{-}\mathbf{2}$ is the additive inverse of 2 because $2 + {}^{-}2 = 0$.

   (b) The additive inverse of ^{-}a can be written as $^{-}(^{-}a) = a$, or the number of the opposite sign. Thus the additive inverse of $^{-}5$ is $^{-}(^{-}5) = \mathbf{5}$.

   (c) The additive inverse of m is $^{-}\boldsymbol{m}$.

   (d) The additive inverse of 0 is **0** because $0 + 0 = 0$.

   (e) The additive inverse of ^{-}m is $^{-}(^{-}m) = \boldsymbol{m}$.

   (f) The additive inverse of $(a + b)$ is $^{-}(a + b) = {}^{-}\boldsymbol{a} + {}^{-}\boldsymbol{b}$.

3. **(a)** Absolute value is the distance on a number line between the origin (0) and a specified number. Distance on the number line between 0 and $^{-}5$ is 5 units, so $|^{-}5| = \mathbf{5}$.

   (b) Distance on the number line between 0 and 10 is 10 units, so $|10| = \mathbf{10}$.

   (c) $^{-}|^{-}5|$ means the additive inverse of the absolute value of $^{-}5$, so $^{-}|^{-}5| = {}^{-}(5) = {}^{-}\mathbf{5}$.

   (d) $^{-}|5| = {}^{-}(5) = {}^{-}\mathbf{5}$.

5. Movement on the number line in addition is always to the right for a positive number and to the left for a negative number.

   (a)

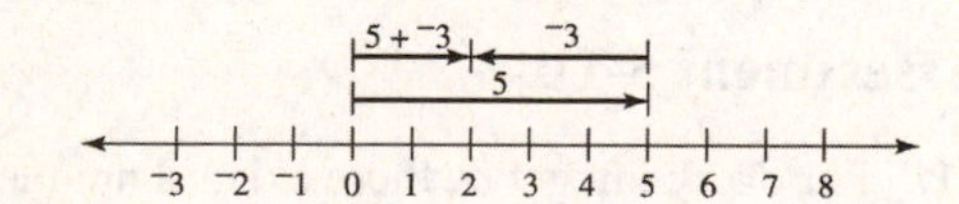

   (b)

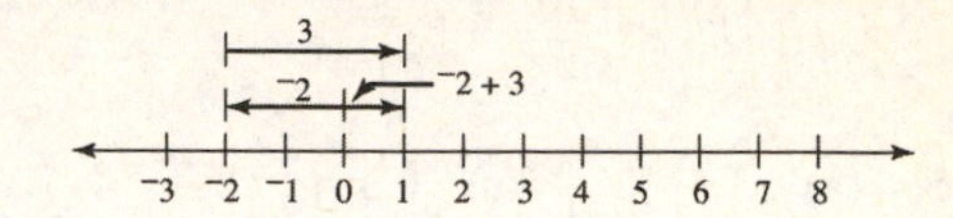

   (c)

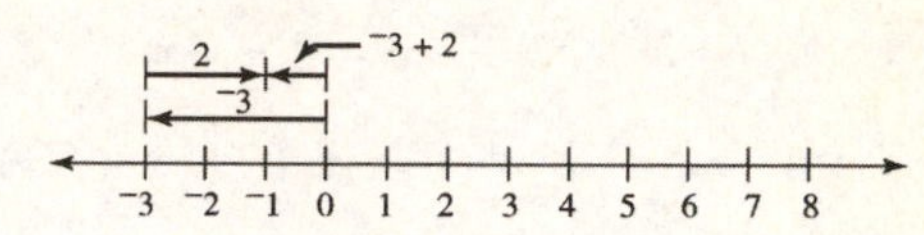

   (d)

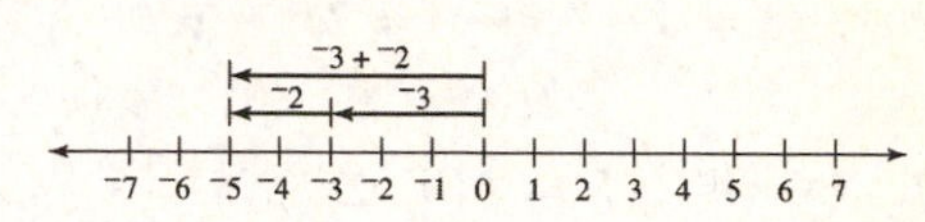

7. **(a)** $3 - (^{-}2) = n$ if and only if $3 = {}^{-}2 + n$. Thus $\boldsymbol{n} = \mathbf{5}$.

   (b) $^{-}3 - 2 = n$ if and only if $^{-}3 = 2 + n$. Thus $\boldsymbol{n} = {}^{-}\mathbf{5}$.

   (c) $^{-}3 - {}^{-}2 = n$ if and only if $^{-}3 = {}^{-}2 + n$. Thus $\boldsymbol{n} = {}^{-}\mathbf{1}$.

9. **(a)** $(^{-}45) + (^{-}55) + (^{-}165) + (^{-}35) + (^{-}100) + 75 + 25 + 400 = {}^{-}400 + 500$, or **\$100**.

   (b) \$300 (beginning balance) + \$100 (in transactions) = **\$400**.

11. **(a)** Starting with subtraction that is already known: $^{-}4 - 2 = {}^{-}6$

    Continuing the pattern:

    $$^{-}4 - 1 = {}^{-}5$$
    $$^{-}4 - 0 = {}^{-}4$$
    $$^{-}4 - {}^{-}1 = {}^{-}3.$$

    (b) Starting with subtraction that is already known:

    $$2 - 1 = 1$$
    $$1 - 1 = 0$$
    $$0 - 1 = {}^{-}1$$
    $$^{-}1 - 1 = {}^{-}2$$
    $$^{-}2 - 1 = {}^{-}3.$$

13. (a) $^-2 + (3 - 10) = {}^-2 + {}^-7 = \mathbf{{}^-9}$.

(b) $[8 - ({}^-5)] - 10 = [8 + {}^-({}^-5)] - 10 =$
$[8 + 5] - 10 = 13 - 10 = \mathbf{3}$.

(c) $({}^-2 - 7) + 10 = ({}^-2 + {}^-7) + 10 =$
$({}^-9) + 10 = \mathbf{1}$.

15. (a) $3 - (2 - 4x) = 3 + {}^-(2 + {}^-4x)$
$= 3 + {}^-(2) + {}^-({}^-4x)$
$= 3 + {}^-2 + 4x = \mathbf{1 + 4x}$.

(b) $x - ({}^-x - y) = x + {}^-({}^-x + {}^-y)$
$= x + {}^-({}^-x) + {}^-({}^-y)$
$= x + x + y = \mathbf{2x + y}$.

17. (a) $W \cup I = \boldsymbol{I}\ (W \subset I)$.

(b) $W \cap I = \boldsymbol{W}\ (W \subset I)$.

(c) $I^+ \cup I^- = \boldsymbol{I} - \{\mathbf{0}\}$. 0 is neither positive nor negative, so it must be removed from the sets of integers.

(d) $I^+ \cap I^- = \emptyset$, the empty set. An integer cannot be both positive and negative simultaneously.

(e) $W - I = \emptyset$. $W \subset I$, so there are no elements in W that are not in I.

19. If $y = {}^-x - 1$:

(a) $y = {}^-({}^-1) - 1 = 1 + {}^-1 = \mathbf{0}$,
when $x = {}^-1$

(b) $y = {}^-(100) + {}^-1 = {}^-100 + {}^-1 = \mathbf{{}^-101}$,
when $x = 100$

(c) $y = {}^-({}^-2) - 1 = 2 + {}^-1 = \mathbf{1}$,
when $x = {}^-2$

(d) $y = {}^-({}^-a) - 1 = \mathbf{a - 1}$,
when $x = {}^-a$

(e) $3 = {}^-x - 1 \Rightarrow 3 + 1 = {}^-x - 1 + 1$
$\Rightarrow 4 = {}^-x \Rightarrow x = \mathbf{{}^-4}$.

21. If $y = |1 - x|$:

(a) $y = |1 - (10)| = |1 + {}^-10| = |{}^-9| = 9$,
when $x = 10$

(b) $y = |1 - ({}^-1)| = |1 + {}^-({}^-1)| = |1 + 1| = 2$,
when $x = {}^-1$

(c) If $1 = |1 - x| \Rightarrow 1 - x = 1$ or $1 - x = {}^-1$:

(*i*) $1 - x = 1 \Rightarrow 1 - 1 - x = 1 - 1$
$\Rightarrow {}^-x = 0 \Rightarrow \mathbf{x = 0}$.

(*ii*) $1 - x = {}^-1 \Rightarrow 1 - 1 - x = {}^-1 - 1 \Rightarrow$
${}^-x = {}^-2 \Rightarrow \mathbf{x = 2}$.

23. For problems such as this, use the arithmetic sequence formula $a_n = a_1 + (n - 1)d$.

(a) $100 = 10 + (n - 1) \cdot 1 \Rightarrow 90 = n - 1$
$\Rightarrow n = 91$. Since there are 91 terms in this sequence, there are then **89 integers** between 10 and 100 (i.e., 11, 12,…, 98, 99).

(b) ${}^-10 = {}^-30 + (n - 1) \cdot 1 \Rightarrow 20 = n - 1 \Rightarrow$
$n = 21$. Since there are 21 terms in this sequence, there are then **19 integers** between $^-30$ and $^-10$.

25. (a) **True**. Absolute value is always non-negative.

(b) **True**. Absolute value of the difference between the two elements is always non-negative.

(c) **True**. ${}^-x + {}^-y = {}^-(x + y)$, and
$|{}^-(x + y)| = |x + y|$.

27. (a) Enter: [1] [2] [+/−] [+] [6] [+/−] [=]
to return **$^-$18**.

(b) Enter: [1] [2] [+/−] [+] [6] [=]
to return **$^-$6**.

(c) Enter: [2] [7] [+] [5] [+/−] [=]
to return **22**.

(d) Enter: [1] [2] [+/−] [−] [6] [=]
to return **$^-$18**.

(e) Enter: [1] [6] [−] [7] [+/−] [=]
to return **23**.

Assessment 5-1B

1. For every integer a, there exists a unique integer ^-a, such that $a + {}^-a = 0 = {}^-a + a$. ^-a is called the additive inverse of a.

(a) The unique integer $^-\mathbf{3}$ is the additive inverse of 3 because $3 + {}^-3 = 0$.

(b) The additive inverse of ^-a can be written as $^-(^-a) = a$, or the number of the opposite sign. Thus the additive inverse of $^-4$ is $^-(^-4) = 4$.

(c) The additive inverse of q is $^-\boldsymbol{q}$.

(d) The additive inverse of 6 is $^-\mathbf{6}$ because $6 + {}^-6 = 0$.

(e) The additive inverse of ^-n is $^-(^-n) = \boldsymbol{n}$.

(f) The additive inverse of $(3 + x)$ is $^-(3 + x) = {}^-\mathbf{3} + {}^-\boldsymbol{x}$.

3. (a) Absolute value is the distance on a number line between the origin (0) and a specified number. Distance on the number line between 0 and $^-3$ is 3 units, so $|^-3| = \mathbf{3}$.

(b) Distance on the number line between 0 and 15 is 15 units, so $|15| = \mathbf{15}$.

(c) $^-|^-3|$ means the additive inverse of the absolute value of $^-3$, so $^-|^-3| = {}^-(3) = {}^-\mathbf{3}$.

(d) $^-|6| = {}^-6$

5. Movement on the number line in addition is always to the right for a positive number and to the left for a negative number.

(a)

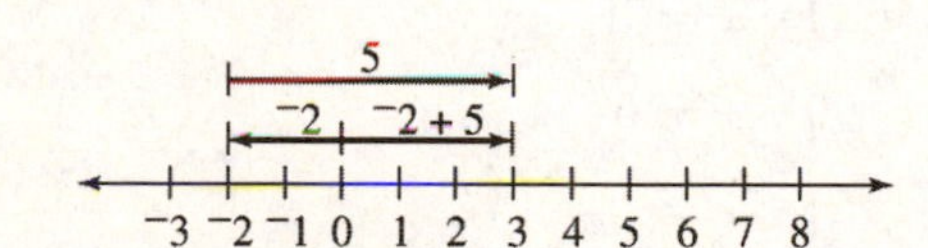

(b)

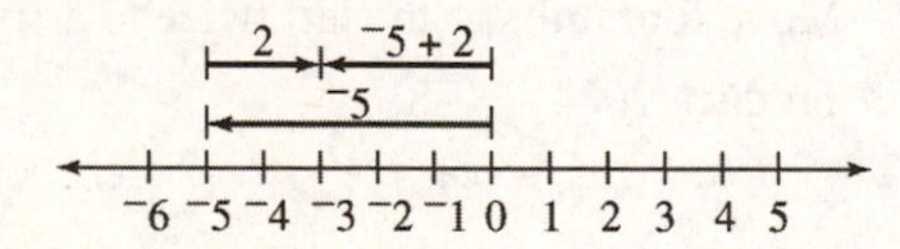

(c)

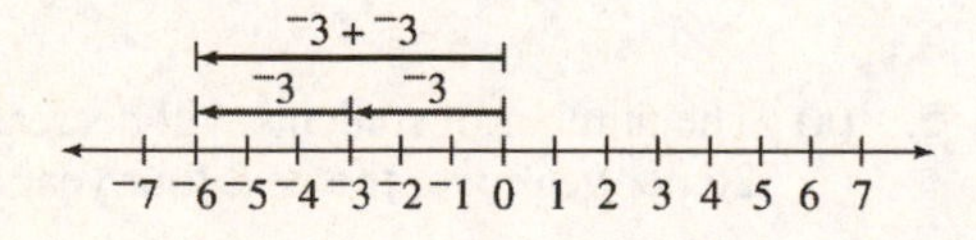

7. (a) $^-3 - 5 = n$ if and only if $^-3 = 5 + n$. Thus $\boldsymbol{n} = {}^-\mathbf{8}$.

(b) $5 - (^-3) = n$ if and only if $5 = {}^-3 + n$. Thus $\boldsymbol{n} = \mathbf{8}$.

(c) $^-2 - (^-3) = n$ if and only if $^-\mathbf{2} = {}^-\mathbf{3} + n$. Thus $\boldsymbol{n} = \mathbf{1}$.

9. (a) $^-247 + {}^-11 + {}^-11 = {}^-258 + {}^-11 = {}^-\mathbf{269°}$ **C**.

(b) $98 - {}^-94 = 98 + {}^-(^-94) = 98 + 94 =$ **192°F**.

11. (a) Starting with subtraction that is already known:

$$^-2 - 1 = {}^-3$$

Continuing the pattern:

$$^-2 - 0 = {}^-2$$
$$^-2 - {}^-1 = {}^-1$$
$$^-2 - {}^-2 = 0$$
$$^-2 - {}^-3 = \mathbf{1}.$$

(b) Starting with subtraction that is already known:

$$3 - 2 = 1$$

Continuing the pattern:

$$2 - 2 = 0$$
$$1 - 2 = {}^-1$$
$$0 - 2 = {}^-2$$
$$^-1 - 2 = {}^-3$$
$$^-2 - 2 = {}^-4$$
$$^-3 - 2 = {}^-\mathbf{5}.$$

13. (a) $^-2 - (7 + 10) = {}^-2 - (17) = {}^-2 + {}^-17 = {}^-\mathbf{19}$.

(b) $8 - 11 - 10 = 8 + {}^-11 + {}^-10 = {}^-3 + {}^-10 = {}^-\mathbf{13}$.

(c) $^-2 - 7 + 3 = {}^-2 + {}^-7 + 3 = {}^-9 + 3 = {}^-\mathbf{6}$.

15. (a) $4x - 2 - 3x = 4x + {}^-(3x) + {}^-(2) = x - 2$.

(b) $4x - (2 - 3x) = 4x + {}^-(2 + {}^-3x) =$

$4x + {}^-(2) + {}^-({}^-3x) = 4x + 3x + {}^-2 =$

$\mathbf{7x - 2}$.

17. (a) $W - I^+ = \{0\}$ $(0 \notin I^+)$.

(b) $W - I^- = \mathbf{W}$. W and I^- are disjoint.

(c) $I \cap I = \mathbf{I}$ $(I \subseteq I)$.

(d) $I - W = I^-$

19. If $y = {}^-3x - 2$:

(a) $y = {}^-3({}^-1) - 2 = 3 + {}^-2 = \mathbf{1}$, when $x = {}^-1$.

(b) $y = {}^-3(100) + {}^-2 = {}^-300 + {}^-2 = \mathbf{{}^-302}$, when $x = 100$.

(c) $y = {}^-3({}^-2) - 2 = 6 + {}^-2 = \mathbf{4}$, when $x = {}^-2$.

(d) $y = {}^-3({}^-a) - 2 = \mathbf{3a - 2}$, when $x = {}^-a$.

(e) ${}^-11 = {}^-3x - 2 \Rightarrow {}^-9 = {}^-3x \Rightarrow \mathbf{x = 3}$.

21. If $y = |x - 5|$:

(a) $y = |(10) - 5| = |10 + {}^-5| = |5| = \mathbf{5}$, when $x = 10$.

(b) $f({}^-1) = |({}^-1) - 5| = |{}^-1 + {}^-5| = |{}^-6| = \mathbf{6}$, when $x = {}^-1$.

(c) If $7 = |x - 5| \Rightarrow x - 5 = 7$ or $x - 5 = {}^-7$:

(*i*) $x - 5 = 7 \Rightarrow x = 7 + 5 \Rightarrow \mathbf{x = 12}$.

(*ii*) $x - 5 = {}^-7 \Rightarrow x = {}^-7 + 5 \Rightarrow \mathbf{x = {}^-2}$.

23. (a) $10 = {}^-10 + (n - 1) \cdot 1 \Rightarrow 20 = n - 1 \Rightarrow$ $n = 21$. Since there 21 terms in this sequence, there are then **19 integers** between ${}^-10$ and 10.

(b) $y = x + (n - 1) \cdot 1 \Rightarrow y - x = n - 1 \Rightarrow$ $n = y - x + 1$. Since there are $y - x + 1$ terms in this sequence, there are then $(y - x + 1) - 2 = \mathbf{y - x - 1}$ **integers** between x and y.

25. (a) **True**. The square of integer is always non-negative.

(b) **False**. Let $x = {}^-2$. Then $|({}^-2)^3| = |{}^-8| = 8 \neq ({}^-2)^3 = {}^-8$.

(c) **True**. $|x^3| = |x^2 \cdot x| = |x^2| \cdot |x| = x^2 \cdot |x|$.

27. Estimates may vary depending upon the method used:

(a) Estimate:

$343 + {}^-42 - 402 \approx 300 - 400 = \mathbf{{}^-100}$

Actual: ${}^-101$.

(b) Estimate:

${}^-1992 + 3005 - 497 \approx$

${}^-2000 + 3000 - 500 = \mathbf{500}$.

Actual: 516.

(c) Estimate: $992 - {}^-10{,}003 - 101 \approx$ $1000 + 10{,}000 - 100 = \mathbf{10{,}900}$.

Actual: 10,894.

(d) Estimate: ${}^-301 - {}^-1303 + 4993 \approx$

${}^-300 + 1300 + 5000 = \mathbf{6000}$.

Actual: 5995.

Assessment 5-2A
Multiplication and Division of Integers

1. $3 \cdot {}^-1 = {}^-1 + {}^-1 + {}^-1 = {}^-3$

$2 \cdot {}^-1 = {}^-1 + {}^-1 = {}^-2$

$1 \cdot {}^-1 = {}^-1$

$0 \cdot {}^-1 = 0$

${}^-1 \cdot {}^-1 = 1$.

3. Move four units to the left twice to arrive at the product $2({}^-4) = {}^-8$:

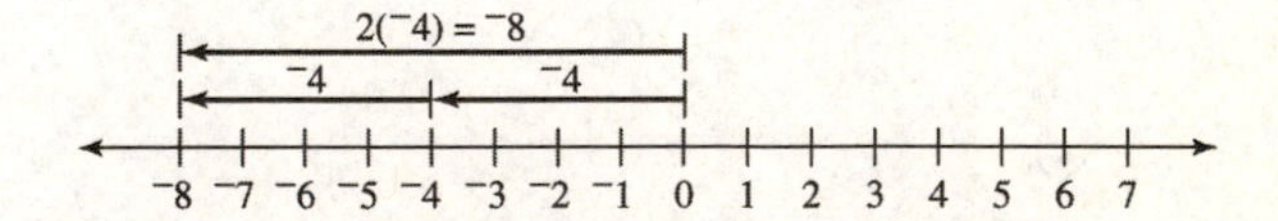

5. (a) The number of students will decrease by 20 per year over the next four years, or $4 \cdot {}^-20 = \mathbf{{}^-80}$.

(b) There were 20 more students per year four years ago, or ${}^-4 \cdot {}^-20 = \mathbf{80}$.

(c) The number of students will decrease by 20 per year over the next n years, or $\boldsymbol{n \cdot {}^{-}20 = {}^{-}20n}$ students.

(d) There were 20 more students per year during the past n years, or $\boldsymbol{{}^{-}n \cdot {}^{-}20 = 20n}$ students.

7. (a) $({}^{-}10 \div {}^{-}2)({}^{-}2) = 5 \cdot {}^{-}2 = \mathbf{{}^{-}10}$.

(b) $({}^{-}10 \cdot 5) \div 5 = {}^{-}50 \div 5 = \mathbf{{}^{-}10}$.

(c) ${}^{-}8 \div ({}^{-}8 + 8) = {}^{-}8 \div 0 \Rightarrow$ **not defined** because there is no integer that would make the multiplication statement true.

(d) $({}^{-}6 + 6) \div ({}^{-}2 + 2) = 0 \div 0 \Rightarrow$ **not defined** because too many integers can make the multiplication equation true.

(e) $|{}^{-}24| \div [4 \cdot (9 - 15)] = 24 \div 4 \cdot {}^{-}6 = {}^{-}1$. Order of operations specifies that multiplication and division be performed from left to right after the operations within the absolute value and parentheses are performed.

9. (a) $4x \div 4 = n$ if and only if $4x = 4n$, where n is an integer. Then $4x = 4n$ if and only if $\boldsymbol{n = x}$, so $4x \div 4 = x$.

(b) $({}^{-}xy) \div y = n$ if and only if $({}^{-}xy) = yn$. Then $({}^{-}xy) = yn$ if and only if $\boldsymbol{n = {}^{-}x}\ (y \neq 0)$.

11. (${}^{-}12{,}000$ acres per year $\times$ 9 years) $=$ ${}^{-}108{,}000$ acres, or **108,000 acres lost**.

13. (a) $({}^{-}2)^3 = {}^{-}2 \cdot {}^{-}2 \cdot {}^{-}2 = ({}^{-}2 \cdot {}^{-}2)\,{}^{-}2 = 4 \cdot {}^{-}2 = \mathbf{{}^{-}8}$.

(b) $({}^{-}2)^4 = ({}^{-}2 \cdot {}^{-}2) \cdot ({}^{-}2 \cdot {}^{-}2) = 4 \cdot 4 = \mathbf{16}$.

(c) $({}^{-}10)^5 \div ({}^{-}10)^2 = {}^{-}100{,}000 \div 100 = \mathbf{{}^{-}1000}$.

(d) $({}^{-}3)^5 \div ({}^{-}3) = {}^{-}243 \div {}^{-}3 = \mathbf{81}$.

(e) $({}^{-}1)^{50} = \mathbf{1}$ (i.e., ${}^{-}1$ taken to an even power).

(f) $({}^{-}1)^{151} = \mathbf{{}^{-}1}$ (i.e., ${}^{-}1$ taken to an odd power).

(g) ${}^{-}2 + 3 \cdot 5 - 1 = {}^{-}2 + 15 - 1 = 13 - 1 = \mathbf{12}$. Note the order of operation; multiplication before addition.

(h) $10 - 3 \cdot 7 - 4({}^{-}2) + 3 = 10 - 21 - {}^{-}8 + 3 = (10 + 8 + 3) - 21 = \mathbf{0}$.

15. (a) Always **negative**. x^2 is always positive, so its additive inverse is always negative.

(b) Always **positive**. Any non-zero number taken to an even power is always positive.

(c) Always **positive**. Any non-zero number taken to an even power is always positive.

(d) **Positive when x is negative; negative when x is positive**. I.e., the additive inverse of x^3, which may be positive or negative.

(e) **Positive when x is negative** (the additive inverse of a negative integer is positive). **Negative when x is positive** (the additive inverse of a positive integer is negative).

17. (a) **Commutative property of multiplication**, or $a \cdot b = b \cdot a$.

(b) **Closure property of addition**. The sum of an integer plus another integer is an integer.

(c) **Associative property of multiplication**, or $a(b \cdot c) = (a \cdot b)c$.

(d) **Distributive property of multiplication over addition**, or $a(b + c) = ab + ac$.

19. (a) ${}^{-}2(x - y) = {}^{-}2x - {}^{-}2y = \mathbf{{}^{-}2x + 2y}$.

(b) $x(x - y) = x \cdot x - x \cdot y = \mathbf{x^2 - xy}$.

(c) ${}^{-}x(x - y) = {}^{-}x^2 - {}^{-}xy = \mathbf{{}^{-}x^2 + xy}$.

(d) ${}^{-}2(x + y - z) = {}^{-}2x + {}^{-}2y - {}^{-}2z = \mathbf{{}^{-}2x - 2y + 2z}$.

21. (a) ${}^{-}3x - 8 = 7 \Rightarrow {}^{-}3x - 8 + 8 = 7 + 8 \Rightarrow {}^{-}3x = 15 \Rightarrow \frac{{}^{-}3}{{}^{-}3}x = \frac{15}{{}^{-}3} \Rightarrow \mathbf{x = {}^{-}5}$.

(b) ${}^{-}2(5x - 3) = 26 \Rightarrow {}^{-}10x + 6 = 26 \Rightarrow {}^{-}10x + 6 - 6 = 26 - 6 \Rightarrow {}^{-}10x = 20 \Rightarrow \frac{{}^{-}10}{{}^{-}10}x = \frac{20}{{}^{-}10} \Rightarrow \mathbf{x = {}^{-}2}$.

(c) $3x - x - 2x = 3 \Rightarrow 0 = 3 \Rightarrow$ **no solution**.

(d) ${}^{-}2(5x - 6) - 30 = {}^{-}x \Rightarrow {}^{-}10x + 12 - 30 = {}^{-}x \Rightarrow {}^{-}10x - 18 = {}^{-}x \Rightarrow {}^{-}18 = 9x \Rightarrow \mathbf{x = {}^{-}2}$.

(e) $x^2 = 4 \Rightarrow \mathbf{x = 2}$ or $\mathbf{x = {}^{-}2}$, since $2^2 = 4$ or $({}^{-}2)^2 = 4$.

(f) $(x - 1)^2 = 9 \Rightarrow$

[Type text] [Type text] [Type text]

(*i*) $(x-1) = 3 \Rightarrow \mathbf{x = 4}$, or

(*ii*) $(x-1) = {}^{-}3 \Rightarrow \mathbf{x = {}^{-}2}$.

(g) $(x-1)^2 = (x+3)^2 \Rightarrow$

(*i*) $(x-1) = (x+3) \Rightarrow 0 = 4 \Rightarrow$ not a viable solution.

(*ii*) $(x-1) = {}^{-}(x+3) \Rightarrow x - 1 = {}^{-}x - 3$ $\Rightarrow 2x = {}^{-}2 \Rightarrow \mathbf{x = {}^{-}1}$.

(h) $(x-1)(x+3) = 0 \Rightarrow$ either $(x-1) = 0$ or $(x+3) = 0$; i.e., the only way that the product of two numbers can be zero is that one or both must equal zero.

(*i*) **If** $(x-1) = 0 \Rightarrow \mathbf{x = 1}$.

(*ii*) **If** $(x+3) = 0 \Rightarrow \mathbf{x = {}^{-}3}$.

23. To *factor* an expression means to find an equivalent expression that is a product; i.e., if $N = ab$, then a and b are factors of N. Factoring may be said to undo the distributive property of multiplication over addition or subtraction.

(a) $3x + 5x = x(3+5) = \mathbf{8x}$. The factor common to both terms, x, divides each and then multiplies their sum.

(b) $xy + x = x \cdot y + x \cdot 1 = \mathbf{x(y+1)}$.

(c) $x^2 + xy = x \cdot x + x \cdot y = \mathbf{x(x+y)}$.

(d) $3xy + 2x - xz = \mathbf{x(3y + 2 - z)}$.

(e) $abc + ab - a = a(bc + b - 1) =$ $\mathbf{a[b(c+1) - 1]}$.

(f) $16 - a^2 = 4^2 - a^2 = \mathbf{(4+a)(4-a)}$; i.e., the factorization of the difference-of-squares formula.

(g) $4x^2 - 25y^2 = (2x)^2 - (5y)^2$ $= \mathbf{(2x + 5y)(2x - 5y)}$.

25. (a) (*i*) Arithmetic sequence; **difference** $(d) =$ ${}^{-}7 - {}^{-}10 = \mathbf{3}$.

(*ii*) The next two terms are: $5 + 3 = \mathbf{8}$ and $8 + 3 = \mathbf{11}$.

(*iii*) $a_n = a_1 + (n-1)d = {}^{-}10 +$ $(n-1) \cdot 3 = {}^{-}10 + 3n - 3 =$ $\mathbf{3n - 13}$.

(b) (*i*) Geometric sequence; **ratio** $(r) =$ ${}^{-}4 \div {}^{-}2 = \mathbf{2}$

(*ii*) The next two terms are: ${}^{-}64 \cdot 2 = \mathbf{{}^{-}128}$ and ${}^{-}128 \cdot 2 = \mathbf{{}^{-}256}$.

(*iii*) $a_n = a_1(r)^{n-1} = {}^{-}2(2)^{n-1} = \mathbf{{}^{-}2^n}$.

(c) (*i*) Geometric sequence; $r = {}^{-}2^2 \div 2 = \mathbf{{}^{-}2}$.

(*ii*) The next two terms are: ${}^{-}2^6 \cdot {}^{-}2 = \mathbf{2^7}$ and $2^7 \cdot {}^{-}2 = \mathbf{{}^{-}2^8}$.

(*iii*) $a_n = 2({}^{-}2)^{n-1} = {}^{-}({}^{-}2)({}^{-}2)^{n-1} =$ ${}^{-}({}^{-}2)^{n-1+1} = \mathbf{{}^{-}({}^{-}2)^n}$.

27. Let t be the temperature after m minutes: In the first reaction the temperature changed $28 - {}^{-}12 = 40°\text{C}$ in 5 minutes, or ${}^{-}8°$ per minute $\Rightarrow t = (28 - 8m)°\text{C}$.

The second reaction $\Rightarrow t = ({}^{-}57 - 3m)°\text{C}$.

(*i*) If temperatures are equal: $28 - 8m =$ ${}^{-}57 - 3m \Rightarrow 85 = 5m \Rightarrow m =$ **17 minutes**.

(*ii*) After 17 minutes of the first reaction:

$$t = 28 - 8(17) = \mathbf{{}^{-}108°C}$$

(using the second reaction to calculate t would have given the same result).

Assessment 5-2B

1. $3 \cdot {}^{-}2 = {}^{-}2 + {}^{-}2 + {}^{-}2 = {}^{-}6$

$2 \cdot {}^{-}2 = {}^{-}2 + {}^{-}2 = {}^{-}4$

$1 \cdot {}^{-}2 = {}^{-}2$

$0 \cdot {}^{-}2 = 0$

${}^{-}1 \cdot {}^{-}2 = 2$

${}^{-}2 \cdot {}^{-}2 = 4$.

3. Move three units to the left twice to arrive at the product $2({}^{-}3) = {}^{-}6$:

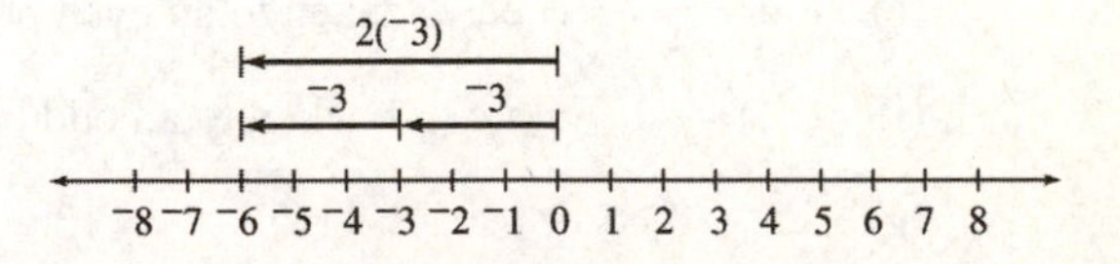

5. (a) $5({}^{-}20) = {}^{-}100$ **students**

(b) $({}^{-}5)({}^{-}20) = 100$ **students**

[Type text] [Type text] [Type text]

(c) $n(^-20) = -20n$ **students**

(d) $(^-n)(^-20) = 20n$ **students**

7. (a) Let $(a \div b)b = x$. Then $a \div b = x \div b \Rightarrow a = x$ and $(a \div b)b = \boldsymbol{a}$ for any integers a and b, $b \neq 0$.

(b) Let $(a \cdot b) \div b = x$. Then $a \cdot b = x \cdot b \Rightarrow a = x$ and $(a \cdot b) \div b = \boldsymbol{a}$ for any integers a and b, $b \neq 0$.

(c) $(^-8 + 8) \div 8 = 0 \div 8 = \mathbf{0}$.

(d) $(^-23 - {^-7}) \div 4 = {^-16} \div 4 = \mathbf{^-4}$. Order of operations specifies that the subtraction within the parentheses be performed first.

(e) $|^-28| \div [2|^-7|] = 28 \div [2(7)] = 28 \div 14 = \mathbf{2}$

9. (a) $(^-4x) \div x = n$ if and only if $(^-4x) = xn$. Then $(^-4x) = xn$ if and only if $\boldsymbol{n} = \mathbf{^-4}$ $(x \neq 0)$.

(b) $(^-10x + 5) \div 5 = n$ if and only if $^-10x + 5 = 5n$. $^-10x + 5 = 5n$ if and only if $5 \cdot (^-2x + 1) = 5n$. $5 \cdot (^-2x + 1) = 5n$ if and only if $\boldsymbol{n} = \mathbf{^-2}\boldsymbol{x} + \mathbf{1}$.

11. (a) $4 \cdot {^-11} = {^-44}$, or **44 yards lost**.

(b) $^-66 \div 11 = {^-6}$, or **6 yards lost on average**.

13. (a) $10 - 3 - 12 = 7 - 12 = \mathbf{^-5}$.

(b) $10 - (3 - 12) = 10 - (^-9) = 10 + 9 = \mathbf{19}$.

(c) $(^-3)^2 = {^-3} \cdot {^-3} = \mathbf{9}$. A negative integer taken to an even power is positive.

(d) $^-3^2 = {^-(3 \cdot 3)} = \mathbf{^-9}$.

(e) $^-5^2 + 3(^-2)^2 = {^-25} + 3 \cdot 4 = {^-25} + 12 = \mathbf{^-13}$.

(f) $^-2^3 = {^-(2 \cdot 2 \cdot 2)} = \mathbf{^-8}$.

(g) $(^-2)^5 = {^-2} \cdot {^-2} \cdot {^-2} \cdot {^-2} \cdot {^-2} = \mathbf{^-32}$. A negative integer taken to an odd power is negative.

(h) $^-2^4 = {^-(2 \cdot 2 \cdot 2 \cdot 2)} = \mathbf{^-16}$.

15. (a) Always **negative**. The additive inverse of a positive integer.

(b) Always **positive**. The product of either four positive or four negative integers is positive; i.e., any non-zero integer taken to an even power is positive.

(c) Always **positive**. Any non-zero integer taken to an even power is positive.

(d) **Positive when *x* is positive; negative when *x* is negative.**

(e) **Positive when *x* is negative; negative when *x* is positive** (i.e., additive inverses).

17. (a) **Closure property of multiplication**. The product of two integers is an integer.

(b) **Zero multiplication property**. For all integers a, $a \cdot 0 = 0 = 0 \cdot a$.

(c) **Distributive property of multiplication over addition**, or $a(b + c) = ab + ac$.

(d) **Commutative property of multiplication**, or $a \cdot b = b \cdot a$.

19. (a) $^-x(x - y - 3) = {^-x^2} - {^-xy} - {^-3x} = \mathbf{^-x^2 + xy + 3x}$.

(b) $(^-5 - x)(5 + x) = {^-5}(5 + x) + {^-x}(5 + x) = {^-25} + {^-5x} + {^-5x} + {^-x^2} = \mathbf{^-25 - 10x - x^2}$.

(c) $(x - y - 1)(x + y + 1) = x(x + y + 1) - y(x + y + 1) - 1(x + y + 1) = x^2 + xy + x + {^-xy} + {^-y^2} + {^-y} + {^-x} + {^-y} + {^-1} = \mathbf{x^2 - y^2 - 2y - 1}$.

(d) $(^-x^2 + 2)(x^2 - 1) = {^-x^2}(x^2 - 1) + 2(x^2 - 1) = {^-x^4} - {^-x^2} + 2x^2 + {^-2} = \mathbf{^-x^4 + 3x^2 - 2}$.

21. (a) $(2x - 1)^2 = (1 - 2x)^2$. True for **all integers**, since regardless of the value of x, the two sides of this equation will be additive inverses, and any number squared is non-negative.

[Type text] [Type text] [Type text]

(b) $x^3 = {}^-2^9 \Rightarrow x^3 = ({}^-2^3)^3$(i.e., ${}^-2^3 \cdot {}^-2^3 \cdot {}^-2^3 = {}^-2^9$) $\Rightarrow \mathbf{x = {}^-2^3 = {}^-8}$.

(c) ${}^-6x > {}^-x + 20 \Rightarrow {}^-5x > 20 \Rightarrow \mathbf{x < {}^-4}$. Note that the direction of the inequality changed under division by a negative number.

(d) ${}^-5(x - 3) > {}^-5 \Rightarrow x - 3 < 1 \Rightarrow \mathbf{x < 4}$.

(e) $x > {}^-2 \Rightarrow 5x > {}^-10 \Rightarrow 3 - 5x < 3 - {}^-10 \Rightarrow \mathbf{3 - 5x < 13}$.

(f) $x < 0 \Rightarrow 7x < 0 \Rightarrow 2 - 7x > 2 - 0 \Rightarrow \mathbf{2 - 7x > 2}$.

23. To *factor* an expression means to find an equivalent expression that is a product; i.e., if $N = ab$, then a and b are factors of N. factoring may be said to undo the distributive property of multiplication over addition or subtraction.

(a) $ax + 2x = \mathbf{x(a + 2)}$.

(b) $ax - 2x = \mathbf{x(a - 2)}$.

(c) $3x - 4x + 7x = x(3 - 4 + 7) = \mathbf{6x}$.

(d) $3x^2 + xy - x = \mathbf{x(3x + y - 1)}$.

(e) $(a + b)(c + 1) - (a + b) = (a + b)[(c + 1) - 1] = \mathbf{(a + b)c}$.

(f) $x^2 - 9y^2 = x^2 - (3y)^2 = \mathbf{(x + 3y)(x - 3y)}$.

(g) $(x^2 - y^2) + x + y = (x + y)(x - y) + (x + y) = (x + y)[(x - y) + 1] = \mathbf{(x + y)(x - y + 1)}$.

25. (a) (*i*) Arithmetic sequence; $\mathbf{d} = 7 - 10 = \mathbf{{}^-3}$.

(*ii*) The next two terms are: ${}^-5 + {}^-3 = \mathbf{{}^-8}$ and ${}^-8 + {}^-3 = \mathbf{{}^-11}$.

(*iii*) $a_n = 10 + (n - 1) \cdot {}^-3 = 10 - 3n + 3 = \mathbf{{}^-3n + 13}$.

(b) (*i*) Geometric sequence; $\mathbf{r} = 4 \div {}^-2 = \mathbf{{}^-2}$.

(*ii*) The next two terms are: $64 \cdot {}^-2 = \mathbf{{}^-128}$ and ${}^-128 \cdot {}^-2 = \mathbf{256}$.

(*iii*) $a_n = {}^-2({}^-2)^{n-1} = \mathbf{({}^-2)^n}$.

27. Let x represent the number of days. The amount in the first account is ${}^-120 + 40x$ and the amount in the second account is $300 - 20x$. Because the amounts are the same, $-120 + 40x = 300 - 20x$. $-120 + 40x = 300 - 20x$. $\Rightarrow 40x = 420 - 20x \Rightarrow 60x = 420$ $\Rightarrow x = \frac{420}{60} = \mathbf{7}$ **days**.

Review Problems

17. Plot **${}^-8$** on the number line then move 5 more units to the left to obtain **${}^-13$**.

19. **(a) 14.**
(b) 21.
(c) 4.
(d) 22.

Chapter 5 Review

1. (a) The additive inverse of 3 of **${}^-3$**.

(b) The additive inverse of ${}^-a$ is ${}^-({}^-a) = \mathbf{a}$.

(c) The additive inverse of ${}^-2 + 3$ is ${}^-({}^-2 + 3) = \mathbf{{}^-1}$.

(d) The additive inverse of $x + y$ is ${}^-(x + y) = {}^-x + {}^-y = \mathbf{{}^-x - y}$.

(e) The additive inverse of ${}^-x + y$ is ${}^-({}^-x + y) = x + {}^-y = \mathbf{x - y}$.

(f) The additive inverse of ${}^-x - y$ is ${}^-({}^-x - y) = x - {}^-y = \mathbf{x + y}$.

(g) $({}^-2)^5 = {}^-32$, thus the additive inverse is **32**.

(h) ${}^-2^5 = {}^-32$, thus the additive inverse is **32**.

3. (a) ${}^-x + 3 = 0 \Rightarrow {}^-x + 3 - 3 = 0 - 3 \Rightarrow {}^-x = {}^-3 \Rightarrow {}^-x({}^-1) = {}^-3({}^-1) \Rightarrow x = 3$.

(b) ${}^-2x = 10 \Rightarrow \frac{{}^-2x}{{}^-2} = \frac{10}{{}^-2} \Rightarrow x = {}^-5$.

(c) $0 \div ({}^-x) = 0 \Rightarrow \frac{0}{{}^-x}({}^-x) = 0({}^-x) \Rightarrow 0 = 0$; i.e., x may be **any integer except 0**.

(d) ${}^-x \div 0 = {}^-1 \Rightarrow$ **no integer solution**. Division by 0 is undefined.

[Type text] [Type text] [Type text]

(e) $3x - 1 = {}^-124 \Rightarrow 3x = {}^-123 \Rightarrow x = \frac{^-123}{3}$

$= \mathbf{{}^-41}$

(f) ${}^-2x + 3x = x \Rightarrow x({}^-2 + 3) = x$

$\Rightarrow x = x$; i.e., x may be **any integer**.

5. **(a)** Removing 5 black chips from $10 \Rightarrow 10 - 5 = \mathbf{5}$.

(b) When the 2 red chips are removed 3 black ones remain, thus $1 - {}^-2 = \mathbf{3}$.

7. **(a)** $x - 3x = x(1 - 3) = \mathbf{{}^-2x}$.

(b) $x^2 + x = \mathbf{x(x + 1)}$.

(c) $x^2 - 36 = x^2 - 6^2 = \mathbf{(x + 6)(x - 6)}$.

(d) $81y^4 - 16x^4 = (9y^2 + 4x^2)(9y^2 - 4x^2)$

$= \mathbf{(9y^2 + 4x^2)(3y + 2x)(3y - 2x)}$.

(e) $5 + 5x = \mathbf{5(1 + x)}$.

(f) $(x - y)(x + 1) - (x - y)$

$= (x - y)[(x + 1) - 1] = \mathbf{(x - y)x}$.

9. Answer may vary:

(a) $2 \div 1 \neq 1 \div 2$.

(b) $3 - (4 - 5) \neq (3 - 4) - 5$.

(c) $1 \div 2 \notin I$.

(d) $8 \div (4 - 2) \neq (8 \div 4) - (8 \div 2)$.

11. **(a)** If $a_n = ({}^-1)^n$:

$a_1 = ({}^-1)^1 = \mathbf{{}^-1}$;

$a_2 = ({}^-1)^2 = \mathbf{1}$;

$a_3 = ({}^-1)^3 = \mathbf{{}^-1}$;

$a_4 = ({}^-1)^4 = \mathbf{1}$;

$a_5 = ({}^-1)^5 = \mathbf{{}^-1}$;

$a_6 = ({}^-1)^6 = \mathbf{1}$.

(b) If $a_n = ({}^-2)^n$:

$a_1 = ({}^-2)^1 = \mathbf{{}^-2}$;

$a_2 = ({}^-2)^2 = \mathbf{4}$;

$a_3 = ({}^-2)^3 = \mathbf{{}^-8}$;

$a_4 = ({}^-2)^4 = \mathbf{16}$;

$a_5 = ({}^-2)^5 = \mathbf{{}^-32}$;

$a_6 = ({}^-2)^6 = \mathbf{64}$.

(c) If $a_n = {}^-2 - 3n$:

$a_1 = {}^-2 - 3 \cdot 1 = \mathbf{{}^-5}$;

$a_2 = {}^-2 - 3 \cdot 2 = \mathbf{{}^-8}$;

$a_3 = {}^-2 - 3 \cdot 3 = \mathbf{{}^-11}$;

$a_4 = {}^-2 - 3 \cdot 4 = \mathbf{{}^-14}$;

$a_5 = {}^-2 - 3 \cdot 5 = \mathbf{{}^-17}$;

$a_6 = {}^-2 - 3 \cdot 6 = \mathbf{{}^-20}$.

13. The questions answered correctly contribute ${}^+4$ points, and the questions answered incorrectly contribute ${}^-7$ points. Terry's score is $87(4) + 46({}^-7) = 348 - 322 =$ **26 points**.

15. **(a)** ${}^-40 - ({}^-62) = -40 + 62 =$ **22 degrees Celsius**.

(b) $8180 - 1100 =$ **7080 ft**.

(c) The pattern above suggest that as one travels South from the North Pole elevations must be higher to achieve negative temperatures. Since ${}^-11 - ({}^-40) = 29$ degrees Celsius, we might guess that since Hawaii is closer to the equator than Alaska, the elevation where ${}^-11°\text{C}$ was recorded was 7000 or 8000 feet higher than the elevation where ${}^-40°\text{C}$ was recorded in Arizona. A reasonable guess is that the elevation in Hawaii is 14,000 to 15,000 ft.

17. If sea level is considered 0 m elevation, then the average depth at the Pacific Ocean is **−3963 m**.

19. **(a)** If we consider the number of forces on the battlefield before the Union and Confederate force arrived as 0, we can describe the change in the number of forces engaged in the battle as ${}^+75{,}000$, ${}^+82{,}289$, ${}^-23{,}049$, and ${}^-28{,}063$.

(b) $^{-}23{,}049 + {}^{-}28{,}063 = {}^{-}51{,}112$. Since the $^{-}$ denotes casualties, there were **51,112** casualties.

CHAPTER 6

RATIONAL NUMBERS AND PROPORTIONAL REASONING

Ongoing Assessment 6-1A: The Set of Rational Numbers

1. **(a)** The solution to $8x = 7$ is $\frac{7}{8}$.

(b) Joe ate seven of the eight apple slices.

(c) The ratio of boys to girls in this math class is seven to eight.

3. The diagrams illustrate the Fundamental Law of Fractions; i.e., the value of a fraction does not change if its numerator and denominator are multiplied by the same nonzero number.

(a) $\frac{2}{3}$. Two of the three parts are shaded.

(b) $\frac{4}{6} = \frac{2}{3}$. Four of the six parts are shaded.

(c) $\frac{6}{9} = \frac{2}{3}$. Six of the nine parts are shaded.

(d) $\frac{8}{12} = \frac{2}{3}$. Eight of the twelve parts are shaded.

5. **(a)** One block of a square sectioned into four equal pieces is the same area as two blocks of a square sectioned into eight equal pieces; i.e., $\frac{1}{4} = \frac{2}{8}$:

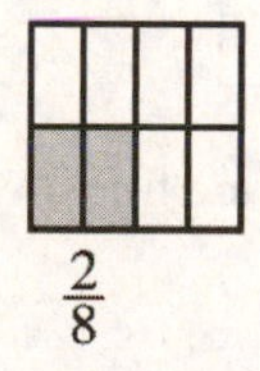

$\frac{2}{8}$

(b) One arc of a circle sectioned into three equal pieces is the same area as three arcs of a circle sectioned into nine equal pieces; i.e., $\frac{1}{3} = \frac{3}{9}$:

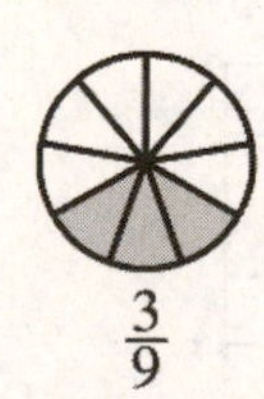

$\frac{3}{9}$

(c) One part of a hexagon sectioned into two equal pieces is the same area as three parts of a hexagon sectioned into six equal pieces; i.e., $\frac{1}{2} = \frac{3}{6}$:

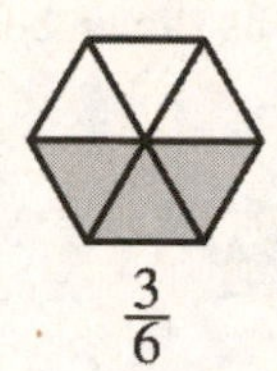

$\frac{3}{6}$

7. Answers may vary (as long as the numerator and denominator of the given fractions are multiplied by the same number); some possibilities are:

(a) $\frac{4}{18}, \frac{6}{27}, \frac{10}{45}$.

(b) $\frac{^-4}{10}, \frac{2}{^-5}, \frac{^-10}{25}$.

(c) $\frac{0}{6}, \frac{0}{9}, \frac{0}{12}$.

(d) $\frac{2a}{4}, \frac{3a}{6}, \frac{4a}{8}$.

9. **(a)** **Undefined.** Division by 0 is undefined.

(b) **Undefined.** Division by 0.

(c) **0.** $\frac{0}{5} = 0$ because $0 \cdot 5 = 0$.

(d) **Cannot be simplified.** $(2 + a)$ and a have no common factors other than 1.

(e) **Cannot be simplified.** $(15 + x)$ and $3x$ have no common factors other than 1.

11. **(a)** **Equal.** $\frac{375}{1000} = \frac{125 \cdot 3}{125 \cdot 8} = \frac{3}{8}$ or $375 \cdot 8 = 1000 \cdot 3$.

(b) **Equal.** $\frac{18}{54} = \frac{18}{3 \cdot 18} = \frac{1}{3}$ and $\frac{23}{69} = \frac{1 \cdot 23}{3 \cdot 23} = \frac{1}{3}$ or $18 \cdot 69 = 54 \cdot 23$.

(b) **Not equal.** $86 = 2 \cdot 43; 215 = 5 \cdot 43 \Rightarrow LCM(86, 215) = 2 \cdot 5 \cdot 43 = 430$. Then $\frac{^-21}{86} = \frac{^-21 \cdot 5}{86 \cdot 5} = \frac{^-105}{430}$ and $\frac{^-51}{215} = \frac{^-51 \cdot 2}{215 \cdot 2} = \frac{^-102}{430}$; $\frac{^-105}{430} \neq \frac{^-102}{430}$.

13. The shaded area takes in three of the four columns and six of the eight small rectangles. Since the area in each case is the same, $\frac{3}{4} = \frac{6}{8}$.

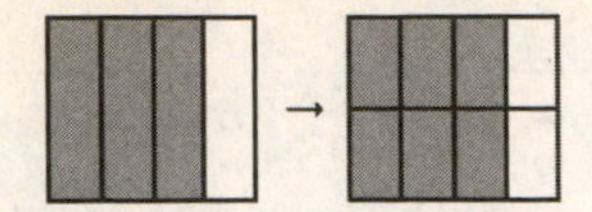

15. Mr. Gomez had $16 - 6 = 10$ gallons left. $\frac{10}{16} = \frac{5}{8}$ of a tank remained; the needle points to the 5th division of 8 as shown below:

17. **(a)** $\frac{7}{8} > \frac{5}{6}$. $LCD(8,6) = 24 \Rightarrow \frac{7}{8} = \frac{21}{24}$ and $\frac{5}{6} = \frac{20}{24}$. Or (noting that if $\frac{a}{b} > \frac{c}{d}$ then $ad > bc$) $7 \cdot 6 > 8 \cdot 5$.

(b) $2\frac{4}{5} > 2\frac{3}{6}$. $LCD(5,6) = 30 \Rightarrow 2\frac{4}{5} = 2\frac{24}{30}$ and $2\frac{3}{6} = 2\frac{15}{30}$, or $4 \cdot 6 > 5 \cdot 3$.

(c) $\frac{^-7}{8} < \frac{^-4}{5}$. $LCD\,(8,5) = 40 \Rightarrow \frac{^-7}{8} = \frac{^-35}{40}$ and $\frac{^-4}{5} = \frac{^-32}{40}$ (note that $^-35 < {}^-32$), or $^-7 \cdot 5 < 8 \cdot {}^-4$.

19. The nth term of the sequence is $\frac{n}{n+2}$, so the $(n + 1)$ th term is $\frac{n+1}{n+3}$. For the sequence to be increasing, then, it must be shown that $\frac{n+1}{n+3} > \frac{n}{n+2}$ for $n \geq 1$. By Theorem 6-3, this is true if, and only if, $(n + 1)(n + 2) > n(n + 3) \Rightarrow$ $n^2 + 2n + n + 2 > n^2 + 3n$, or $2 > 0$. Since 2 is in fact greater than 0, the statement $\frac{n+1}{n+3} > \frac{n}{n+2}$ is true.

21.

Number of miles	12	x
Number of inches	1	38

$\Rightarrow \frac{12 \text{ miles}}{1 \text{ inch}} = \frac{x \text{ miles}}{38 \text{ inches}} \Rightarrow 12 \cdot 38 = 1 \cdot x \Rightarrow$ **456 miles** $= x$.

23. Even though the figure is divided in an inconsistent manner, we could redraw the figure as follows.

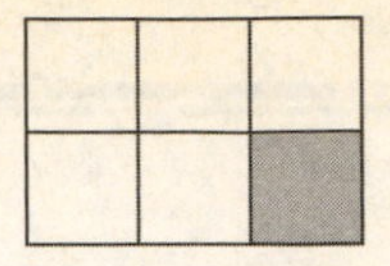

The shades region is $\frac{1}{6}$ of the entire figure.

25. Since $5 \cdot 31 > 6 \cdot 23$, $\frac{5}{23} > \frac{6}{31}$. Thus, Bren's class had a higher rate of As.

Assessment 6-1B

1. **(a)** The solution to $10x = 7$ is $\frac{7}{10}$.

(b) Joe ate seven of the ten apple slices.

(c) The ratio of boys to girls in this math class is seven to ten.

3. **(a)** The figure highlights three of the five parts of the bar:

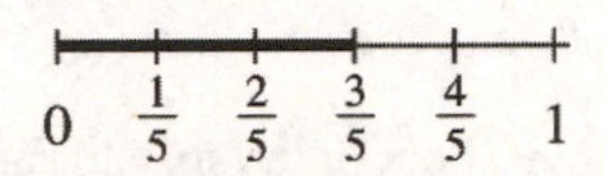

(b) Three of the five dots are filled:

(c) Nine of the fifteen blocks are shaded $= \frac{3}{5}$ of the area:

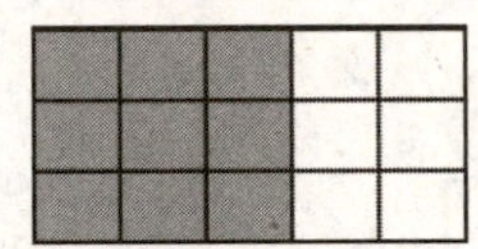

(d) Six of the ten dots are filled $= \frac{3}{5}$ of the dots:

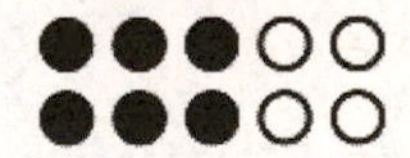

(e) Fifteen of the 25 blocks are shaded $= \frac{3}{5}$ of the area:

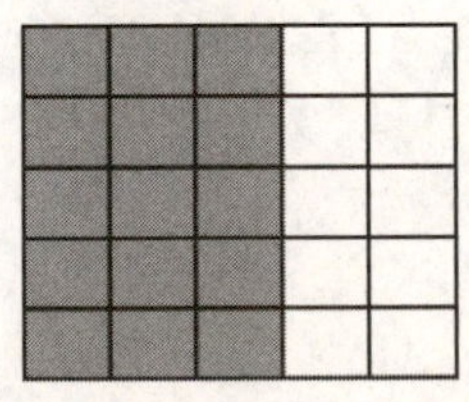

(f) $\frac{3}{5}$ of the volume is shaded:

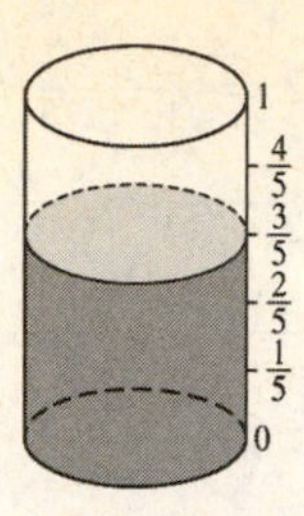

5. **(a)** Divide the figure into three equal parts, then add another equal part. The shaded area is then the whole:

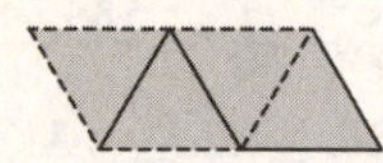

(b) Divide the figure into four equal parts, then remove one of the equal parts. The shaded area is then the whole:

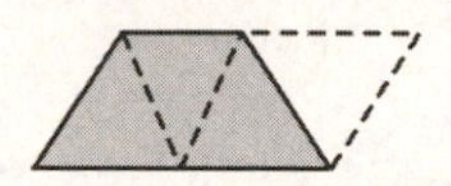

(c) If two circles are $\frac{1}{5}$ or $\frac{2}{10}$ of the whole, then the whole is ten shaded circles:

(d) If the rectangle is $\frac{1}{4}$ of the whole, then add three other equal rectangles:

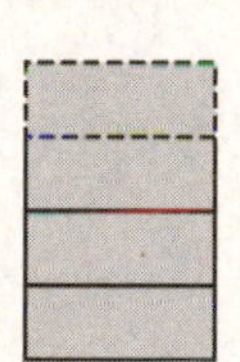

7. Answers may vary.

(a) $\frac{2}{6}, \frac{3}{9}, \frac{4}{12}$.

(b) $\frac{8}{10}, \frac{12}{15}, \frac{16}{20}$.

(c) $\frac{-6}{14}, \frac{-9}{21}, \frac{-12}{28}$.

(d) $\frac{2a}{6}, \frac{3a}{9}, \frac{4a}{12}$.

9. **(a)** **Cannot be simplified.** $\frac{6+x}{3x} \neq \frac{2+x}{x}$ because there are no factors common to each term.

(b) $\frac{2^6+2^5}{2^4+2^7} = \frac{2^5(2^1+1)}{2^4(1+2^3)} = \frac{2^5(3)}{2^4(9)} = \frac{32 \cdot 3}{16 \cdot 9} = \frac{2 \cdot 16 \cdot 3}{16 \cdot 3 \cdot 3} = \mathbf{\frac{2}{3}}$.

(c) $\frac{2^{100}+2^{98}}{2^{100}-2^{98}} = \frac{2^{98}(2^2+1)}{2^{98}(2^2-1)} = \frac{2^{98} \cdot 5}{2^{98} \cdot 3} = \mathbf{\frac{5}{3}}$.

11. **(a)** $\frac{6}{16} = \frac{6}{4^2} = \frac{6}{4^2} \cdot \frac{25}{25} \cdot \frac{25}{25} = \frac{6 \cdot 25 \cdot 25}{4 \cdot 25 \cdot 4 \cdot 25} = \frac{3{,}750}{10{,}000}$. The fractions are **equal**.

(b) **Not equal**. $\frac{17}{27}$ is in its reduced form and $\frac{25}{45} = \frac{5 \cdot 5}{5 \cdot 9} = \frac{5}{9}$ or $17 \cdot 45 \neq 27 \cdot 25$.

13.

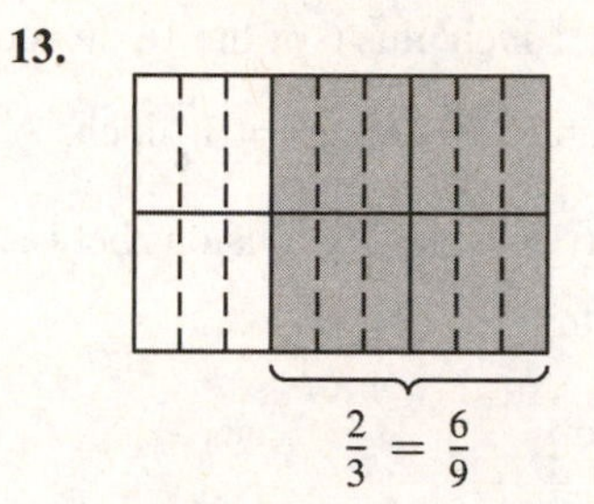

Alternative:

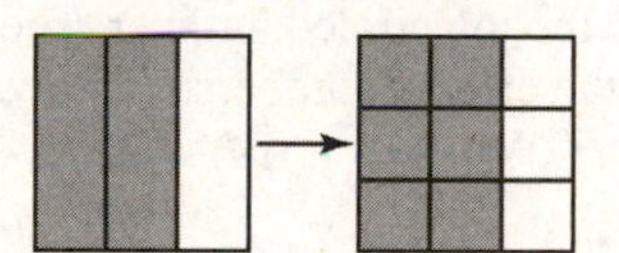

15. **Meter A by 3 minutes**. Meter A has $\frac{4}{10}$ of one hour $=$ 24 minutes remaining; meter B has $\frac{7}{10}$ of $\frac{1}{2}$ hour $= \frac{7}{20}$ of one hour $=$ 21 minutes remaining.

17. **(a)** $\frac{1}{^-7} < \frac{1}{^-8}$. $LCD(7, 8) = 56 \Rightarrow$ $\frac{1}{^-7} = \frac{^-1}{7} = \frac{^-8}{56}$ and $\frac{1}{^-8} = \frac{^-1}{8} = \frac{^-7}{56}$, or $1 \cdot {}^-8 < {}^-7 \cdot 1$.

(b) $\frac{2}{5} = \frac{4}{10} \cdot 2 \cdot 10 = 5 \cdot 4$.

(c) $\frac{0}{7} = \frac{0}{17} \cdot \frac{0}{7} = 0 = \frac{0}{17}$.

19. $\frac{a}{b} < 1$ and $\frac{c}{d} > 0 \Rightarrow \frac{a}{b} \cdot \frac{c}{d} < 1 \cdot \frac{c}{d} \Rightarrow \frac{a}{b} \cdot \frac{c}{d} < \frac{c}{d}$.

21.

Number of miles	120	x
Number of inches	1	3 / 4

$\Rightarrow \frac{120 \text{ miles}}{1 \text{ inch}} = \frac{x \text{ miles}}{3/4 \text{ inches}} \Rightarrow 120 \cdot 3/4 = 1 \cdot x \Rightarrow$ **90 miles** $= x$.

23. **(a)** $\mathbf{2\frac{7}{8}}$ **inch**, or 2 inches plus 14 of the 16 divisions between 2 and 3; $2\frac{14}{16}$ inch $= 2\frac{7}{8}$ inch.

(b) $\mathbf{2\frac{3}{8}}$ **inch**, or 2 inches plus 6 of the 16 divisions between 2 and 3; $2\frac{6}{16}$ inch $= 2\frac{3}{8}$ inch.

(c) $\mathbf{1\frac{3}{8}}$ **inch**, or 1 inch plus 6 of the 16 divisions between 1 and 2; $1\frac{6}{16}$ inch $= 1\frac{3}{8}$ inch.

(d) $\mathbf{\frac{15}{16}}$ **inch**, or 15 of the 16 divisions between 0 and 1 inch.

25. **(a)** $\frac{A}{B} = \frac{94}{94} = \mathbf{1}$.

(b) $\frac{A}{B} = \frac{86}{86} = \mathbf{1}$.

(c) Let x be the top circled number. The sum of the circled numbers is always $x + (x + 12) + (x + 19) + (x + 31) = 4x + 62$.

The sum of the interior numbers is always $(x + 10) + (x + 11) + (x + 20) + (x + 21) = 4x + 62$; thus the ratio of circled to interior numbers is always 1.

Assessment 6-2A

Addition and Subtraction of Rational Numbers

1. **(a)** Three possible methods are illustrated below:

(*i*) $\frac{1}{2} + \frac{2}{3} = \frac{1 \cdot 3 + 2 \cdot 2}{2 \cdot 3} = \frac{3+4}{6} = \frac{7}{6}$; or

(*ii*) $LCD(2,3) = 6 \Rightarrow \frac{1}{2} + \frac{2}{3} = \frac{1}{2} \cdot \frac{3}{3} + \frac{2}{3} \cdot \frac{2}{2} = \frac{3}{6} + \frac{4}{6} = \frac{7}{6}$; or

(*iii*)

$\frac{1}{2} + \frac{2}{3} = \frac{14}{12}$ or $\frac{7}{6}$

$\frac{1}{2} = \frac{6}{12}$ $\quad \frac{2}{3} = \frac{8}{12}$

0 $\quad$ 1 $\quad \frac{14}{12}$

(b) $LCD(12,3) = 12 \Rightarrow \frac{4}{12} - \frac{2}{3} = \frac{4}{12} - \frac{2}{3} \cdot \frac{4}{4} = \frac{4}{12} - \frac{8}{12} = \frac{^-4}{12} = \mathbf{\frac{^-1}{3}}$.

(c) $\frac{5}{x} + \frac{^-3}{y} = \frac{5 \cdot y - x \cdot 3}{x \cdot y} = \mathbf{\frac{5y - 3x}{xy}}$.

(d) $LCD(2x^2y, 2xy^2, x^2) = 2x^2y^2 \Rightarrow \frac{^-3}{2x^2y} + \frac{5}{6xy^2} + \frac{7}{x^2} = \frac{^-3}{2x^2y} \cdot \frac{y}{y} + \frac{5}{2xy^2} \cdot \frac{x}{x} + \frac{7}{x^2} \cdot \frac{2y^2}{2y^2} = \mathbf{\frac{^-3y + 5x + 14y^2}{2x^2y^2}}$.

(e) $\frac{5}{6} + 2\frac{1}{8} = \frac{5}{6} + \frac{17}{8} = \frac{5 \cdot 8 + 6 \cdot 17}{6 \cdot 8} = \frac{40 + 102}{48} = \frac{142}{48} = \frac{71}{24} = \mathbf{2\frac{23}{24}}$.

(f) $^-4\frac{1}{2} - 3\frac{1}{6} = {^-4}\frac{3}{6} - 3\frac{1}{6} = {^-7}\frac{4}{6} = \mathbf{^-7\frac{2}{3}}$ $\left(\text{or } \mathbf{\frac{^-23}{3}}\right)$.

3. **(a)** $6\frac{3}{4} = \frac{6}{1} + \frac{3}{4} = \frac{6 \cdot 4 + 1 \cdot 3}{1 \cdot 4} = \frac{24 + 3}{4} = \mathbf{\frac{27}{4}}$.

(b) $^-3\frac{5}{8} = {^-}\left(\frac{3}{1} + \frac{5}{8}\right) = {^-}\left(\frac{3 \cdot 8 + 1 \cdot 5}{1 \cdot 8}\right) = {^-}\left(\frac{24+5}{8}\right) = \mathbf{\frac{^-29}{8}}$.

5. Two possible methods are illustrated:

(*i*) $7\frac{1}{4} + 3\frac{5}{12} - 2\frac{1}{3}$

$= 7 + \frac{1}{4} + 3 + \frac{5}{12} - \left(2 + \frac{1}{3}\right)$

$= 7 + 3 - 2 + \frac{1}{4} + \frac{5}{12} - \frac{1}{3}$

$= 8 + \frac{3}{12} + \frac{5}{12} - \frac{4}{12}$

$= 8 + \frac{4}{12} = \mathbf{8\frac{1}{3}}$

(*ii*)

$$\begin{array}{rcr} 7\frac{1}{4} & = & 7\frac{3}{12} \\ +3\frac{5}{12} & = & +3\frac{5}{12} \\ \hline & & 10\frac{8}{12} \end{array}$$

$$\begin{array}{rcr} 10\frac{8}{12} & = & 10\frac{2}{3} \\ -2\frac{1}{3} & = & -2\frac{1}{3} \\ \hline & & \mathbf{8\frac{1}{3}} \end{array}$$

7.

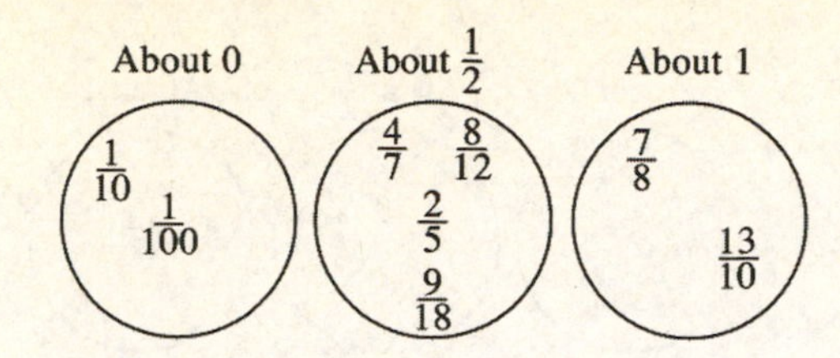

Note that $\frac{8}{12} = \frac{3}{4}$ so it could be placed in either the "about $\frac{1}{2}$" or the "about 1" oval.

9. **(a)** **2**. Each addend is about $\frac{1}{2}$; thus the best approximation would be $4 \cdot \frac{1}{2} = 2$.

(b) $\frac{3}{4}$. $\frac{30}{41}$ is about $\frac{3}{4}$; the other two addends are negligible compared to $\frac{3}{4}$.

11. **(a)** Region *A*. $\frac{20}{8} = \frac{10}{4}$ is between 2 and 3.

(b) Region *H*. $\frac{36}{8} = \frac{18}{4}$ is between 4 and 5.

(c) Region *T*. $\frac{60}{16} = \frac{15}{4}$ is between 3 and 4.

(d) Region *H*. $\frac{18}{4}$ is between 4 and 5.

13. The entire student population is represented by 1; subtract to obtain the senior's fraction; i.e., seniors make up $1 - \frac{2}{5} - \frac{1}{4} - \frac{1}{10}$ of the class. Thus $1 - \frac{2}{5} - \frac{1}{4} - \frac{1}{10} = \frac{20}{20} - \frac{8}{20} - \frac{5}{20} - \frac{2}{20} = \frac{5}{20} = \mathbf{\frac{1}{4}}$.

15. The amount of fabric to be used is $1\frac{7}{8} + 2\frac{3}{8} + 1\frac{2}{3} = 1\frac{21}{24} + 2\frac{9}{24} + 1\frac{16}{24} = 4\frac{46}{24} = 5\frac{22}{24}$ yards. The amount left over is $8\frac{3}{4} - 5\frac{22}{24} = 8\frac{18}{24} - 5\frac{22}{24} = 7\frac{42}{24} - 5\frac{22}{24} = 2\frac{20}{24} = \mathbf{2\frac{5}{6}}$ **yards.**

17. Assume in each case that the pattern continues:

(a) **(*i*)** $\mathbf{\frac{3}{2}, \frac{7}{4}, 2}$.

(*ii*) **Arithmetic.**

(*iii*) $\frac{1}{2} - \frac{1}{4} = \frac{3}{4} - \frac{1}{2} = 1 - \frac{3}{4} = \frac{5}{4} - 1 = \frac{1}{4}$. The sequence has a common difference of $\frac{1}{4}$; the next three terms are thus $\frac{5}{4} + \frac{1}{4} = \frac{3}{2}$, $\frac{3}{2} + \frac{1}{4} = \frac{7}{4}$, and $\frac{7}{4} + \frac{1}{4} = 2$.

(b) **(*i*)** $\mathbf{\frac{6}{7}, \frac{7}{8}, \frac{8}{9}}$.

(*ii*) **Not arithmetic.**

(*iii*) $\frac{2}{3} - \frac{1}{2} \neq \frac{3}{4} - \frac{2}{3}$. The sequence has no common difference; the general term is $\frac{n}{n+1}$ thus the next three terms are $\frac{6}{7}, \frac{7}{8}$, and $\frac{8}{9}$.

19. **(a)** **(*i*)** $\frac{1}{4} + \frac{1}{3 \cdot 4} = \frac{1}{4} + \frac{1}{12} = \frac{3}{12} + \frac{1}{12} = \frac{4}{12} = \frac{1}{3}$.

(*ii*) $\frac{1}{5} + \frac{1}{4 \cdot 5} = \frac{1}{5} + \frac{1}{20} = \frac{4}{20} + \frac{1}{20} = \frac{5}{20} = \frac{1}{4}$.

(*iii*) $\frac{1}{6} + \frac{1}{5 \cdot 6} = \frac{1}{6} + \frac{1}{30} = \frac{5}{30} + \frac{1}{30} = \frac{6}{30} = \frac{1}{5}$.

(b) $\mathbf{\frac{1}{n} = \frac{1}{n+1} + \frac{1}{n(n+1)}}$.

21. **(a)** $\frac{3x}{xy^2} + \frac{y}{x^2} = \frac{3x}{xy^2} \cdot \frac{x}{x} + \frac{y}{x^2} \cdot \frac{y^2}{y^2} = \mathbf{\frac{3x^2 + y^3}{x^2y^2}}$.

(b) $\frac{a}{xy^2} - \frac{b}{xyz} = \frac{a}{xy^2} \cdot \frac{z}{z} - \frac{b}{xyz} \cdot \frac{y}{y} = \mathbf{\frac{az - by}{xy^2z}}$.

(c) $\frac{a^2}{a^2 - b^2} - \frac{a-b}{a+b} = \frac{a^2}{(a+b)(a-b)} - \frac{a-b}{a+b} \cdot \frac{a-b}{a-b} = \frac{a^2 - (a^2 - 2ab + b^2)}{(a+b)(a-b)} = \mathbf{\frac{2ab - b^2}{(a^2 - b^2)}}$.

23. **No**. To make both recipes, you need $1\frac{3}{4} + 1\frac{1}{2} = 1 + \frac{3}{4} + 1 + \frac{1}{2} = 2 + \frac{3}{4} + \frac{1}{2} = 2 + 1\frac{1}{4} = 3\frac{1}{4}$ cups. $3\frac{1}{4} - 3 = \frac{1}{4}$ cups less than what is needed.

Assessment 6-2B

1. Various methods may be used:

(a) $\frac{^-1}{2} + \frac{2}{3} \Rightarrow \text{LCD}(2,3) = 6 \Rightarrow \frac{^-1}{2} \cdot \frac{3}{3} + \frac{2}{3} \cdot \frac{2}{2} = \frac{^-3}{6} + \frac{4}{6} = \mathbf{\frac{1}{6}}$.

(b) $\frac{5}{12} - \frac{2}{3} = \frac{5 \cdot 3 - 2 \cdot 12}{12 \cdot 3} = \frac{15 - 24}{36} =$
$\frac{^-9}{36} = \mathbf{\frac{^-1}{4}}$.

(c) $\frac{5}{4x} + \frac{^-3}{2y} = \frac{5 \cdot 2y + {^-3} \cdot 4x}{4x \cdot 2y} =$
$\frac{10y - 12x}{8xy} = \mathbf{\frac{5y - 6x}{4xy}}$.

(d) $\frac{^-3}{2x^2y^2} + \frac{5}{2xy^2} + \frac{7}{x^2y} \Rightarrow$
$\text{LCD}(2x^2y^2, 2xy^2, x^2y) =$
$2x^2y^2 \Rightarrow \frac{^-3}{2x^2y^2} + \frac{5}{2xy^2} \cdot \frac{x}{x} +$
$\frac{7}{x^2y} \cdot \frac{2y}{2y} = \mathbf{\frac{^-3 + 5x + 14y}{2x^2y^2}}$.

(e) $\frac{5}{6} - 2\frac{1}{8} = \frac{5}{6} - \frac{17}{8} = \frac{5 \cdot 8 - 6 \cdot 17}{6 \cdot 8} =$
$\frac{40 - 102}{48} = \frac{^-62}{48} = \mathbf{\frac{^-31}{24}} = \mathbf{^-1\frac{7}{24}}$.

(f) $^-4\frac{1}{2} + 3\frac{1}{6} \Rightarrow \text{LCD}(2,6) = 6 \Rightarrow$
$^-4\frac{1}{2} \cdot \frac{3}{3} + 3\frac{1}{6} = {^-4}\frac{3}{6} + 3\frac{1}{6} =$
$^-1\frac{2}{6} = \mathbf{^-1\frac{1}{3}} = \mathbf{\frac{^-4}{3}}$.

3. **(a)** $7\frac{1}{2} = \frac{7}{1} + \frac{1}{2} = \frac{7 \cdot 2 + 1 \cdot 1}{1 \cdot 2} = \frac{14 + 1}{2} = \mathbf{\frac{15}{2}}$.

(b) $^-4\frac{2}{3} = {^-}\left(\frac{4}{1} + \frac{2}{3}\right) = {^-}\left(\frac{4 \cdot 3 + 1 \cdot 2}{1 \cdot 3}\right) =$
$^-\left(\frac{12 + 2}{3}\right) = \mathbf{\frac{^-14}{3}}$.

5. $5\frac{1}{3} + 5\frac{5}{6} - 3\frac{1}{9} = 5 + 5 - 3 + \frac{1}{3} + \frac{5}{6} - \frac{1}{9}$
$= 7 + \frac{6}{18} + \frac{15}{18} - \frac{2}{18}$
$= 7 + \frac{19}{18} = 7 + 1\frac{1}{18} = \mathbf{8\frac{1}{18}}$

Alternative:

$$\begin{array}{rcr} 5\frac{1}{3} & = & 5\frac{2}{6} \\ +5\frac{5}{6} & = & +5\frac{5}{6} \\ \hline & & 10\frac{7}{6} \end{array}$$

$10\frac{7}{6} = 11\frac{1}{6}$.

$$\begin{array}{rcr} 11\frac{1}{6} & = & 11\frac{3}{18} \\ -3\frac{1}{9} & = & -3\frac{2}{18} \\ \hline & & \mathbf{8\frac{1}{18}} \end{array}$$

7.

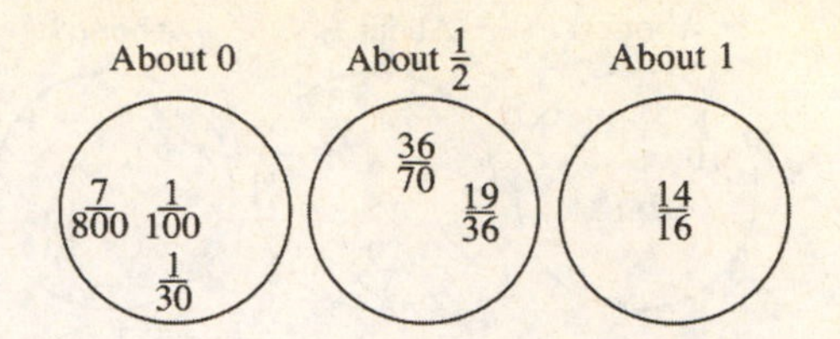

9. **(a)** The first addend is about 0, the next two addends are about $\frac{1}{2}$, and the third is about 1. **2** best approximates the sum.

(b) The third addend is about 0 and the second is close to one-fourth. The first is estimated at about three-forths. **1** is the best approximation.

11. **(a)** Region ***M***. $\frac{9}{8} = 1\frac{1}{8}$ is between 1 and 2.

(b) Region ***A***. $\frac{18}{8} = \frac{9}{4} = 2\frac{1}{4}$ is between 2 and 3.

(c) Region ***T***. $\frac{50}{16} = \frac{25}{8} = 3\frac{1}{8}$ is between 3 and 4.

(d) Region ***H***. $\frac{17}{4} = 4\frac{1}{4}$ is between 4 and 5.

13. **(a)** $\frac{1}{5}$ (Japan) $- \frac{1}{6}$ (Canada) $= \frac{6}{30} - \frac{5}{30} = \mathbf{\frac{1}{30}}$.

(b) $\frac{7}{20}$ (United States) $-\frac{1}{4}$ (England) $=$
$\frac{7}{20} - \frac{5}{20} = \frac{2}{20} = \mathbf{\frac{1}{10}}$.

(c) $\frac{7}{20}(2012) - \frac{1}{3}(2009) = \frac{21}{60} - \frac{20}{60} = \mathbf{\frac{1}{60}}$.

(d) **No**. The total number of dollars might have been greater in 2009 than in 2012, but the fraction of the total dollars $\left(\text{i.e., } \frac{1}{10} \text{ versus } \frac{1}{20}\right)$ was greater in 2012.

15. $38\frac{1}{4} - \left(15\frac{3}{4} + \frac{3}{8}\right) = 38\frac{2}{8} - \left(15\frac{6}{8} + \frac{3}{8}\right) =$
$38\frac{2}{8} - 16\frac{1}{8} = \mathbf{22\frac{1}{8}}$ **inches.**

17. **(a)** **(*i*)** $\mathbf{\frac{17}{3}, \frac{20}{3}, \frac{23}{3}}$.

(*ii*) **Arithmetic**.

(*iii*) $\frac{5}{3} - \frac{2}{3} = \frac{8}{3} - \frac{5}{3} = \frac{11}{3} - \frac{8}{3} = \frac{14}{3} =$
$\frac{11}{3} = \frac{3}{3}$. The sequence has a common difference of $\frac{3}{3}$ so the next three terms

are $\frac{14}{3} + \frac{3}{3} = \frac{17}{3}, \frac{17}{3} + \frac{3}{3} = \frac{20}{3}$, and $\frac{20}{3} + \frac{3}{3} = \frac{23}{3}$.

(b) ***(i)*** $\frac{^-5}{4}, \frac{^-7}{4}, \frac{^-9}{4}$.

(ii) **Arithmetic.**

(iii) $\frac{3}{4} - \frac{5}{4} = \frac{1}{4} - \frac{3}{4} = \frac{^-1}{4} - \frac{1}{4} = \frac{^-3}{4} - \left(\frac{^-1}{4}\right) = \frac{^-2}{4}$. The sequence has a common difference of $\frac{^-2}{4}$ so the next three terms are $\frac{^-3}{4} + \left(\frac{^-2}{4}\right) = \frac{^-5}{4}$, $\frac{^-5}{4} + \left(\frac{^-2}{4}\right) = \frac{^-7}{4}$, and $\frac{^-7}{4} + \left(\frac{^-2}{4}\right) = \frac{^-9}{4}$.

19. **(a)** $x - \frac{5}{6} = \frac{2}{3} \Rightarrow x = \frac{5}{6} + \frac{2}{3} = \frac{5}{6} + \frac{4}{6} = \frac{9}{6} = \frac{3}{2}$. $\mathbf{x = 1\frac{1}{2}}$.

(b) $x - \frac{7}{2^3 \cdot 3^2} = \frac{5}{2^2 \cdot 3^2} \Rightarrow x = \frac{5}{2^2 \cdot 3^2} + \frac{7}{2^3 \cdot 3^2} = \frac{5}{2^2 \cdot 3^2} \cdot \frac{2}{2} + \frac{7}{2^3 \cdot 3^2} = \frac{10+7}{2^3 \cdot 3^2} \Rightarrow \mathbf{x = \frac{17}{2^3 \cdot 3^2} = \frac{17}{72}}$.

21. Write the equivalent operations of using (take away) paint and adding paint:

$\frac{3}{4} - \frac{1}{3} + \frac{1}{2} = \frac{9}{12} - \frac{4}{12} + \frac{6}{12} = \mathbf{\frac{11}{12}}$ **cup.**

Review Problems

17. **(a)** $\frac{14}{21} = \frac{7 \cdot 2}{7 \cdot 3} = \mathbf{\frac{2}{3}}$.

(b) $\frac{117}{153} = \frac{9 \cdot 13}{9 \cdot 17} = \mathbf{\frac{13}{17}}$.

(c) $\frac{5^2}{7^2} = \mathbf{\frac{25}{49}}$, which cannot be further simplified.

(d) $\frac{a^2 + a}{1 + a} = \frac{a(a+1)}{a+1} = \frac{a}{1} = \mathbf{a}$.

(e) $\frac{a^2 + 1}{a + 1}$ **cannot be further simplified.** There are no factors common to both the numerator and denominator.

(f) $\frac{a^2 - b^2}{a - b} = \frac{(a+b)(a-b)}{a-b} = \mathbf{a + b}$ (if $a \neq b$).

Assessment 6-3A Multiplication and Division of Rational Numbers

1. **(a)** The blue-shaded vertical region represents $\frac{1}{3}$ of the total area; the yellow-shaded horizontal region represents $\frac{1}{4}$ of the total area. The blue-yellow region therefore represents $\frac{1}{4}$ of $\frac{1}{3}$, or the product of the two fractions. Since one of the twelve blocks is blue-yellow it represents $\mathbf{\frac{1}{4} \cdot \frac{1}{3} = \frac{1}{12}}$.

(b) The blue-shaded vertical region represents $\frac{3}{5}$ of the total area; the yellow-shaded horizontal region represents $\frac{2}{4}$ of the total area. The blue-yellow region therefore represents $\frac{2}{4}$ of $\frac{3}{5}$, or the product of the two fractions. Since six of the twenty blocks are blue-yellow it represents $\mathbf{\frac{2}{4} \cdot \frac{3}{5} = \frac{6}{20}}$.

3. **(a)** $\frac{49}{65} \cdot \frac{26}{98} = \frac{1274}{6370} = \frac{1 \cdot 1274}{5 \cdot 1274} = \mathbf{\frac{1}{5}}$.

(b) $\frac{a}{b} \cdot \frac{b^2}{a^2} = \frac{ab^2}{a^2 b} = \frac{b \cdot ab}{a \cdot ab} = \mathbf{\frac{b}{a}}$.

(c) $\frac{xy}{z} \cdot \frac{z^2 a}{x^3 y^2} = \frac{axyz^2}{x^3 y^2 z} = \frac{az \cdot xyz}{x^2 y \cdot xyz} = \mathbf{\frac{az}{x^2 y}}$.

5. **(a)** Multiplicative inverse of $\frac{^-1}{3}$ is $\frac{3}{^-1} = \mathbf{^-3}$.

(b) Multiplicative inverse of $3\frac{1}{3} = \frac{10}{3}$ is $\mathbf{\frac{3}{10}}$.

(c) Multiplicative inverse of $\frac{x}{y}$ is $\mathbf{\frac{y}{x}}$$(x, y \neq 0)$.

(d) Multiplicative inverse of $^-7 = \frac{^-7}{1}$ is $\frac{1}{^-7} = \mathbf{\frac{^-1}{7}}$.

7. Answers may vary; e.g.,

(a) $\frac{1}{2} \div \frac{1}{4} \neq \frac{1}{4} \div \frac{1}{2}$.

(b) $\left(\frac{2}{3} \div \frac{1}{2}\right) \div \frac{3}{5} \neq \frac{2}{3} \div \left(\frac{1}{2} \div \frac{3}{5}\right)$

9. **(a)** **20.** $3\frac{11}{12} \cdot 5\frac{3}{100} \approx 4 \cdot 5 = 20$.

(b) **16.** $2\frac{1}{10} \cdot 7\frac{7}{8} \approx 2 \cdot 8 = 16$.

(c) **1.** $\frac{1}{101}$ and $\frac{1}{103}$ are approximately equal.

11. **(c)**, between \$6 and \$8. Estimation gives a cost of approximately $6 \cdot 60¢ + 3 \cdot 80¢ = \$6.00$. All rounding was down, so the estimate is low.

13. Alberto has $\frac{5}{9}$ of the stock; Renatta has $\frac{1}{2} \cdot \frac{5}{9} = \frac{5}{18}$ of the stock. Thus $1 - \frac{5}{9} - \frac{5}{18} = \mathbf{\frac{1}{6}}$ **of the stock** is not owned by them.

15. Let a be the amount of money in the account. After spending \$50 there was $a - 50$ remaining. He spent $\frac{3}{5}$ of that, or $\frac{3}{5}(a - 50)$, leaving him $\frac{2}{5}(a - 50)$; half goes back into the bank, or $\frac{1}{2} \cdot \frac{2}{5}(a - 50) = \frac{1}{5}(a - 50)$. The other half was \$35; or $\frac{1}{5}(a - 50) = 35 \Rightarrow \frac{1}{5}a - 10 = 35 \Rightarrow \frac{1}{5}a = 45 \Rightarrow a = 5 \cdot 45 = \mathbf{\$225}$.

17. If a is any number and m and n are natural numbers, then $a^m \cdot a^n = a^{m+n}, \frac{a^m}{a^n} = a^{m-n}$, and $a^{-m} = \frac{1}{a^m}$:

(a) $3^{^-7} \cdot 3^{^-6} = 3^{^-7+^-6} = 3^{^-13} = \frac{1}{3^{13}} = \left(\frac{1}{3}\right)^{\mathbf{13}}$.

(b) $3^7 \cdot 3^6 = 3^{7+6} = \mathbf{3^{13}}$.

(c) $5^{15} \div 5^4 = 5^{15-4} = \mathbf{5^{11}}$.

(d) $5^{15} \div 5^{^-4} = 5^{15-^-4} = 5^{15+4} = \mathbf{5^{19}}$.

(e) $(^-5)^{^-2} = \frac{1}{(^-5)^2} = \frac{1}{\mathbf{5^2}} = \left(\frac{1}{5}\right)^{\mathbf{2}}$.

(f) $\frac{a^2}{a^{^-3}} = a^{2-^-3} = a^{2+3} = \mathbf{a^5}$.

19. Counterexamples may vary:

(a) **False**. $2^3 \cdot 3^2 = 8 \cdot 9 = 72 \neq (2 \cdot 3)^{3+2} = 6^5 = 7776$.

(b) **False**. $2^3 \cdot 3^2 = 72 \neq (2 \cdot 3)^{2 \cdot 3} = 6^6 = 46{,}656$.

(c) **False**. $2^3 \cdot 3^3 = (2 \cdot 3)^3 = 216 \neq (2 \cdot 3)^{2 \cdot 3} = 46{,}656$.

(d) **False**. Any number (except 0) to the 0 power $= 1$; 0^0 is undefined.

(e) **False**. $(2 + 3)^2 = (2 + 3)(2 + 3) = 25 \neq 2^2 + 3^2 = 13$.

(f) **False**. $(2 + 3)^{^-2} = \frac{1}{(2+3)^2} = \frac{1}{25} \neq \frac{1}{2^2} + \frac{1}{3^2} = \frac{13}{36}$.

21. **(a)** $3^x \leq 9 \Rightarrow 3^x \leq 3^2 \Rightarrow \mathbf{x \leq 2}$, where x is an integer.

(b) $25^x < 125 \Rightarrow (5^2)^x < 5^3 \Rightarrow 5^{2x} < 5^3 \Rightarrow 2x < 3 \Rightarrow \mathbf{x < \frac{3}{2}}$, where x is an integer.

(c) $3^{2x} > 27 \Rightarrow 3^{2x} > 3^3 \Rightarrow 2x > 3 \Rightarrow \mathbf{x > \frac{3}{2}}$, where x is an integer.

(d) $4^x \geq 1 \Rightarrow 4^x \geq 4^0 \Rightarrow \mathbf{x \geq 0}$, where x is an integer.

23. **(a)** $2S = 2\left(\frac{1}{2} + \frac{1}{2^2} + \cdots + \frac{1}{2^{64}}\right) = \frac{2}{2} + \frac{2}{2^2} + \cdots + \frac{2}{2^{64}} = \mathbf{1 + \frac{1}{2} + \frac{1}{2^2} + \cdots + \frac{1}{2^{63}}}$.

(b) $2S = 1 + S - \frac{1}{2^{64}} \Rightarrow 2S - S = 1 + S - \frac{1}{2^{64}} - S = \mathbf{1 - \frac{1}{2^{64}}}$.

(c) $S = \frac{1}{2} + \frac{1}{2^2} + \frac{1}{2^3} + \cdots + \frac{1}{2^n} = \mathbf{1 - \frac{1}{2^n}}$.

25. **(a)** $\mathbf{32^{50}}. 32^{50} = (2^5)^{50} = 2^{250}$, while $4^{100} = (2^2)^{100} = 2^{200}$.

(b) $\mathbf{(^-3)^{^-75}}. (^-27)^{^-15} = [(^-3)^3]^{^-15} = (^-3)^{^-45} = \frac{1}{(^-3)^{45}} = {^-}\left(\frac{1}{3^{45}}\right)$, while $(^-3)^{^-75} = \frac{1}{(^-3)^{75}} = {^-}\left(\frac{1}{3^{75}}\right)$ $\Rightarrow (^-3)^{^-75} > (^-27)^{^-45}\Big(\frac{1}{3^{45}} > \frac{1}{3^{75}}$; but their negatives reverse the direction of the inequality).

Assessment 6-3B

1. **(a)**

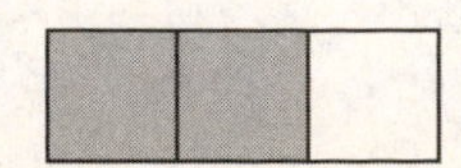

The figure above represents $\frac{2}{3}$. Divide this figure horizontally into four equal parts. This divides $\frac{2}{3}$ into 4 pieces of 6. Only 2 of the 6 have both shadings. This represents $\mathbf{\frac{1}{2} \cdot \frac{2}{3} = \frac{2}{6}}$.

(b) The figure shows $\frac{1}{3}$ divided into two equal parts. $\mathbf{\frac{1}{2} \cdot \frac{1}{3} = \frac{1}{6}}$.

3. **(a)** $2\frac{1}{3} \cdot 3\frac{3}{4} = \frac{7}{3} \cdot \frac{15}{4} = \frac{105}{12} = \frac{35 \cdot 3}{4 \cdot 3} = \frac{35}{4} = \mathbf{8\frac{3}{4}}$.

(b) $\frac{22}{7} \cdot 4\frac{2}{3} = \frac{22}{7} \cdot \frac{14}{3} = \frac{308}{21} = \frac{44 \cdot 7}{3 \cdot 7} = \frac{44}{3} = \mathbf{14\frac{2}{3}}$.

(c) $\frac{^-5}{2} \cdot 2\frac{1}{2} = \frac{^-5}{2} \cdot \frac{5}{2} = \frac{^-25}{4} = \mathbf{^-6\frac{1}{4}}$.

(d) $2\frac{3}{4} \cdot 2\frac{1}{3} = \frac{11}{4} \cdot \frac{7}{3} = \frac{77}{12} = \mathbf{6\frac{5}{12}}$.

(e) $\frac{a^2}{b^3} \cdot \frac{b^2}{a^3} = \frac{a^2b^2}{a^3b^3} = \frac{a^2b^2 \cdot 1}{a^2b^2 \cdot ab} = \mathbf{\frac{1}{ab}}$.

(f) $\frac{x^3y^2}{z} \cdot \frac{z}{x^2y} = \frac{x^3y^2z}{x^2yz} = \frac{x^2yzxy}{x^2yz} = \boldsymbol{xy}$.

5. **(a)** $\mathbf{\frac{7}{6}}$. $\frac{6}{7} \cdot \frac{7}{6} = 1$.

(b) $\mathbf{\frac{1}{8}}$. $8 \cdot \frac{1}{8} = 1$.

(c) $4\frac{1}{5} = \frac{21}{5}$. $\frac{21}{5} \cdot \mathbf{\frac{5}{21}} = 1$.

(d) $^-1\frac{1}{2} = \frac{^-3}{2}$. $\frac{^-3}{2} \cdot \left(\mathbf{\frac{^-2}{3}}\right) = 1$.

7. Let the fraction be $\frac{a}{b}$. Then $\frac{a+b}{b} = 3\left(\frac{a}{b}\right)$ or $\frac{a+b}{b} = \frac{3a}{b}$. The denominators are equal, so $a + b = 3a \Rightarrow b = 2a$. The fraction then becomes $\frac{a}{2a} = \mathbf{\frac{1}{2}}$.

9. **(a)** **2.** $20\frac{2}{3} \div 9\frac{7}{8} \approx 20 \div 10 = 2$.

(b) **24.** $3\frac{1}{20} \cdot 7\frac{77}{100} \approx 3 \cdot 8 = 24$.

(c) **1.** $\frac{1}{10^3} \div \frac{1}{1001} \approx \frac{1}{1000} \div \frac{1}{1000} = 1$.

11. If n is the number, then $3n - \frac{7}{18} = 2n + \frac{5}{12} \Rightarrow n = \frac{5}{12} + \frac{7}{18} = \frac{5 \cdot 18 + 12 \cdot 7}{12 \cdot 18} = \mathbf{\frac{29}{36}}$.

13. **(a)** Let u be the number of uniforms to be made. Assuming no waste, $u = 29\frac{1}{2} \div \frac{3}{4} = \frac{59}{2} \cdot \frac{4}{3} = \frac{236}{6} = \frac{118}{3} = 39\frac{1}{3}$, or **39 uniforms** can be made.

(b) Enough material for $\frac{1}{3}$ uniform will be left over. Each uniform needs $\frac{3}{4}$ yard material, so $\frac{1}{3} \cdot \frac{3}{4} = \frac{3}{12} = \mathbf{\frac{1}{4}}$ **yard** will remain.

15. Let b be the number of pages in the book. Jasmine has read $\frac{3}{4}b$, so she has $b - \frac{3}{4}b = \frac{1}{4}b$ yet to read. If 82 pages $= \frac{1}{4}b$ then multiply each side of the equation by 3 to yield $246 = \frac{3}{4}b$, or she has read **246 pages**.

17. **(a)** $\left(\frac{1}{3}\right)^{^-1} = \left(\frac{3}{1}\right)^1 = \mathbf{3}$.

(b) $\frac{a^{^-3}}{a} = \frac{\left(\frac{1}{a}\right)^3}{a} = \frac{1}{a^3} \cdot \frac{1}{a} = \mathbf{\frac{1}{a^4}}$.

(c) $\frac{(a^{^-4})^3}{a^{^-4}} = \frac{a^{^-12}}{a^{^-4}} = a^{^-12-(^-4)} = a^{^-8} = \mathbf{\frac{1}{a^8}}$.

(d) $\frac{a}{a^{^-1}} = \frac{a}{\left(\frac{1}{a}\right)} = \frac{a}{1} \cdot \frac{a}{1} = \boldsymbol{a^2}$.

(e) $\frac{a^{^-3}}{a^{^-2}} = \frac{a^2}{a^3} = \boldsymbol{\frac{1}{a}}$.

19. False counterexamples may vary:

(a) **False.** $\frac{2^3}{3^2} = \frac{8}{9} \neq \left(\frac{2}{3}\right)^{3-2} = \frac{2}{3}$.

(b) **True.**
$(ab)^{^-m} = \frac{1}{(ab)^m} = \frac{1}{a^mb^m} = \frac{1}{a^m} \cdot \frac{1}{b^m}$.

(c) **False.** $\left(\frac{2}{2^{^-1}+3^{^-1}}\right)^{^-1} = \frac{\left(\frac{1}{2}+\frac{1}{3}\right)}{2} = \frac{\left(\frac{5}{6}\right)}{2} = \frac{5}{12} \neq \frac{1}{2} \cdot \frac{1}{2+3} = \frac{1}{2} \cdot \frac{1}{5} = \frac{1}{10}$.

(d) **True.** $2(a^{^-1} + b^{^-1})^{^-1} = 2 \cdot \frac{1}{a^{^-1}+b^{^-1}} = 2 \cdot \frac{1}{\frac{1}{a}+\frac{1}{b}} = 2 \cdot \frac{1}{\left(\frac{a+b}{ab}\right)} = \frac{2ab}{a+b}$.

(e) **False.** $3^{2 \cdot 3} = 3^6 = 729 \neq 3^2 \cdot 3^3 = 243$.

(f) **True.** $\left(\frac{a}{b}\right)^{^-1} = \frac{1}{\left(\frac{a}{b}\right)} = \frac{b}{a}$.

21. (a) $3^x \geq 81 \Rightarrow 3^x \geq 3^4 \Rightarrow \mathbf{x \geq 4}$.

(b) $4^x \geq 8 \Rightarrow (2^2)^x \geq 2^3 \Rightarrow 2x \geq 3$
$\Rightarrow x \geq \frac{3}{2} \Rightarrow \mathbf{x \geq 2}$, where x is an integer.

(c) $3^{2x} \leq 27 \Rightarrow 3^{2x} \leq 3^3 \Rightarrow 2x \leq 3$
$\Rightarrow x \leq \frac{3}{2} \Rightarrow \mathbf{x \leq 1}$, where x is an integer.

(d) $2^x < 1 \Rightarrow 2^x < 2^0 \Rightarrow$
$\mathbf{x < 0 \Rightarrow x \leq {}^{-}1}$.

23. (a) $3S = 3\left(\frac{1}{3} + \frac{1}{3^2} + \cdots + \frac{1}{3^{61}}\right)$
$= \frac{3}{3} + \frac{3}{3^2} + \cdots + \frac{3}{3^{61}}$
$= \mathbf{1 + \frac{1}{3} + \frac{1}{3^2} + \cdots + \frac{1}{3^{63}}}$.

(b) $3S = 1 + S - \frac{1}{3^{64}} \Rightarrow 3S - S =$
$1 + S - \frac{1}{2^{64}} - S = \mathbf{1 - \frac{1}{3^{64}}}$.

(c) $2S = 1 - \frac{1}{3^n} \Rightarrow S = \mathbf{\frac{1}{2}\left(1 - \frac{1}{3^n}\right)}$.

25. (a) $\mathbf{32^{100}}$. $32^{100} = (2^5)^{100} = 2^{500}$, while
$4^{200} = (2^2)^{200} = 2^{400}$.

(b) $\mathbf{({}^{-}3)^{{}^{-}50}}$. $({}^{-}27)^{{}^{-}15} = [({}^{-}3)^3]^{{}^{-}15} =$
$({}^{-}3)^{{}^{-}45} = \frac{1}{({}^{-}3)^{45}} = {}^{-}\left(\frac{1}{3^{45}}\right)$, while
$({}^{-}3)^{{}^{-}50} = \frac{1}{({}^{-}3)^{50}} = \left(\frac{1}{3^{50}}\right)$, a positive number.
So $({}^{-}3)^{{}^{-}50} > ({}^{-}27)^{{}^{-}15}$
($\frac{1}{3^{45}} > \frac{1}{3^{50}}$, but the negative reverses the direction of the inequality).

27. Let $x =$ the amount Brandy spent on the horse's keep. Counting the amount she paid for the horse, Brandy spent $270 + x$. Her profit or loss can be found by $270 + x - 540$. In the second sentence we are told Brandy lost $\frac{1}{2}(270) + \frac{1}{4}x$. Thus,

$$\frac{1}{2}(270) + \frac{1}{4}x = 270 + x - 540$$
$$135 + \frac{1}{4}x = x - 270$$
$$135 + 270 = x - \frac{1}{4}x$$
$$405 = \frac{3}{4}x$$
$$x = \frac{4}{3} \cdot 405 = 540 \text{ dollars.}$$

Since x represents the money Brandy spent on keep, her loss is $270 + 540 - 540 =$ **270 dollars**.

Review Problems

21. (a) $\frac{{}^{-}3}{16} + \frac{7}{4} = \frac{{}^{-}3}{16} + \frac{28}{16} = \frac{25}{16} = \mathbf{1\frac{9}{16}}$.

(b) $\frac{1}{6} + \frac{{}^{-}4}{9} + \frac{5}{3} = \frac{3}{18} + \frac{{}^{-}8}{18} + \frac{30}{18} =$
$\frac{25}{18} = \mathbf{1\frac{7}{18}}$.

(c) $\frac{{}^{-}5}{2^3 \cdot 3^2} - \frac{{}^{-}5}{2 \cdot 3^3} = \frac{{}^{-}5 \cdot 3}{2^3 \cdot 3^3} - \frac{{}^{-}5 \cdot 2^2}{2^3 \cdot 3^3} = \frac{{}^{-}15}{216} +$
$\frac{20}{216} = \mathbf{\frac{5}{216}}$.

(d) $3\frac{4}{5} + 4\frac{5}{6} = 3\frac{24}{30} + 4\frac{25}{30} = 7\frac{49}{30} = \mathbf{8\frac{19}{30}}$.

(e) $5\frac{1}{6} - 3\frac{5}{8} = 5\frac{4}{24} - 3\frac{15}{24} = 4\frac{28}{24} - 3\frac{15}{24} =$
$\mathbf{1\frac{13}{24}}$.

(f) ${}^{-}4\frac{1}{3} - 5\frac{5}{12} = \frac{{}^{-}13}{3} - \frac{65}{12} = \frac{{}^{-}52}{12} - \frac{65}{12} =$
$\frac{{}^{-}117}{12} = \frac{{}^{-}39}{4} = \mathbf{{}^{-}9\frac{3}{4}}$.

Assessment 6-4A: Proportional Reasoning

1. (a) There are five vowels and 21 consonants. Their ratio is $\frac{5 \text{ vowels}}{21 \text{ consonants}} = \frac{5}{21}$, or **5:21**.

(b) $\frac{21 \text{ consonants}}{5 \text{ vowels}} = \frac{21}{5}$, or **21:5**.

(c) $\frac{21 \text{ consonants}}{26 \text{ alphabet letters}} = \frac{21}{26}$, or **21:26**.

(d) Answers may vary. "Break" (2 vowels, 3 consonants) or "minor" (2 vowels, 3 consonants) are two.

3. (a) Because the ratio is 2:3, there are $2x$ boys and $3x$ girls. The ratio of boys to all students is then $\frac{2x}{2x+3x} = \frac{2x}{5x} = \frac{2}{5} =$ **2:5**.

(b) $\mathbf{m:(m+n)}$. See part (a) above.

(c) Because the ratio of girls to all students is $\frac{3}{5}$, there are 3 girls to every 2 boys, or a ratio of girls to boys of **3:2**.

5. $\frac{4 \text{ grapefruit}}{80¢} = \frac{12 \text{ grapefruit}}{x¢} \Rightarrow 4 \cdot x = 80 \cdot 12 \Rightarrow$ $x = \frac{80 \cdot 12}{4} = 240¢$, so **$1.80 for 12** is a better buy.

7. $\frac{40 \text{ pages}}{50 \text{ minutes}} = \frac{x \text{ pages}}{80 \text{ minutes}} \Rightarrow 50 \cdot x = 40 \cdot 80 \Rightarrow$ $x = \frac{40 \cdot 80}{50} = \mathbf{64 \text{ pages}}$.

9. (*i*) There are a total of $2 + 3 + 5 = 10$ shares. Using a proportion, and letting G represent the amount of money Gary would receive, we write: $\frac{2}{10} = \frac{G}{82{,}000} \Rightarrow 10G =$ $2 \cdot 82{,}000 \Rightarrow G = \mathbf{\$16{,}400}$.

Alternative thinking: Common Core State Standards suggests we consider using rates to solve this type of problem. Gary's rate is 2 to 10. Thus,

$\frac{2}{10} \cdot 82{,}000 = \frac{2 \cdot 82{,}000}{10} = \mathbf{\$16{,}400}$.

(*ii*) $\frac{3}{10} = \frac{\text{Bill's amount}}{82{,}000} \Rightarrow 10(\text{Bill's}) =$ $3 \cdot 82{,}000 \Rightarrow \text{Bill's amount} = \mathbf{\$24{,}600}$.

(*iii*) $\frac{5}{10} = \frac{\text{Carmella's amount}}{82{,}000} \Rightarrow$ $10 \cdot \text{Carmella's} = 5 \cdot 82{,}000 \Rightarrow$ $\text{Carmella's amount} = \mathbf{\$41{,}000}$.

11. Success:failure $= 5{:}4 \Rightarrow \frac{5 \text{ successes}}{4 \text{ failures}} =$ $\frac{75 \text{ successes}}{x \text{ failures}} \Rightarrow 5 \cdot x = 4 \cdot 75 \Rightarrow x = \frac{4 \cdot 75}{5} =$ 60 failures. 75 successes + 60 failures = **135 attempts**.

13. The proportion implies $12¢ \cdot 48 \text{ oz} =$ $16¢ \cdot 36 \text{ oz}$. Other equivalent proportions are thus:

(*i*) $\frac{12¢}{16¢} = \frac{36 \text{ ounces}}{48 \text{ ounces}}$.

(*ii*) $\frac{48 \text{ ounces}}{16¢} = \frac{36 \text{ ounces}}{12¢}$.

(*iii*) $\frac{16¢}{12¢} = \frac{48 \text{ oz}}{36 \text{ oz}}$.

15. (a) $\frac{4 \text{ rpm on large gear}}{6 \text{ rpm on small gear}} = \frac{18 \text{ teeth on small gear}}{x \text{ teeth on large gear}} \Rightarrow$ $4 \cdot x = 6 \cdot 18 \Rightarrow x = \frac{6 \cdot 18}{4} = \mathbf{27 \text{ teeth}}$.

(b) $\frac{200 \text{ rpm on large gear}}{600 \text{ rpm on small gear}} = \frac{x \text{ teeth on small gear}}{60 \text{ teeth on large gear}} \Rightarrow$ $600 \cdot x = 200 \cdot 60 \Rightarrow x = \frac{200 \cdot 60}{600} =$ **20 teeth**.

17. $\frac{160 \text{ pounds on Earth}}{416 \text{ pounds on Jupiter}} = \frac{120 \text{ pounds on Earth}}{x \text{ pounds on Jupiter}} \Rightarrow$ $160 \cdot x = 416 \cdot 120 \Rightarrow x = \frac{416 \cdot 120}{160} =$ **312 pounds**.

19. $\frac{4.2 \text{ ohms}}{5 \text{ feet}} = \frac{x \text{ ohms}}{18 \text{ feet}} \Rightarrow 5 \cdot x = 4.2 \cdot 18 \Rightarrow$ $x = \frac{4.2 \cdot 18}{5} = \mathbf{15.12 \text{ ohms}}$.

21. The ratio between the mass of the gold in the ring to the mass of the ring is 18:24. If x is the number of ounces of pure gold in a ring that weighs 0.4 ounces, then $\frac{18 \text{ ounces of gold}}{24 \text{ ounces of ring}} =$ $\frac{x \text{ ounces of gold}}{0.4 \text{ ounces of ring}} \Rightarrow 24x = 18 \cdot 0.4 \Rightarrow$ $x = \frac{18 \cdot 0.4}{24} = 0.3$ ounces of gold in the ring. 0.3 ounces at \$300 per ounce of = **\$90**.

23. (a) The total number of men in all three rooms is $1 + 2 + 5 = 8$; the total number of women in all three rooms is $2 + 4 + 10 = 16$. The ratio of men to women is $\frac{8}{16} = \frac{1}{2}$, or **1:2**.

(b) Let $\frac{a}{b} = \frac{c}{d} = \frac{e}{f} = r$.

Then $a = br$;

$c = dr$;

$e = fr$.

So $a + c + e = br + dr + fr \Rightarrow$

$a + c + e = r(b + d + f) \Rightarrow$

$r = \frac{a+c+e}{b+d+f}$

Thus $r = \frac{a}{b} = \frac{c}{d} = \frac{e}{f} = \frac{a+c+e}{b+d+f}$.

Assessment 6-4B

1. **(a)** There are four vowels and seven consonants in "Mississippi." Their ratio is $\frac{4 \text{ vowels}}{7 \text{ consonants}} = \frac{4}{7}$, or **4:7**.

(b) $\frac{7 \text{ consonants}}{4 \text{ vowels}} = \frac{7}{4}$, or **7:4**.

(c) There are seven consonants in the eleven letters, so $\frac{7 \text{ consonants}}{11 \text{ letters}} = \frac{7}{11}$, or **7:11**.

3. $\frac{5 \text{ adults}}{1 \text{ teen}} = \frac{12{,}345 \text{ adults}}{x \text{ teens}} \Rightarrow 5 \cdot x = 1 \cdot 12{,}345 \Rightarrow x = \frac{1 \cdot 12{,}345}{5} = $ **2469 teen drivers.**

5. Let $5x$ represent width and $9x$ represent length. The perimeter of the rectangle is $2 \cdot 5x + 2 \cdot 9x = 2800 \Rightarrow 28x = 2800 \Rightarrow x = 100$. Thus width $= 5 \cdot 100 =$ **500 feet** and length $= 9 \cdot 100 =$ **900 feet.**

7. $\frac{9 \text{ months}}{6 \text{ vocation days}} = \frac{12 \text{ months}}{x \text{ vocation days}} \Rightarrow 9 \cdot x = 6 \cdot 12 \Rightarrow x = \frac{6 \cdot 12}{9} =$ **8 days.**

9. $\frac{30 \text{ feet tall}}{12 \text{ foot shadow}} = \frac{x \text{ feet tall}}{14 \text{ foot shadow}} \Rightarrow 12 \cdot x = 30 \cdot 14 \Rightarrow x = \frac{30 \cdot 14}{12} =$ **35 feet.**

11. Solutions may vary. The ratio of 4 to 20 is the same in each case; x (the number of tickets) and y (the cost) could therefore be $\frac{8 \text{ tickets}}{\$40}, \frac{12 \text{ tickets}}{\$60}, \frac{16 \text{ tickets}}{\$80}, \ldots$.

13. **(a)** $\frac{\text{Length of hand}}{\text{Length of big toe}} = \frac{14}{3} \Rightarrow \frac{14 \text{ cm}}{3 \text{ cm}} = \frac{x \text{ cm}}{6 \text{ cm}} \Rightarrow 3x = 14 \cdot 6 \Rightarrow x = \frac{14 \cdot 6}{3} =$ **28 cm.**

(b) $\frac{\text{Length of hand}}{\text{Length of foot}} = \frac{7}{9} \Rightarrow \frac{7 \text{ cm}}{9 \text{ cm}} = \frac{21 \text{ cm}}{x \text{ cm}} \Rightarrow 7x = 21 \cdot 9 \Rightarrow x = \frac{21 \cdot 9}{7} =$ **27 cm.**

(c) $\frac{\text{Length of elbow to end of hand}}{\text{Shoulder to elbow}} = \frac{8}{5} \Rightarrow \frac{8 \text{ in}}{5 \text{ in}} = \frac{20 \text{ in}}{x \text{ in}} \Rightarrow 8x = 20 \cdot 5 \Rightarrow x = \frac{20 \cdot 5}{8} =$ **12.5 in.**

15. $\frac{240 \text{ mi}}{15 \text{ gallons}} = \frac{x \text{ mi}}{3 \text{ gallons}} \Rightarrow 15x = 240 \cdot 3 \Rightarrow x = \frac{240 \cdot 3}{15} =$ **48 mi.**

17. **(a)** There are 50 stars and 13 stripes in the flag, or a ratio of $\frac{50 \text{ stars}}{13 \text{ stripes}} =$ **50:13.**

(b) $\frac{13 \text{ stripes}}{50 \text{ stars}} =$ **13:50.**

19. $\frac{x}{y} = \frac{a}{b} \Rightarrow xb = ya \Rightarrow \frac{b}{a} = \frac{y}{x}, \frac{a}{x} = \frac{b}{y}$, and $\frac{x}{a} = \frac{y}{b}$ $(x, y, a, b \neq 0)$.

21. **(a)** $\frac{a}{b} = \frac{c}{d} \Rightarrow \frac{a}{b} + 1 = \frac{c}{d} + 1 \Rightarrow \frac{a}{b} + \frac{b}{b} = \frac{c}{d} + \frac{d}{d} \Rightarrow \frac{a+b}{b} = \frac{c+d}{d}$.

(b) $\frac{a}{b} = \frac{c}{d} \Rightarrow ad = bc \Rightarrow ac + ad = ac + bc \Rightarrow a(c+d) = c(a+b) \Rightarrow \frac{a}{a+b} = \frac{c}{c+d} (a \neq {}^{-}b; c \neq {}^{-}d)$.

(c) $\frac{a}{b} = \frac{c}{d} \Rightarrow \frac{a}{a+b} = \frac{c}{c+d} \Rightarrow \frac{2a}{a+b} = \frac{2c}{c+d} \Rightarrow \frac{2a}{a+b} - 1 = \frac{2c}{c+d} - 1 \Rightarrow \frac{2a-(a+b)}{a+b} = \frac{2c-(c+d)}{c+d} \Rightarrow \frac{a-b}{a+b} = \frac{c-d}{c+d}$.

Chapter 6 Review

1. **(a)** Answers may vary. Shade three of four parts:

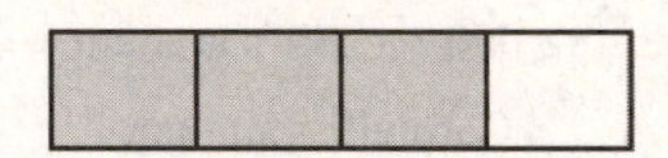

(b) Two of three blocks shaded:

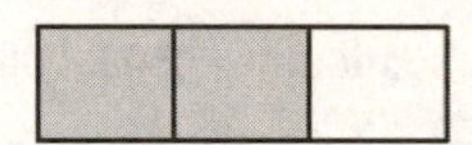

(c) Three of four horizontal bars are shaded, meshed with two of three vertical bars. Six of the twelve bars are dark-shaded; i.e., $\frac{3}{4} \cdot \frac{2}{3} = \frac{6}{12}$.

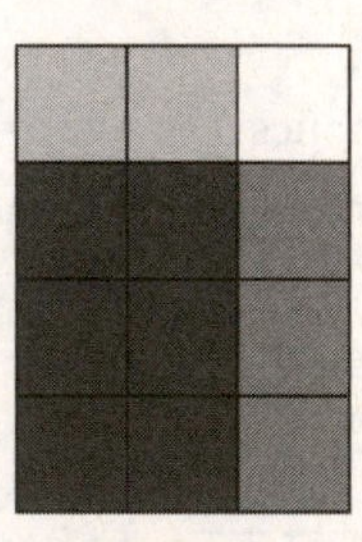

3. **(a)** $\frac{24}{28} = \frac{6 \cdot 4}{7 \cdot 4} = \mathbf{\frac{6}{7}}$.

(b) $\frac{ax^2}{bx} = \frac{ax \cdot x}{b \cdot x} = \mathbf{\frac{ax}{b}}$.

(c) $\frac{0}{17} = \frac{0 \cdot 17}{1 \cdot 17} = \frac{\mathbf{0}}{\mathbf{1}}$.

(d) $\frac{45}{81} = \frac{5 \cdot 9}{9 \cdot 9} = \frac{\mathbf{5}}{\mathbf{9}}$.

(e) $\frac{bx^2 + bx}{b + x} = \frac{b \cdot (b + x)}{1 \cdot (b + x)} = \frac{\boldsymbol{b}}{\mathbf{1}}$.

(f) $\frac{16}{216} = \frac{2 \cdot 8}{27 \cdot 8} = \frac{\mathbf{2}}{\mathbf{27}}$.

(g) $\frac{x+a}{x-a}$ **cannot be further reduced**. There are no factors common to both numerator and denominator (x and a are terms).

(h) $\frac{xa}{x+a}$ **cannot be further reduced**.

5. Additive inverse: $n +$ inverse $= 0$.
Multiplicative inverse: $n \cdot$ inverse $= 1$.

	n	Additive Inverse	Multiplicative Inverse
(a)	3	$^-3$	$\frac{1}{3}$
(b)	$3\frac{1}{7} = \frac{22}{7}$	$^-3\frac{1}{7}$	$\frac{7}{22}$
(c)	$\frac{5}{6}$	$\frac{^-5}{6}$	$\frac{6}{5}$
(d)	$\frac{^-3}{4}$	$\frac{3}{4}$	$\frac{^-4}{3}$

7. **Yes**. Apply the laws of multiplication and the commutative and associative laws of multiplication to find: $\frac{4}{5} \cdot \frac{7}{8} \cdot \frac{5}{14} = \frac{4 \cdot 7 \cdot 5}{5 \cdot 8 \cdot 14} = \frac{4 \cdot 7 \cdot 5}{8 \cdot 14 \cdot 5} = \frac{4}{8} \cdot \frac{7}{14} \cdot \frac{5}{5}$.

9. (*a*) Assuming no waste, $54\frac{1}{4} \div 3\frac{1}{12} = 17\frac{22}{37}$, so **17 pieces** can be cut.

(b) $\frac{22}{37} \cdot 3\frac{1}{12} = \frac{\mathbf{11}}{\mathbf{6}} = \mathbf{1\frac{5}{6}}$ **yards left over**.

11. (*i*) $\frac{a}{b} \div \frac{c}{d} = \frac{\left(\frac{a}{b}\right)}{\left(\frac{c}{d}\right)} = \frac{\left(\frac{a}{b} \cdot \frac{d}{c}\right)}{\left(\frac{c}{d} \cdot \frac{d}{c}\right)} = \frac{a}{b} \cdot \frac{d}{c}$.

(*ii*) $\frac{a}{b} \div \frac{c}{d} = \frac{ad}{bd} \div \frac{bc}{bd} = ad \div bc = \frac{a \cdot d}{b \cdot c} = \frac{a}{b} \cdot \frac{d}{c}$.

13. Answers may vary. $\frac{3}{4} = \frac{60}{80}$ and $\frac{4}{5} = \frac{64}{80}$ so $\frac{61}{80}$ and $\frac{62}{80}$ are between $\frac{3}{4}$ and $\frac{4}{5}$.

15. Jim ate $\frac{1}{3} \cdot \frac{1}{2} = \frac{1}{6}$ pizza. $\frac{1}{6} \cdot 2000 = \mathbf{333\frac{1}{3}}$ **calories**.

17. $\frac{240 \text{ heads}}{1000 \text{ flips}} = \frac{6 \cdot 40}{25 \cdot 40} = \frac{\mathbf{6}}{\mathbf{25}}$ **of the time**.

19. We could convert to the same units (for example inches to feet); however, this is not necessary when using proportional reasoning provide corresponding measures are in the same units.

Let $x =$ the distance between pupils on the carring of George Washington's head.

$\frac{2\frac{1}{2} \text{ in}}{9 \text{ in}} = \frac{x \text{ ft}}{60 \text{ ft}} \Rightarrow 9x = 2\frac{1}{2} \cdot 60 \Rightarrow x = \frac{2\frac{1}{2} \cdot 60}{9} = \frac{\frac{5}{2} \cdot 60}{9} = \frac{150}{9} = \frac{50}{3} = \mathbf{16\frac{2}{3}}$ **ft.**

21. 12 acres $\cdot\, 9\frac{1}{3}$ bags per acre $= 108 + 4 =$ **112 bags**.

23. $\frac{^-12}{10} - \frac{^-11}{9} = \frac{^-108}{90} - \frac{^-110}{90} = \frac{2}{90}$, which is a positive number, therefore $\frac{^-\mathbf{12}}{\mathbf{10}} > \frac{^-\mathbf{11}}{\mathbf{9}}$.

Alternatively, $\frac{^-12}{10} \square \frac{^-11}{9} \Rightarrow {}^-12 \cdot 9 \square {}^-11 \cdot 10 \Rightarrow {}^-108 \square {}^-110$. Since $^-108 > {}^-110$, then $\frac{^-\mathbf{12}}{\mathbf{10}} > \frac{^-\mathbf{11}}{\mathbf{9}}$.

25. (a) $2x - \frac{5}{3} = \frac{5}{6} \Rightarrow 2x = \frac{5}{6} + \frac{5}{3} = \frac{15}{6} \Rightarrow x = \frac{15}{6} \div 2 = \frac{15}{6} \cdot \frac{1}{2} \Rightarrow \boldsymbol{x} = \frac{\mathbf{5}}{\mathbf{4}}$.

(b) $x + 2\frac{1}{2} = 5\frac{2}{3} \Rightarrow x = 5\frac{2}{3} - 2\frac{1}{2} \Rightarrow \boldsymbol{x} = \mathbf{3\frac{1}{6}}$.

(c) $\frac{20+x}{x} = \frac{4}{5} \Rightarrow 5(20 + x) = 4x \Rightarrow 100 + 5x = 4x \Rightarrow 100 = {}^-x \Rightarrow \boldsymbol{x} = {}^-\mathbf{100}$.

(d) $2x + 4 = 3x - \frac{1}{3} \Rightarrow 4 + \frac{1}{3} = 3x - 2x \Rightarrow \boldsymbol{x} = \mathbf{4\frac{1}{3}}$.

27. (a) $\frac{3a}{xy^2} + \frac{b}{x^2y^2} = \frac{3a}{xy^2} \cdot \frac{x}{x} + \frac{b}{x^2y^2} = \frac{\mathbf{3ax + b}}{\boldsymbol{x^2y^2}}$.

(b) $\frac{5}{xy^2} - \frac{2}{3x} = \frac{5}{xy^2} \cdot \frac{3}{3} - \frac{2}{3x} \cdot \frac{y^2}{y^2} = \frac{\mathbf{15 - 2y^2}}{\mathbf{3xy^2}}$.

(c) $\frac{a}{x^3y^2z} - \frac{b}{xyz} = \frac{a}{x^3y^2z} - \frac{b}{xyz} \cdot \frac{x^2y}{x^2y} =$

$\frac{\mathbf{a - bx^2y}}{\mathbf{x^3y^2z}}$.

(d) $\frac{7}{2^3 3^2} + \frac{5}{2^2 3^3} = \frac{7}{2^3 3^2} \cdot \frac{3}{3} + \frac{5}{2^2 3^3} \cdot \frac{2}{2} =$

$\frac{21+10}{2^3 3^3} = \frac{\mathbf{31}}{\mathbf{216}}$.

29. $\left(\frac{a^{-1}+b^{-1}+c^{-1}}{2}\right)^{-1} = \frac{2}{\frac{1}{a}+\frac{1}{b}+\frac{1}{c}} =$

$\frac{2}{\frac{1}{a}\cdot\frac{bc}{bc}+\frac{1}{b}\cdot\frac{ac}{ac}+\frac{1}{c}\cdot\frac{ab}{ab}} = \frac{2}{\left(\frac{bc+ac+ab}{abc}\right)} = \frac{\mathbf{2abc}}{\mathbf{bc+ac+ab}}$.

31. **(*i*)** 48 fl oz for \$3.05 $\Rightarrow \frac{305¢}{48\text{ fl oz}} \approx$ 6.35¢ per fl oz;

(*ii*) 64 fl oz for \$3.60 $\Rightarrow \frac{360¢}{64\text{ fl oz}} \approx$ 5.63¢ per fl oz;

Based on cost per ounce, the **64 fl oz** size is the **better buy**.

33. **(a)** $\frac{15}{12} = \frac{21}{x} \Rightarrow 15x = 21 \cdot 12 \Rightarrow x = \frac{21\cdot12}{15} = \mathbf{16.8}$.

(b) $\frac{20}{35} = \frac{110}{x} \Rightarrow 20x = 110 \cdot 35$

$\Rightarrow x = \frac{110\cdot35}{20} = \mathbf{192.5}$.

(c) $\frac{\left(\frac{1}{2}\right)}{\left(\frac{1}{3}\right)} = \frac{\left(\frac{3}{2}\right)}{x} \Rightarrow \frac{1}{2}x = \frac{3}{2} \cdot \frac{1}{3} = \frac{1}{2}$

$\Rightarrow x = \frac{1}{2} \div \frac{1}{2} = \mathbf{1}$.

35. $\frac{1\text{ cm}}{2.5\text{ m}} = \frac{3\text{ cm}}{x\text{ m}} \Rightarrow x = 3 \cdot 2.5 = \mathbf{7.5\,m}$.

37. The ratio **cannot be determined exactly**, but it will always be between 12:100 and 15:100 as a ratio of defective to non-defective chips. 12:100 is $\frac{12}{112} = 10\frac{5}{7}\%$ defective; 15:100 is $\frac{15}{115} = 13\frac{1}{23}\%$ defective. If the observed defective percentage is closer to $13\frac{1}{23}\%$ then more chips came from the first plant. If the percentage is closer to $10\frac{5}{7}\%$, then more chips came from the second.

39. **(a)** Games won to games lost was $\frac{18}{7}$ or **18:7**.

(b) 25 games were played. Games won to games played was $\frac{18}{25}$ or **18:25**.

41. $\frac{\text{boys}}{\text{girls}} = \frac{3}{5} = \frac{x\text{ boys}}{15\text{ girls}} \Rightarrow 5 \cdot x = 3 \cdot 15 \Rightarrow$

$x = \frac{3\cdot15}{5} = \mathbf{9\ boys}$.

CHAPTER 7

DECIMALS: RATIONAL NUMBERS AND PERCENTS

Assessment 7-1A: Introduction to Decimals

1. (a) $0.023 = \mathbf{0 \cdot 10^0 + 0 \cdot 10^{-1} + 2 \cdot 10^{-2} + 3 \cdot 10^{-3}}$.

   (b) $206.06 = \mathbf{2 \cdot 10^2 + 0 \cdot 10^1 + 6 \cdot 10^0 + 0 \cdot 10^{-1} + 6 \cdot 10^{-2}}$.

   (c) $312.0103 = \mathbf{3 \cdot 10^2 + 1 \cdot 10^1 + 2 \cdot 10^0 + 0 \cdot 10^{-1} + 1 \cdot 10^{-2} + 0 \cdot 10^{-3} + 3 \cdot 10^{-4}}$.

   (d) $0.000132 = \mathbf{0 \cdot 10^0 + 0 \cdot 10^{-1} + 0 \cdot 10^{-2} + 0 \cdot 10^{-3} + 1 \cdot 10^{-4} + 3 \cdot 10^{-5} + 2 \cdot 10^{-6}}$.

3. (a) **536.0076**; i.e., seventy-six ten-thousandths $= 7 \cdot 10^{-3} + 6 \cdot 10^{-4}$.

   (b) **3.008**; i.e., eight thousandths $= 8 \cdot 10^{-3}$.

   (c) **0.000436**; i.e., four hundred thirty-six millionths $= 4 \cdot 10^{-4} + 3 \cdot 10^{-5} + 6 \cdot 10^{-6}$.

   (d) **5,000,000.2**.

5. (a) $0.436 = \frac{436}{1000} = \frac{109 \cdot 4}{250 \cdot 4} = \mathbf{\frac{109}{250}}$.

   (b) $25.16 = 25\frac{16}{100} = 25 + \frac{16}{100} = \frac{2500}{100} + \frac{16}{100} = \frac{2516}{100} = \mathbf{\frac{629}{25}}$.

   (c) $^{-}316.027 = {}^{-}316\frac{27}{1000} = \mathbf{\frac{^{-}316{,}027}{1000}}$.

   (d) $28.1902 = 28\frac{1902}{10{,}000} = \frac{281{,}902}{10{,}000} = \mathbf{\frac{140{,}951}{5000}}$.

   (e) $^{-}4.3 = {}^{-}4\frac{3}{10} = \mathbf{\frac{^{-}43}{10}}$.

   (f) $^{-}62.01 = {}^{-}62\frac{1}{100} = \mathbf{\frac{^{-}6201}{100}}$.

7. (a) $\frac{4}{5} = \frac{4}{5} \cdot \frac{2}{2} = \frac{8}{10} = \mathbf{0.8}$.

   (b) $\frac{61}{2^2 \cdot 5} = \frac{61}{2^2 \cdot 5} \cdot \frac{5}{5} = \frac{305}{2^2 \cdot 5^2} = \frac{305}{100} = \mathbf{3.05}$.

   (c) $\frac{3}{6} = \frac{1}{2} = \frac{1}{2} \cdot \frac{5}{5} = \frac{5}{10} = \mathbf{0.5}$.

   (d) $\frac{1}{2^5} = \frac{1}{2^5} \cdot \frac{5^5}{5^5} = \frac{3125}{10^5} = \frac{3125}{100{,}000} = \mathbf{0.03125}$.

   (e) $\frac{36}{5^5} = \frac{36}{5^5} \cdot \frac{2^5}{2^5} = \frac{1152}{10^5} = \frac{1152}{100{,}000} = \mathbf{0.01152}$.

   (f) $\frac{133}{625} = \frac{133}{5^4} \cdot \frac{2^4}{2^4} = \frac{2128}{10^4} = \frac{2128}{10{,}000} = \mathbf{0.2128}$.

9. Answers may vary. Many values between 0 and 100 are composed of whole-number powers of 2 and 5, thus dividing 100 and being capable of being expressed as a two-digit decimal. These numbers include the coin designations of 1, 5, 10, 25, and 50 cents, but there are others which could have been used, such as 2¢ or 20¢.

11. (a) Fourteen thousandths inch $= \frac{14}{1000}$ inch $=$ **0.014 inch**.

    (b) Twenty-four hundredths $= \frac{24}{100} = 0.24$ days. The rotational period is thus **365.24 days**.

13. The largest number is furthest to the right on the number line. Thus $0.804 < 0.8399 <$ **0.84**.

15. (a) Answers may vary. One method could be to find the difference, not matter how slight, between the two decimal numbers and add some fraction of that to the smaller.

    (b) Part (a) is a recursive process; no matter how small the difference between the terminating decimals, that differcence can be divided.

17. **Rhonda, Martha, Kathy, Molly, Emily,** because 63.54 (Rhonda) < 63.59 (Martha) < 64.02 (Kathy) < 64.46 (Molly) < 64.54 (Emily).

Assessment 7-1B

1. (a) $0.045 = \mathbf{0 \cdot 10^0 + 0 \cdot 10^{-1} + 4 \cdot 10^{-2} + 5 \cdot 10^{-3}}$.

   (b) $103.03 = \mathbf{1 \cdot 10^2 + 0 \cdot 10^1 + 3 \cdot 10^0 + 0 \cdot 10^{-1} + 3 \cdot 10^{-2}}$.

   (c) $245.6701 = \mathbf{2 \cdot 10^2 + 4 \cdot 10^1 + 5 \cdot 10^0 + 6 \cdot 10^{-1} + 7 \cdot 10^{-2} + 0 \cdot 10^{-3} + 1 \cdot 10^{-4}}$.

   (d) $0.00034 = \mathbf{0 \cdot 10^0 + 0 \cdot 10^{-1} + 0 \cdot 10^{-2} + 0 \cdot 10^{-3} + 3 \cdot 10^{-4} + 4 \cdot 10^{-5}}$.

3. (a) **2.027**. Two thousand thousandths = 2; twenty-seven thousandths = 0.027.

   (b) **2000.027**. The difference between this value and that of part (a) is the word *and*.

   (c) **2020.007**.

   (d) **0.00004**. Four hundred-thousandths $= 4 \cdot 10^{-5}$.

5. (a) $28.32 = 28 + \frac{32}{100} = \frac{2800}{100} + \frac{32}{100} = \frac{2832}{100} = \frac{708 \cdot 4}{25 \cdot 4} = \mathbf{\frac{708}{25}}$.

   (b) $34.1736 = 34 + \frac{1736}{10{,}000} = \frac{340{,}000}{10{,}000} + \frac{1736}{10{,}000} = \frac{341{,}736}{10{,}000} = \frac{42{,}717 \cdot 8}{1250 \cdot 8} = \mathbf{\frac{42{,}717}{1250}}$.

   (c) $^{-}27.32 = {}^{-}\left(27 + \frac{32}{100}\right) = {}^{-}\left(\frac{2700}{100} + \frac{32}{100}\right) = {}^{-}\left(\frac{2732}{100}\right) = {}^{-}\left(\mathbf{\frac{683}{25}}\right)$.

7. (a) $\frac{4}{8} = \frac{1}{2} = \frac{1}{2} \cdot \frac{5}{5} = \frac{5}{10} = \mathbf{0.5}$.

   (b) $\frac{1}{2^6} = \frac{1}{2^6} \cdot \frac{5^6}{5^6} = \frac{15{,}625}{10^6} = \frac{15{,}625}{1{,}000{,}000} = \mathbf{0.015625}$.

   (c) $\frac{137}{625} = \frac{137}{5^4} = \frac{137}{5^4} \cdot \frac{2^4}{2^4} = \frac{2192}{10^4} = \frac{2192}{10{,}000} = \mathbf{0.2192}$.

   (d) Nonterminating.

   (e) $\frac{3}{25} = \frac{3}{5^2} = \frac{3}{5^2} \cdot \frac{2^2}{2^2} = \frac{12}{10^2} = \frac{12}{100} = \mathbf{0.12}$.

   (f) $\frac{14}{35} = \frac{2}{5} = \frac{4}{10} = \mathbf{0.4}$.

9. (a) If a block is represented by $\frac{1}{10}$, there are 1000 cubes in a base-ten block, so each cube would have a value of $\frac{1}{10} \div 1000 = \frac{1}{10{,}000}$, or **one ten-thousandth**.

   (b) If each cube has a value of $\frac{1}{10{,}000}$, and there are 100 cubes in a flat, each flat would have a value of $100 \cdot \frac{1}{10{,}000} = \frac{1}{100}$, or **one hundredth**.

   (c) One long is composed of 10 cubes, thus a long would have a value of $10 \cdot \frac{1}{10{,}000} = \frac{1}{1000}$, or **one thousandth**.

   (d) Three blocks, 1 long, and 4 cubes $= 3 \cdot \frac{1}{10} + 1 \cdot \frac{1}{1000} + 4 \cdot \frac{1}{10{,}000} = 0.3 + 0.001 + 0.0004 = \mathbf{0.3014}$, or **three thousand fourteen ten-thousandths.**

   (The least numbers are those furthest from 0.)

11. (a) $\frac{1}{16} = \frac{1}{2^4} = \frac{5^4}{2^4 \cdot 5^4} = \frac{625}{10{,}000} = \mathbf{0.0625}$.

    (b) $224 + \frac{7006}{10{,}000} = 224.7006$.

13. $0.8114 = \frac{8114}{10{,}000}$, $0.8119 = \frac{8119}{10{,}000}$, and $0.82 = \frac{8200}{10{,}000}$. Since $8114 < 8119 < 8200$, $\frac{8200}{10{,}000} =$ **0.82** is furthest to the right. Alternetively, .82 has two hundredths while the other two fractions have one hundredth.

15. Answers may vary, but for example just as there are infinitely many rational numbers of the form $\frac{a}{b}$ where $a, b \in I$ and $b \neq 0$, there are infinitely many terminating decimals. E.g., just as between

0.0625 and 0.125. (representing $\frac{1}{16}$ and $\frac{1}{8}$, respectively), $\frac{1}{16} = \frac{2}{32}$ and $\frac{1}{8} = \frac{2}{16} = \frac{4}{32}$, so $\frac{3}{32} = 0.09375$ is between 0.0625 and 0.125. This process can continue indefinitely.

17. Sort first by the tens values, then by the ones values, then by the tenths, and finally by the hundredths. **Ricky, Michael, Karl, Marius, Eddie.**

Assessment 7-2A
Operations on Decimals

1. She bought a total of:

$$\begin{array}{r} \$17.95 \\ 13.59 \\ 14.86 \\ 179.98 \\ 2.43 \\ 2.43 \\ \hline \mathbf{\$231.24} \end{array}$$

in her shopping (excluding sales tax)

3. Let $x =$ the price per pound of the other type of fruit. The total cost of the fruit is $25 \cdot 4 + 15 \cdot 2 + 10\,x = 130 + 10x$ dollars. The total weight of the fruit is $25 + 15 + 10 = 50$ lbs.

To find the average cost per pound divide the total cost by the total weight. This will equal $3.50 per lb.

$$\frac{130 + 10x}{50} = 3.50 \Rightarrow \frac{130 + 10x}{50} = \frac{350}{100}$$

$$\Rightarrow 130 + 10x = \frac{350}{100} \cdot 50 \Rightarrow 130 + 10x = 175$$

$$\Rightarrow 10x = 45 \Rightarrow x = \mathbf{4.5\ lbs.}$$

5. Let p be the price of the stock. Then $p + 0.24 = 73.245 \Rightarrow p = 73.245 - 0.24$, or

$$\begin{array}{r} 73.245 \\ -0.240 \\ \hline \mathbf{73.005} \end{array}$$

7. **(a)** $\frac{3 \text{ heaters}}{1} \cdot \frac{1200 \text{ w}}{1 \text{ heater}} \cdot \frac{24 \text{ hrs}}{1 \text{ day}} \cdot \frac{1 \text{ kw}}{1000 \text{ w}} \cdot \frac{\$0.06715}{\text{kw-hr}} = \$5.80176$ which rounds to **$5.80**. (The actual bill would be lumped with other appliance usage.)

(b) 75 watts $= 0.075$ kilowatts. $0.075 \text{ kw} \times 1 \text{ hour} \times \$0.06715$ per kw-hr $= \$0.00503625$ to operate one bulb for one hour. $\$1.00 \div \$0.00503625 =$ **199 hours** (rounded to the nearest hour).

9. **(a)** There is a difference of 0.9 between each element of the sequence, thus it is arithmetic. The next three elements are: $4.5 + 0.9 = \mathbf{5.4}$, $5.4 + 0.9 = \mathbf{6.3}$, $6.3 + 0.9 = \mathbf{7.2}$.

(b) There is a difference of 0.2 between each element of the sequence, thus it is arithmetic. The next three elements are: $1.1 + 0.2 = \mathbf{1.3}$, $1.3 + 0.2 = \mathbf{1.5}$, $1.5 + 0.2 = \mathbf{1.7}$.

11. A finite geometric sequence is one with a constant ratio between terms and a finite number of terms. 0.2222 can be expressed as a sum: $\frac{2}{10} + \frac{2}{100} + \frac{2}{1000} + \frac{2}{10{,}000}$; note that the denominators are powers of ten. Thus to make it a geometric sequence with constant ratio $\frac{1}{10}$, it could be written as $\frac{2}{10} + \left(\frac{2}{10}\right)\left(\frac{1}{10}\right) + \left(\frac{2}{10}\right)\left(\frac{1}{10}\right)^2 + \left(\frac{2}{10}\right)\left(\frac{1}{10}\right)^3$, or $\mathbf{0.2 + (0.2)(0.1) + (0.2)(0.1)^2 + (0.2)(0.1)^3}$.

13. The bank is **not correct**. The total of outstanding checks is:

$$\begin{array}{r} \$3.21 \\ 14.56 \\ 12.44 \\ 6.98 \\ 9.51 \\ 7.49 \\ \hline \$54.19 \end{array}$$

Adding the total of outstanding checks to the checkbook balance gives $\$54.19 + 21.69 =$

$75.88, which differs from the bank statement by $7.74. The bank balance is too high.

15. **(a)** Move the decimal point 7 places to the left to obtain the product of a number between 1 and 10 and an integer power of 10. I.e., 1.27 multiplied by 10^7, or about $\mathbf{1.27 \cdot 10^7}$ **m**.

(b) Move the decimal point 9 places to the left and multiply the result by 10^9 to obtain about $\mathbf{4.486 \cdot 10^9}$ **km**.

(c) Move the decimal point 7 places to the left and multiply the result by 10^7 to obtain about $\mathbf{5 \cdot 10^7}$ **cans**.

17. **(a)** Move the decimal point ten places to the left to obtain **0.000000000753 g**.

(b) Move the decimal point five places to the right to obtain **298,000 km per sec**.

(c) Move the decimal point eight places to the right to obtain **778,570,000 km**.

19. **(a)** $\underline{2}03.651$ is between 200 and 300 and is closer to 200, so round down to **200**.

(b) $2\underline{0}3.651$ is between 200 and 210 and is closer to 200, so round down to **200**.

(c) $20\underline{3}.651$ is between 203 and 204 and is closer to 204, so round up to **204**.

(d) $203.\underline{6}51$ is between 203.6 and 203.7 and is closer to 203.7, so round up to **203.7**.

(e) $203.6\underline{5}1$ is between 203.65 and 203.66 and is closer to 203.65, so round down to **203.65**.

21. Camera rounds to $55, film rounds to $5, and case rounds to $18. Total estimated cost is 55 + 5 + 18 = **$78**.

23. To find the greatest products use the largest digits in the highest place values:

(*i*) Least: $\boxed{2}.\boxed{3} \times \boxed{1} = \mathbf{2.3}$.

(*ii*) Greatest: $\boxed{8}.\boxed{7} \times \boxed{9} = \mathbf{78.3}$.

25. Answers may vary. E.g., $40 \cdot \$8 + 40 \cdot \$\left(\frac{1}{4}\right) = \$320 + \$10 = \mathbf{\$330}$.

27. $\frac{7}{0.25} = \frac{70}{2.5} = \frac{700}{25}$. Thus, ***a*, *b*, and *d*** have equal quotients.

Assessment 7-2B

1. He bought a total of:

$$\begin{array}{r} \$4.99 \\ 0.79 \\ 49.99 \\ 1.49 \\ \hline \$57.26 \end{array}$$

in his shopping (excluding sales tex).

3. There would be a total of 30 + 20 + 10 = 60 pounds of nuts; at an average price per pound of $4.50 he would pay a total of $60 \cdot 4.50 = \$270$. He has already paid $\$3.00 \cdot 30 + \$5.00 \cdot 20 = \$190.00$, so he has $\$270.00 - 190.00 =$ **$80.00** to pay for 10 pounds.

5.

$$\begin{array}{r} \$63.28 \\ -27.45 \\ \hline \end{array}$$

$35.83 was the loss.

7. The computer plus the monitor uses 45 + 35 = 80 watts. $\frac{80\text{ w}}{1} \cdot \frac{3.4\text{ hr}}{1} \cdot \frac{1\text{ kw}}{1000\text{ w}} \cdot \frac{\$0.07104}{\text{kw}-\text{hr}} =$ **$0.01932288**, or about 2¢.

9. **(a)** Each element of the sequence is 0.5 times the previous element, thus it is geometric. The next three elements are: $0.125 \cdot 0.5 = \mathbf{0.0625}$, $0.0625 \cdot 0.5 = \mathbf{0.03125}$, $0.03125 \cdot 0.5 = \mathbf{0.015625}$.

(b) There is a difference of 1.3 between each element of the sequence, thus it is arithmetic. The next three elements are: 5.4 + 1.3 = **6.7**, 6.7 + 1.3 = **8.0**, 8.0 + 1.3 = **9.3**.

11. $0.3333333 = 0.3 + 0.03 + 0.003 + \cdots + 0.0000003$ or $\frac{3}{10} + \frac{3}{10}\left(\frac{1}{10}\right) + \frac{3}{10}\left(\frac{1}{10}\right)^2 + \cdots + \frac{3}{10}\left(\frac{1}{10}\right)^6$. This is the finite sum of a geometric sequence with $a_1 = 0.3$, $n = 7$, $a_7 = 0.0000003$, and $r = \frac{0.03}{0.3} = \frac{1}{10} = 0.1$.

13. 18 shares · \$61.48 = \$1106.64 net return on first transaction.
350 shares · \$85.35 − \$495 commission = \$29,377.50 net return on second transaction.

Total net return = \$1106.64 + 29377.50 = \$30,484.14.

Total cost = \$964.00 + 27,422.50 = \$28,386.50.

Profit = \$30,484.14 − 28,386.85 = \$2097.64, which rounds to **\$2098**.

15. **(a)** Move the decimal point five places to the left to obtain the product of a number between 1 and 10 and an integer power of 10. I.e., 9.99347 multiplied by 10^5, or $\mathbf{9.99347 \cdot 10^5}$ **people**.

(b) Move the decimal point seven places to the left and multiply the result by 10^7 to obtain $\mathbf{2.449 \cdot 10^7}$ $\mathbf{mi^2}$.

(c) Multiply 1 by 10^{-15} to obtain about $\mathbf{1.0 \cdot 10^{-15}}$**m**.

17. **(a)** Move the decimal point 6 places to the left to obtain **0.0000044 sec**.

(b) Move the decimal point 4 places to the right to obtain about **19,900 km**.

(c) Move the decimal point 9 places to the right to obtain approximately **3,000,000,000 years**.

19. **(a)** $\underline{7}15.04$ is between 700 and 800 and is closer to 700, so round down to **700**.

(b) $715.\underline{0}4$ is between 715.0 and 715.1 and is closer to 715.0, so round down to **715.0**.

(c) $71\underline{5}.04$ is between 715 and 716 and is closer to 715, so round down to **715**.

(d) $7\underline{1}5.04$ is between 710 and 720 and is closer to 720, so round up to **720**.

(e) $\underline{0}715.04$ is between 0 and 1000, and is closer to 1000, so round up to **1000**.

21. Answers may vary. E.g., **10.3** because $9 \cdot 10 = 90$ and $9 \cdot 0.3$ is greater than 2 and less than 4.

23. To find the least products use the smallest digits in the highest place values. To find the greatest products use the largest digits in the highest place values:

(i) Least: $\boxed{1}.\boxed{3} \times \boxed{2}.\boxed{4} =$ **3.12**.

(ii) Greatest: $\boxed{8}.\boxed{7} \times \boxed{9}.\boxed{6} =$ **83.52**.

25. Answers may vary. E.g., $40 \cdot \$6 + 40 \cdot \$\left(\frac{1}{4}\right) = \$240 + \$10 =$ **\$250**.

27. $9 \div 0.35 = \dfrac{9}{0.35} = \dfrac{90}{3.5} = \dfrac{900}{35} = \dfrac{.9 \cdot 1000}{.35 \cdot 1000} = \dfrac{.9}{.35}$.

Thus, a, b, c and d have the same quotient.

Review Problems

15. $14.0479 = 1 \cdot 10^1 + 4 \cdot 10^0 + 0 \cdot 10^{-1} + 4 \cdot 10^{-2} + 7 \cdot 10^{-3} + 9 \cdot 10^{-4}$.

17. **Yes**, if the numerator is a multiple of 13. E.g., $\frac{13}{26} = \frac{1}{2}$ in simplest form, and $\frac{1}{2}$ can be written as a terminating decimal.

Assessment 7-3A Nonterminating Decimals

1. In each of the following, divide the numerator by the denominator either with a calculator or by use of the division algorithm. The overline in the quotient indicates that the block of digits underneath is repeated an infinite number of times.

(a) $\frac{4}{9} = 4 \div 9 = \mathbf{0.\overline{4}}$.

(b) $\frac{2}{7} = 2 \div 7 = \mathbf{0.\overline{285714}}$.

(c) $\frac{3}{11} = 3 \div 11 = \mathbf{0.\overline{27}}$.

(d) $\frac{1}{15} = 1 \div 15 = \mathbf{0.0\overline{6}}$.

(e) $\frac{2}{75} = 2 \div 75 = \mathbf{0.02\overline{6}}$.

(f) $\frac{1}{99} = 1 \div 99 = \mathbf{0.\overline{01}}$.

(g) $\frac{5}{6} = 5 \div 6 = \mathbf{0.8\overline{3}}$.

(h) $\frac{1}{13} = 1 \div 13 = \mathbf{0.\overline{076923}}$.

(i) $\frac{1}{21} = 1 \div 21 = \mathbf{0.\overline{047619}}$.

(j) $\frac{3}{19} = 3 \div 19 = \mathbf{0.\overline{157894736842105236}}$.

The repetend has more digits than most calculator's display capability. Use the division algorithm, assuming an 8-digit display:

```
     0. 1 5 7 8 9 4 7 3 6 8 4 2 1 0 5 2 6 3
1 9 |3.0 0 0 0  0 0 0 0 0 0 0 0 0 0 0 0 0 0
     2.9 9 9 9 8 6
     -------------
            1 4  0 0  0 0 0 0
            1 3  9 9 9 9 9 8
            ----------------
                          2 0  0 0 0 0 0
                          1 9 9 9 9 9 7
                          -------------
                                      3
```

where the last remainder, 3, again begins the repeating pattern. Note that since the denominator is 19, the most digits that can be in the repetend is 18.

3. 1 minute $= \frac{1}{60}$ hour, and $1 \div 60 = \mathbf{0.01\overline{6}}$ **hour.**

5. Each element could be a decimal representation of $\frac{n}{n+1}$, beginning at $n = 0$. If so, then the next three elements would be:

$\frac{6}{7} = \mathbf{0.\overline{857142}}$,

$\frac{7}{8} = \mathbf{0.875}$, and

$\frac{8}{9} = \mathbf{0.\overline{8}}$.

7. The repeating decimal part is determined by the 7 in the denominator of $3\frac{1}{7} = \frac{22}{7}$. Seven has no prime factors of 2 or 5 and so is a repeating decimal.

9. **Yes.** Zeros after the last non-zero digit of the decimal can be repeated.

11. To find a number halfway between any two others, find their average; i.e., add the original numbers and divide by 2.

$(0.5 + 0.\overline{4}) \div 2 = (0.9444\ldots) \div 2 =$
$\mathbf{0.47222\cdots} = \mathbf{0.47\overline{2}}$.

13. **(a)** $\frac{3}{7} = 0.\overline{428571}$, a six-digit repetend. 21 when divided by 6 has a remainder of 3. Thus the 21st digit is the third in the repetend $= \mathbf{8}$.

(b) $17^{-1} = 0.\overline{0588235294117647}$, a sixteen-digit repetend. 5280 when divided by 16 has a remainder of 0. Thus the 5280th digit is the sixteenth in the repetend $= \mathbf{7}$.

15. **(a)** Since $0.\overline{9} = 1, 0.0\overline{9} = \frac{1}{10} \cdot (0.\overline{9}) =$
$\frac{1}{10}(1) = \mathbf{\frac{1}{10}} = \mathbf{0.1}$.

(b) $0.3\overline{9} = 0.3 + 0.0\overline{9} = 0.3 + 0.1 = \mathbf{0.4}$.

(c) $9.\overline{9} = 9 + 0.\overline{9} = 9 + 1 = \mathbf{10}$.

17. See problem 14(a)(*i*) above:

(a) $0.\overline{05} = \mathbf{\frac{5}{99}}$. In the fraction $\frac{a}{b}, a = 5$ and b is two 9's.

(b) $0.\overline{003} = \frac{3}{999} = \mathbf{\frac{1}{333}}$. In the fraction $\frac{a}{b}$, $a = 3$ and b is three 9's.

19. To find the sum of an infinite geometric series with ratio $0 < r < 1$ (if r were greater than 1, the series could not have a finite sum; if $r = 0$ the series would be just the first term):

$$S\infty = a + ar + ar^2 + ar^3 + \cdots$$
$$^{-}rS_\infty = ^{-}(ar + ar^2 + ar^3 + \cdots)$$
$$S_\infty - rS\infty = a$$
$$S_\infty(1 - r) = a$$
$$S\infty = \frac{a}{1 - r}$$

(a) $0.2\overline{9} = 0.29999\ldots = 0.2 + 0.9999\ldots =$
$0.2 + 0.09 + 0.09(0.1) + 0.09(0.01) + \cdots$.

Thus $0.2\overline{9}$ is 0.2 plus an infinite geometric sequence where $a_1 = 0.09$ and $r = 0.1$.

$S_\infty = 0.2 + \frac{0.09}{1-0.1} = 0.2 + \frac{0.09}{0.9} = 0.2 + 0.1 = \frac{3}{10}$.

(b) $2.0\overline{29} = 2 + 0.029 + 0.00029 + \cdots$.
Thus $2.0\overline{29}$ is 2 plus an infinite geometric sequence where $a_1 = 0.029$ and $r = 0.01$.

$S_\infty = 2 + \frac{0.029}{1-0.01} = 2\frac{29}{990} = \mathbf{\frac{2009}{990}}$.

21. **(a)** $1 - 3x = 8 \Rightarrow -3x = 7 \Rightarrow x = \frac{^-7}{3} = {}^-2.\overline{3}$.

(b) $1 = 3x + 8 \Rightarrow {}^-7 = 3x \Rightarrow 3x = {}^-7 \Rightarrow x = \frac{^-7}{3} = {}^-2.\overline{3}$.

(c) $1 = 8 - 3x \Rightarrow {}^-7 = -3x \Rightarrow 3x = 7 \Rightarrow x = \frac{7}{3} = 2.\overline{3}$.

Assessment 7-3B

1. In each of the following, divide the numerator by the denominator either with a calculator or by use of the division algorithm. The overline in the quotient indicates that the block of digits underneath is repeated an infinite number of times.

(a) $\frac{2}{3} = 2 \div 3 = \mathbf{0.\overline{6}}$.

(b) $\frac{7}{9} = 7 \div 9 = \mathbf{0.\overline{7}}$.

(c) $\frac{1}{24} = 1 \div 24 = \mathbf{0.041\overline{6}}$.

(d) $\frac{3}{60} = \frac{1}{20} = 1 \div 20 = \mathbf{0.05}$.

(e) $\frac{2}{99} = 2 \div 99 = \mathbf{0.\overline{02}}$.

(f) $\frac{7}{6} = 7 \div 6 = \mathbf{1.1\overline{6}}$.

(g) $\frac{2}{21} = 2 \div 21 = \mathbf{0.\overline{095238}}$.

(h) $\frac{4}{19} = 4 \div 19 = \mathbf{0.\overline{210526315789473684}}$.

The repetend has more digits than most calculators' display capability. Use the division algorithm, assuming an 8-digit display:

```
        0. 2 1 0 5 2 6 3 1 5 7 8 9 4 7 3 6 8 4
   1 9 |4.0 0 0 0 0 0 0 0 0 0 0 0 0 0 0 0 0 0
        3.9 9 9 9 9 4
                    6 0 0 0 0 0 0
                    5 9 9 9 9 9 1
                                9 0 0 0 0 0 0
                                8 9 9 9 9 9 6
                                            4
```

where the last remainder, 4, again begins the repeating pattern. Note that since the denominator is 19, the most digits that can be in the repetend is 18.

3. One second $= \frac{1}{3600}$ hour; $1 \div 3600 =$ $\mathbf{0.0002\overline{7}}$ **hour**.

5. If this is an arithmetic sequence, then add 0.3 to each element to obtain the following element. The next three elements thus could be:

$1.\overline{3} + 0.\overline{3} = \mathbf{1.\overline{6}}$;

$1.\overline{6} + 0.\overline{3} = \mathbf{1.\overline{9}} = \mathbf{2.0}$;

$2.0 + 0.\overline{3} = \mathbf{2.\overline{3}}$.

7. $\frac{2}{26} = \frac{1}{13}$. Since the denominator has a prime factor other than 2 or 5, the decimal will not terminate. Because $\frac{2}{26}$ is equal to the reduced fraction $\frac{1}{13}$ and the number of digits in the repetend for a proper fraction cannot exceed the denominator minus one, the maximum possible length of the repetend is **12**.

9. **Yes**. If, for example, the repeating digits are zeros, then the repeating decimal could be written as a terminating decimal.

11. To find a number halfway between any two others, find their average; i.e., add the original numbers and divide by 2. (Note that $0.\overline{9} = 1.0$.)

$(0.\overline{9} + 1.1) \div 2 = (2.1) \div 2 = \mathbf{1.05}$.

13. $17^{-1} = 0.\overline{0588235294117647}$, a sixteen-digit repetend. 23 when divided by 16 has a remainder of 7, so the 23rd digit is the 7th in the repetend, or **5**.

15. **(a)** $1.\overline{9} = 1 + 0.\overline{9} = 1 + 1 = 2$.

(b) $0.00\overline{9} = \frac{1}{100} \cdot 0.\overline{9} = \frac{1}{100} \cdot 1 = \mathbf{0.01}$.

(c) $0.3\overline{9} = 0.3 + 0.0\overline{9} = 0.3 + \frac{1}{10} \cdot 0.\overline{9}$

$= 0.3 + 0.1 = \mathbf{0.4}$.

17. **(a)** First observe that $25(0.010101\ldots) = 0.252525\ldots$. Therefore, $3.\overline{25} = 3 + 25(0.\overline{01}) = 3 + \frac{25}{99} = \mathbf{\frac{322}{99}}$.

(b) $3.\overline{125} = 3 + 125(0.\overline{001}) = 3 + \frac{125}{999} = \mathbf{\frac{3122}{999}}$.

19. To find the sum of an infinite geometric series with ratio $0 < r < 1$ (if r were greater than l, the series could not have a finite sum; if $r = 0$ the series would be just the first term):

$$S\infty = a + ar + ar^2 + ar^3 + \cdots$$
$$^{-}rS_\infty = {}^{-}(ar + ar^2 + ar^3 + \cdots)$$
$$S_\infty - rS\infty = a$$
$$S_\infty(1 - r) = a$$
$$S\infty = \frac{a}{1 - r}$$

(a) $0.\overline{29} = 0.29 + 0.0029 + \cdots = 0.29 + 0.29(0.01) + 0.29(0.0001) + \cdots$.
Thus $a_1 = 0.29$ and $r = 0.01$.

$S_\infty = \frac{0.29}{1-0.01} = \frac{0.29}{0.99} = \mathbf{\frac{29}{99}}$.

(b) $0.000\overline{29} = 0.00029 + 0.0000029 + \cdots$.
Thus $0.000\overline{29}$ is an infinite geometric sequence where $a_1 = 0.00029$ and $r = 0.01$.

$S = \frac{0.00029}{1-0.01} = \frac{0.00029}{0.99} = \mathbf{\frac{29}{99{,}000}}$.

21. **(a)** $3x = 8 \Rightarrow x = \frac{8}{3} = \mathbf{2\frac{2}{3}} = \mathbf{2.\overline{6}}$.

(b) $3x + 1 = 8 \Rightarrow 3x = 7 \Rightarrow x = \frac{7}{3} = \mathbf{2\frac{1}{3}} = \mathbf{2.\overline{3}}$.

(c) $3x - 1 = 8 \Rightarrow 3x = 9 \Rightarrow x = \mathbf{3.0}$.

Review Problems

13. Total deductions were \$1520.63 + \$723.30 + \$2843.62 = \$5087.55. Gross pay less deductions was \$27,849.50 − \$5087.55 = **\$22,761.95**.

15. The rule states that decimal point placement should be four places, or 0.0770. Because 0.077 as found on the calculator equals 0.0770; i.e., trailing zeros add no value; the rule still applies.

7.4 Percents and Interest

1. **(a)** $7.89 = 100\left[\frac{7.89}{100}\right] = \frac{789}{100} = \mathbf{789\%}$.

(b) $193.1 = 100\left[\frac{193.1}{100}\right] = \frac{19310}{100} = \mathbf{19310\%}$.

(c) $\frac{5}{6} = 100\left[\frac{\frac{5}{6}}{100}\right] = \frac{100\left[\frac{5}{6}\right]}{100}$.

$= \frac{\frac{50 \cdot 5}{1 \cdot 3}}{100} = \frac{\frac{250}{3}}{100} = \frac{83\frac{1}{3}}{100} = \mathbf{83\frac{1}{3}\%}$.

(d) $\frac{1}{8} = 100\left[\frac{\left(\frac{1}{8}\right)}{100}\right] = \frac{\left(\frac{100}{8}\right)}{100} = \frac{12.5}{100} = \mathbf{12.5\%}$.

(e) $\frac{5}{8} = 100\left[\frac{\left(\frac{5}{8}\right)}{100}\right] = \frac{\left(\frac{500}{8}\right)}{100} = \frac{62.5}{100} = \mathbf{62.5\%}$.

(f) $\frac{4}{5} = 100\left[\frac{\left(\frac{4}{5}\right)}{100}\right] = \frac{\left(\frac{400}{5}\right)}{100} = \frac{80}{100} = \mathbf{80\%}$.

3. **(a)** $\underline{4}$ for every 100. $4\% = \frac{4}{100}$.

(b) $\underline{2}$ for every 50. $4\% = \frac{4}{100} = \frac{2}{50}$.

(c) 1 for every $\underline{25}$. $4\% = \frac{4}{100} = \frac{1}{25}$.

(d) 8 for every $\underline{200}$. $4\% = \frac{4}{100} = \frac{8}{200}$.

(e) 0.5 for every $\underline{12.5}$. $4\% = \frac{4}{100} = \frac{0.5}{12.5}$.

5. **(a)** 5% of $x \Rightarrow \frac{5}{100}x$, or $\frac{5x}{100} = \frac{x}{20}$.

(b) 10% of amount $= a \Rightarrow \frac{10}{100} \cdot$ amount $= a$

$\Rightarrow$ amount $= \mathbf{10}\boldsymbol{a}$.

7. Let s be her last salary. 7% of s + 100% of s = \$27,285 $\Rightarrow$ 107% $\cdot\, s$ = \$27,285 $\Rightarrow 1.07 \cdot s =$ 27,285. Thus $s = \frac{27{,}285}{1.07} =$ **\$25,500**.

9. The amount of the discount is \$35 − \$28 = \$7. The percent of the discount is \$7 as percent of \$35 $\Rightarrow \frac{7}{35} = 0.2 = 100\left(\frac{0.2}{100}\right) = \frac{20}{100} =$ **20%**.

11. Sale price was regular price − 20% of regular price. \$28.00 − 20% of \$28.00 = \$28 − (0.2 · \$28) = \$28.00 − \$5.60 = **\$22.40**.

13. Bill answered 80 − 52 = 28 questions incorrectly. 28 as percentage of 80 is $\frac{28}{80} = 0.35 = 100\left(\frac{0.35}{100}\right) = \frac{35}{100} =$ **35%**.

15. $\frac{325}{500} \cdot \frac{325}{500} = \frac{650}{1000} = \frac{65}{100} = 65\%$; $\frac{600}{1000} = \frac{60}{100} = 60\%$.

17. The amount of John's 20% profit is 0.20 · \$330 = \$66. Sale price of the bike after a 10% discount must be \$330 + \$66 = \$396.

Let p be the list price; then p − 10% of p is the sale price. p − 10% of p = \$396 $\Rightarrow 0.9p =$ 396 $\Rightarrow p = \frac{396}{0.9} =$ \$440. Thus if John prices the bike at **\$440** he can offer a 10% discount of \$44 and still realize his \$66 profit.

19. A journeyman makes 200% of an apprentice's pay and a master makes 150% of a journeyman's pay. 150% of 200% = 1.5 · 2 = 3 = 300% of an apprentice's pay.

The \$4200 must have 1 + 2 + 3 = 6 shares, or $\frac{\$4200}{6}$ = \$700 per share. The **apprentice earns \$700**. The **journeyman** earns 200% of \$700 = 2 · \$700 = **\$1400**. The **master** earns 300% of \$700 = 3 · \$700 = **\$2100**.

21. Let s be the salary of the previous year. 100% of s + 10% of s = 110% of s = $1.1 \cdot s$ is current salary. $1.1s$ = \$100,000 (this year), so

$s = \frac{\$100.000}{1.1} \approx$ \$90,909.09 (last year).

$1.1s$ = \$90,909.09 (last year), so

$s = \frac{\$90{,}909{,}09}{1.1}$, or about **\$82,644.63** (two years ago).

23. **No**. Since 56% is more then double 25%, \$950 should be more than double \$500, but it is not.

25. **(a)** 15% of \$30 = 0.15 · 30 = **\$4.50**.

(b) $100\left[\frac{\left(\frac{1}{2}\right)}{100}\right] =$ **50%**.

(c) $100\left(\frac{1}{100}\right) =$ **100%**.

27. In the table below:

(i) Interest rate per period is found by dividing the annual rate by the number of compounding periods per year.

(ii) Number of periods: Semiannually is twice per year, quarterly is four times per year, monthly is twelve times per year, daily is 365 times per year.

(iii) Amount of interest paid: Where

A = compound amount;
P = principal (\$1000 in this example);
i = interest rate per period expressed as a decimal;
n = number of periods per year;
I = compound interest;

then $A = P(1 + i)^n$ and $I = A - P$.

Interest Rate per Period (i)	Number of Periods (n)	Amount of Interest Paid (I)	Compound Amount (A)

(a)	$\frac{6\%}{2} = 3\% = 0.03$	4	\$125.51	\$1125.51
(b)	$\frac{8\%}{4} = 2\% = 0.02$	12	\$268.24	\$1268.24
(c)	$\frac{10\%}{2} = 0.8\overline{3}\%$ $= 0.008\overline{3}$	60	\$645.31	\$1645.31
(d)	$\frac{12\%}{365} = \frac{0.12}{365}$	1460	\$615.95	\$1615.95

29. Look for the principal to be invested.
If $A = P(1+i)^n$ then $P = \frac{A}{(1+i)^n}$, where
$A = \$50{,}000$, $i = \frac{3\%}{4} = 0.75\% = 0.0075$,
and $n = 5 \text{ years} \cdot 4 \text{ periods per year} = 20$.
$P = \frac{\$50{,}000}{(1+0.0075)^{20}} \approx$ **\$43,059.49** rounded to the nearest penny.

31. Prices starting at \$1.35 are compounding at $11\% = 0.11$ annually for six years.
$A = 1.35 \cdot (1+0.11)^6 \approx$ **\$2.53.**

33. **Geometric**. To see this, let P be the original price of the house.

Month	Price
1	$P - .002P = .998P$
2	$.998P - .002(.998P)$ $= (1-.002).998P = .998^2P$
⋮	
k	$.998^kP$

35. Suppose \$1 is invested in each bank for 1 year.

(*i*) At New Age, $i = \frac{0.04}{365}$ and $n = 365$.
$A = \$1\left(1 + \frac{0.04}{365}\right)^{365} \approx \$1.041.$
Subtracting the \$1 invested, interest is 4.1¢ which corresponds to an effective annual rate of $\frac{4.1¢}{100¢} = 4.1\%$.

(*ii*) At Pay More, $i = \frac{0.052}{1}$ and $n = 1 \cdot 1 = 1$. $A = \$1(1+0.052)^1 =$ \$1.052. Subtracting the \$1 invested, interest is 5.2¢ which corresponds to an effective annual rate of 5.2%. The **Pay More bank offers the better rate**.

37. The preincipal remain constant, so the situation is represented by an **arithmetical sequence** with $d = 0$ or by a **geometric sequence with $r = 1$**.

39. In each case, interest as a percentage of saving is represented by $\frac{\text{interest earned}}{\text{amount invested}}$.

(a) Passbook savings account: $\frac{\$53.90}{\$980} = 0.055 = 5.5\%$ interest.

(b) Certificate of deposit: $\frac{\$55.20}{\$600} = 0.092$ $=$ **9.2%** interest.

(c) Money-marked certificates: $0.132 =$ **13.2%** interest. **This was best rate.**

7-4(B) Percents and Interests

1. (a) $0.032 = 100\left[\frac{0.032}{100}\right] = \frac{3.2}{100} = \mathbf{3.2\%}.$

(b) $0.2 = 100\left[\frac{0.2}{100}\right] = \frac{20}{100} = \mathbf{20\%}.$

(c) $\frac{3}{20} = 100\left[\frac{\left(\frac{3}{20}\right)}{100}\right] = \frac{\left(\frac{300}{20}\right)}{100} = \frac{15}{100} = \mathbf{15\%}.$

(d) $\frac{13}{8} = 100\left[\frac{\left(\frac{13}{8}\right)}{100}\right] = \frac{\left(\frac{1300}{8}\right)}{100} = \frac{162.5}{100}$ $= \mathbf{162.5\%}.$

(e) $\frac{1}{6} = 100\left[\frac{\left(\frac{1}{6}\right)}{100}\right] = \frac{\left(\frac{100}{6}\right)}{100} = \frac{16.\overline{6}}{100} = \mathbf{16.\overline{6}\%}$ $= \mathbf{16\frac{2}{3}\%}.$

(f) $\frac{1}{40} = 100\left[\frac{\left(\frac{1}{40}\right)}{100}\right] = \frac{\left(\frac{100}{40}\right)}{100} = \frac{2.5}{100} = \mathbf{2.5\%}.$

3. (a) $\underline{5}$ for every 100. $5\% = \frac{5}{100}$.

(b) $\underline{\mathbf{2.5}}$ for every 50. $5\% = \frac{5}{100} = \frac{2.5}{50}$.

(c) 1 for every $\underline{\mathbf{20}}$. $5\% = \frac{5}{100} = \frac{1}{20}$.

(d) 8 for every $\underline{\mathbf{160}}$. $5\% = \frac{5}{100} \cdot \frac{1.6}{1.6} = \frac{8}{160}$.

(e) 0.5 for every $\underline{\mathbf{10}}$. $5\% = \frac{5}{100} = \frac{0.5}{10}$.

5. (a) $.005x = \frac{5}{1000}x = \frac{x}{\mathbf{200}}$.

(b) One percent of x (an amount) is $0.001x$. If this equals a, then $0.001x = a \Rightarrow x = \frac{a}{0.001} \Rightarrow x = \mathbf{1000a}$.

7. Let x represent her previous salary.
$x + .1x = 60,000 \Rightarrow 1.1x = 60,000 \Rightarrow$
$x = \frac{60,000}{1.1} \approx \$54,545.45.$

9. The amount of depreciation was $\$1700 - \$1400 = \$300$. The depreciation as a percentage of $1700 is given by $\frac{300}{1700} \approx 0.1765 =$
$100\left[\frac{0.1765}{100}\right] = \frac{17.65}{100} = \mathbf{17.65\%}.$

11. Let t be the total income for the month. Then
$t = \$900 + 4\%$ of $\$1800 = 900 + \frac{4}{100} \cdot 1800$
$= \mathbf{\$972}.$

13. $\frac{4500 \text{ mammals}}{1{,}700{,}000 \text{ species}} \approx 0.0026 = 100\left(\frac{0.0026}{100}\right)$
$= \frac{0.26}{100}$, or **about 0.26%**.

15. **0.625%** is greater. $0.625\% = \frac{62.5}{100}$. On the other hand $(0.625\%)^2 = \frac{62.5}{100} \cdot \frac{62.5}{100} = \frac{62.5 \cdot \frac{62.5}{100}}{100} =$
$\frac{.625(62.5)}{100}$. Notice that 62.5 is multiplied by a number less than one and thus, it decreases.

17. The reduced price of the suit is
200 – .25(200) = 150. Let x be the percent increase. Then
$150 + x(150) = 200 \Rightarrow x = \frac{50}{150} \Rightarrow$
$x = \frac{1}{3} = \mathbf{33\frac{1}{3}\%}.$

19. Set up a proportion: $\frac{\frac{1}{4}\text{cup}}{x \text{ cups}} = \frac{0.5\%}{100\%} \Rightarrow$
$\frac{0.25 \text{ cup}}{x \text{ cups}} = \frac{0.005}{1} \Rightarrow x \cdot 0.005 = 0.25 \cdot 1 \Rightarrow$
$x = \frac{0.25}{0.005}$. $x =$ **50 cups**.

21. Let n be the original number and let N be 20% more than $n \Rightarrow 120\%$ of $n = 1.2n$. Then n as a percentage of N is $\frac{n}{N} = \frac{n}{1.2n} = \frac{1}{1.2} = 0.83\frac{1}{3}$;
i.e., $n = 83\frac{1}{3}\%$ of N. 100% of $N - 83\frac{1}{3}\%$ of $N = \mathbf{16\frac{2}{3}\%}$ of N.

23. Let g be the number of girls in the class, and let b be the number of boys. Then the problem may be stated as:

$0.7g + 0.6b = 0.5(g + b)$

$0.7g + 0.6b = 0.5g + 0.5b$

$0.2g = -0.1b.$

That is, it is **impossible**.

25. Let g be the number of girls in the class, and let b be the number of boys. Then the problem may be stated as:

$0.7g + 0.4b = 0.5(g + b)$

$0.7g + 0.4b = 0.5g + 0.5b$

$0.2g = 0.1b$, or $2g = b$.

That is, if the number of boys in the class is twice the number of girls, the **yes**, it is possible.

27. In the table below:

(i) Interest rate per period is found by dividing the annual rate by the number of compounding periods per year.

(ii) Number of periods: Semiannually is twice per year, quarterly is four times per year, monthly is twelve times per year, daily is 365 times per year.

(iii) Amount of interest paid: Where
A = compound amount;
P = principal ($1000 in this example);
i = interest rate per period expressed as a decimal;
n = number of periods per year;
I = compound interest;
then $A = P(1 + i)^n$ and $I = A - P$.

	Interest Rate per Period (i)	Number of Periods (n)	Amount of Interest Paid (I)	Compound Amount (A)
(a)	$\frac{4\%}{2} = 2\% = 0.02$	4	$82.43	$1082.43
(b)	$\frac{6\%}{4} = 1.5\% = 0.015$	12	$195.62	$1195.62
(c)	$\frac{18\%}{12} = 1.5\% = 0.015$	60	$1443.22	$2443.22
(d)	$\frac{18\%}{365} = \frac{0.18}{365}$	1460	$1054.07	$2054.07

29. An interest rate of 3% annually is a rate of $\frac{3}{4}\%$ each quarter. Over 5 years, there will be $5 \cdot 4 = 20$ compoundings.
$100,000 = P(1 + .0075)^{20} \Rightarrow 100,000 =$
$P(1.161184142) \Rightarrow P \approx \mathbf{\$86,119}.$

31. $10,000 compounded daily at 2% annual interest for 15 years is calculated as

$$A = \$10{,}000\left(1 + \frac{0.02}{365}\right)^{365\cdot 15} \approx \mathbf{\$13{,}498.48}.$$

33. Assume the interest earned each year is reinvested at the same simple interest rate. Then the interest from the first year is $I = prt$, where $t = 1$, so the amount at the end of the year is $p + I = p + pr = p(1 + r)$. If $r = 0.04$, then the amount at the end of the year is $1.04p$. This effect repeats each year. Since there is a constant ratio of 1.04 each year, the situation is represented by a **geometric** sequence.

35. Al's yearly return is given by $P\left(1 + \frac{.06}{365}\right)^{365}$. Betty's is given by $1000(1 + .07)$.

Al's Investment		Betty's Investment
Year 1	1061.83	1070
Year 2	1127.49	1140
Year 3	1197.20	1210
Year 4	1271.22	1280
Year 5	1349.83	1350
Year 6	1433.29	1420

Review Problems 7-4

27. **(a)** $16.72 = \frac{16.72\cdot 100}{100} = \frac{1672}{100} = \mathbf{\frac{418}{25}}$.

(b) $0.003 = \mathbf{\frac{3}{1000}}$.

(c) $^-5.07 = \mathbf{\frac{^-507}{100}}$.

(d) $0.123 = \frac{.123\cdot 1000}{1000} = \mathbf{\frac{123}{1000}}$.

29. $0.00024 = \frac{0.00024\cdot 100{,}000}{100{,}000} = \frac{24}{100{,}000}$

$= \frac{2^3\cdot 3}{10^5} = \frac{2^3\cdot 3}{2^5\cdot 5^5} = \frac{3}{2^2 5^5} = \mathbf{\frac{3}{12500}}$.

31. **(a)** $2.08 \cdot 10^5 = \mathbf{208{,}000}$.

(b) $3.8 \cdot 10^{^-4} = \mathbf{0.00038}$.

Chapter 7 Review

1. Each division on the number line corresponds to 0.01.

(a) *(i)* Point A is two divisions to the right of point 0, or **0.02**.

(ii) Point B is five divisions to the right of point 0, or **0.05**.

(iii) Point C is one division to the right of point 0.1, or **0.11**.

(b) *(i)* Point D (0.09) is one division to the left of point 0.1; i.e., $0.1 - 0.01 = 0.09$.

(ii) Point E (0.15) is five divisions to the right of point 0.1; i.e., $0.1 + 0.05 = 0.15$.

0 0.1 0.2
A B D C E

3. A fraction in simplest from, $\frac{a}{b}$, can be written as a terminating decimal if and only if the prime factorization of the denominator contains no primes other than 2 or 5. If only 2's and 5's are present, then a power of 10 can be obtained using the Fundamental Law of Fractions (which shows that the fraction can be written directly as a terminating decimal to the same number of decimal places as the smallest power of 10 in the denominator).

5. **(a)** $\frac{4}{7} = 4 \div 7 = \mathbf{0.\overline{571428}}$.

(b) $\frac{1}{8} = 1 \div 8 = \mathbf{0.125}$.

(c) $\frac{2}{3} = 2 \div 3 = \mathbf{0.\overline{6}}$.

(d) $\frac{5}{8} = 5 \div 8 = \mathbf{0.625}$.

7. **(a)** **307.63**. $307.6\underline{2}5$ is between 307.62 and 307.63; when the number to be rounded is at the mid-point, by convention it is commonly rounded up.

(b) **307.6**. $307.\underline{6}25$ is between 307.6 and 307.7 and is closer to 307.6.

(c) **308**. $30\underline{7}.625$ is between 307 and 308 and is closer to 308.

(d) **300**. $\underline{3}07.625$ is between 300 and 400 and is closer to 300.

9. Line up the decimal points:

$1.\overline{4519} = 1.4519519519...$

$1.451\overline{9} = 1.4519999999...$

$1.4519 = 1.4519000000...$

$1.45\overline{19} = 1.4519191919...$

$^-0.134 = {}^-0.1340000000...$

$^-0.\overline{1340}1 = {}^-0.1340140140...$

$0.1\overline{3401} = 0.1340140140...$

Ordering from greatest to least:

$\mathbf{1.451\overline{9} > 1.4\overline{519} > 1.45\overline{19} > 1.4519 >}$
$\mathbf{0.13\overline{401} > {}^{-}0.134 > {}^{-}0.13\overline{401}}$.

11. **(a)** Answers may vary. E.g., $0.11 > 0.105 > 0.104 > 0.103 > 0.102 > 0.101 > 0.1$.

(b) Begin with 0.1 and find half of each succeeding element, or $0 < 0.00625 < 0.0125 < 0.025 < 0.05 < 0.1$.

(c) $0.1 < 0.15 < 0.175 < 0.1875 < 0.19375 < 0.2$.

13. **(a)** $\frac{1}{8} = 0.125 = 100\left(\frac{0.125}{100}\right) = \frac{12.5}{100} = \mathbf{12.5\%}$.

(b) $\frac{3}{40} = 0.075 = 100\left(\frac{0.075}{100}\right) = \frac{7.5}{100} = \mathbf{7.5\%}$.

(c) $6.27 = 100\left(\frac{6.27}{100}\right) = \frac{627}{100} = \mathbf{627\%}$.

(d) $0.0123 = 100\left(\frac{0.0123}{100}\right) = \frac{1.23}{100} = \mathbf{1.23\%}$.

(e) $\frac{3}{2} = 1.5 = 100\left(\frac{1.5}{100}\right) = \frac{150}{100} = \mathbf{150\%}$.

15. $11\% \cdot \text{investment} = \$1020.80$
$\Rightarrow \text{investment} = \frac{\$1020.80}{0.11} = \mathbf{\$9280}$.

17. Assume "nearest tenth" refers to the nearest tenth of a percent. Then percent correct
$= 100\left(\frac{\frac{62}{70}}{100}\right) = \frac{\left(\frac{6200}{70}\right)}{100} \approx \frac{88.6}{100} = \mathbf{88.6\%}$.

19. A discount of $d\%$ means the customer pays $1 - \frac{d}{100}$ for the purchase. Discounts of 5%, 10%, and 20% mean the customer pays 0.95, 0.90, and 0.80, respectively, of cost. Their product, 0.648, is the same in any order, therefore there is **no difference**. The customer would pay 0.684 times the purchase price, for a discount of $1 - 0.684 = 0.316$, or 31.6%.

21. The price difference was $\$89.95 - \$62.00 = \$27.95$. The percentage difference was
$100\left[\frac{\left(\frac{27.95}{89.95}\right)}{100}\right] \approx \mathbf{31.1\%}$.

Chapter 8

Real Numbers and Algebraic Thinking

Assessment 8-1A: Real Numbers

1. Answers may vary. E.g., one such number could be 0.232233222333..., continuing the pattern of adding a 2 and a 3 to each succeeding group.

3. Line up the decimal points:

$0.9 = 0.90000000\ldots$

$0.\overline{9} = 0.99999999\ldots$

$0.\overline{98} = 0.98989898\ldots$

$0.9\overline{8} = 0.98888888\ldots$

$0.\overline{998} = 0.99898989\ldots$

$0.\overline{898} = 0.89889889\ldots$

$\sqrt{0.98} = 0.98994949\ldots$

Ordering from greatest to least:

$0.\overline{9} > 0.\overline{998} > \sqrt{0.98} > 0.\overline{98} > 0.9\overline{8} > 0.9 > 0.\overline{898}.$

5. (a) $15 \cdot 15 = 225 \Rightarrow \sqrt{225} = \mathbf{15}.$

(b) $13 \cdot 13 = 169 \Rightarrow \sqrt{169} = \mathbf{13}.$

(c) **Impossible**. There is no real number n such that $n^2 = {}^-81$.

(d) $25 \cdot 25 = 625 \Rightarrow \sqrt{625} = \mathbf{25}.$

7. Answers may vary.

(a) **False.** $0 + \sqrt{2}$ is irrational.

(b) **False.** ${}^-\sqrt{2} + \sqrt{2} = 0$, which is rational.

(c) **False.** $\sqrt{2} \cdot \sqrt{2} = 2$, which is rational.

(d) **True.** $\sqrt{2} - \sqrt{2} = 0$, which is rational.

9. Consider 8(c). We form an irrational number between 0.5 and 0.6 by placing digits in lesser place-values so that the decimal does not terminate or repeat, e.g., 0.5151151115.... A similar approach can be taken for any two rational numbers that are 0.1 a part. For example, 1.7171171117... is between 1.7 and 1.8.

11. In the tables of 14. and 15. below: N is the set of natural (or counting) numbers; I is the set of integers; Q is the set of rational numbers; R is the set of real numbers; and S is the set of irrational numbers. $N \subset I \subset Q \subset R; R = Q \cup S.$

		N	I	Q	R	S
(a)	6.7			√	√	
(b)	5	√	√	√	√	
(c)	$\sqrt{2}$				√	√
(d)	${}^-5$		√	√	√	
(e)	$3\frac{1}{7}$			√	√	

13. (a) $x = \mathbf{64}.$ $\sqrt{64} = 8.$

(b) **No real values.** $\sqrt{x}$ is the principal square root of x.

(c) $x = \mathbf{{}^-64}.$ $\sqrt{{}^-({}^-64)} = \sqrt{64} = 8.$

(d) **No real values.** $\sqrt{{}^-x}$ is the principal square root of x, if $x < 0$.

(e) All real numbers > 0. If $x = 0$ then $\sqrt{x} \not> 0$.

(f) **No real values.** $\sqrt{x}$ is the principal square root of x.

15. (a) $\sqrt[3]{{}^-54} = \sqrt[3]{{}^-27 \cdot 2} = \sqrt[3]{{}^-27} \cdot \sqrt[3]{2} = \mathbf{{}^-3\sqrt[3]{2}}.$

(b) $\sqrt[5]{96} = \sqrt[5]{32 \cdot 3} = \mathbf{2\sqrt[5]{3}}.$

(c) $\sqrt[3]{250} = \sqrt[3]{125 \cdot 2} = \mathbf{5\sqrt[3]{2}}.$

(d) $\sqrt[5]{{}^-243} = \mathbf{{}^-3\sqrt[5]{1} = {}^-3 \cdot 1 = {}^-3}.$

17. The sides of the gate form the sides of a right triangle with the diagonal brace as the hypotenuse. If c is the length of the hypotenuse, $c^2 = 4^2 + 5^2 = 16 + 25 = 41.$ $\sqrt{41} \approx$ **6.4 feet.**

19. (a) $3^x = 81 \Rightarrow 3^x = 3^4 \Rightarrow \mathbf{x = 4}.$

(b) $4^x = 8 \Rightarrow (2^2)^x = 2^3 \Rightarrow 2^{2x} = 2^3$
$\Rightarrow 2x = 3 \Rightarrow \boldsymbol{x = \frac{3}{2}}.$

(c) $128^{^-x} = 16 \Rightarrow (2^7)^{^-x} = 2^4 \Rightarrow {}^-7x = 4$
$\Rightarrow \boldsymbol{x = \frac{^-4}{7}}.$

(d) $\left(\frac{4}{9}\right)^{3x} = \frac{32}{243} \Rightarrow \left[\left(\frac{2}{3}\right)^2\right]^{3x} = \left(\frac{2}{3}\right)^5$
$\Rightarrow 6x = 5 \Rightarrow \boldsymbol{x = \frac{5}{6}}.$

(e) $2\sqrt{3} = \sqrt{x} \Rightarrow x = \left(\sqrt{x}\right)^2 = \left(2\sqrt{3}\right)^2$
$\Rightarrow x = 2^2 \cdot 3 = \mathbf{12}.$

21. Construct a right triangle with one side starting at 0 and ending at 18, then extending up approximately $4\frac{1}{4}$ units at a right angle. The resulting hypotenuse will have a length of $\sqrt{18^2 + \left(4\frac{1}{4}\right)^2} \approx \sqrt{342.1}$ units $\left(\text{close to } \sqrt{342}\right)$ and would lie directly on the number line from 0 to about 18.5 units if an arc with radius 18.5 were to be constructed. Thus $\sqrt{342} \approx 18.5$, about as close as possible with whole number units.

Assessment 8-1B

1. Answers may vary; one such could be 0.454455444555..., and continuing the pattern of adding a 4 and a 5 to each succeeding group.

3. Line up the decimal points:

$$0.8 = 0.80000\ldots$$
$$0.\overline{8} = 0.88888\ldots$$
$$0.\overline{89} = 0.89898\ldots$$
$$0.8\overline{89} = 0.88989\ldots$$
$$\sqrt{0.7744} = 0.88000\ldots$$

Ordering from greatest to least:
$0.\overline{89} > 0.8\overline{89} > 0.\overline{8} > \sqrt{0.7744} > 0.8.$

5. **(a)** $16 \cdot 16 = 256 \Rightarrow \sqrt{256} = \mathbf{16}.$

(b) $18 \cdot 18 = 324 \Rightarrow \sqrt{324} = \mathbf{18}.$

(c) **Impossible**. There is no real number n such that $n^2 = {}^-25$ (or any other negative number).

(d) $32 \cdot 32 = 1024 \Rightarrow \sqrt{1024} = \mathbf{32}.$

7. **(a)** **True**. The rationals are closed under addition.

(b) **False**. For example, $\sqrt{2} - \sqrt{2} = 0$, which is rational.

(c) **False**. If the rational number is 0, any product will be 0, which is rational.

9. Suppose $\frac{\sqrt{2}}{2}$ is a rational number which can be written in the form $\frac{a}{b}$, where a and b are integers and $b \neq 0$. Then $\frac{\sqrt{2}}{2} = \frac{a}{b} \Rightarrow \sqrt{2} = \frac{2a}{b}$, which is a rational number. But $\sqrt{2}$ is an irrational number which cannot equal a rational number, so a contradiction exists and therefore $\frac{\sqrt{2}}{2}$ must be irrational.

11. In tables 12. and 13. below: N is the set of natural (or counting) numbers; I is the set of integers; Q is the set of rational numbers; R is the set of real numbers; and S is the set of irrational numbers.
$N \subset I \subset Q \subset R; R = Q \cup S.$

		N	I	Q	R	S
(a)	$\sqrt{3}$				√	√
(b)	$4\frac{1}{2}$			√	√	
(c)	${}^-3\frac{1}{7}$			√	√	

13. **(a)** $\boldsymbol{x = 49}.\ \sqrt{49} = 7.$

(b) **No real numbers.** $\sqrt{x}$ is the principal square root of x.

(c) $\boldsymbol{x = {}^-49}.\ \sqrt{^-(^-49)} = 7.$

(d) **No real numbers.** $\sqrt{x}$ is always positive, so its additive inverse would be negative.

(e) **49**. The additive inverse of a positive number is negative.

15. **(a)** $\sqrt[3]{^-102} = \boldsymbol{{}^-1\sqrt[3]{102}}$. There is no number n such that $n^3 = 102$.

(b) $\sqrt[6]{64} = \sqrt[6]{2^6} = \mathbf{2}.$

(c) $\sqrt[3]{64} = \sqrt[3]{(2^2)^3} = 2^2 = \mathbf{4}.$

17. $h^2 = 5^2 + 12^2 = 169 \Rightarrow h = \sqrt{169} = \mathbf{13}.$

19. **(a)** $2^x = 64 \Rightarrow 2^x = 2^6 \Rightarrow x = \mathbf{6}$, (if the bases are equal, the exponents must be equal).

(b) $4^x = 64 \Rightarrow (2^2)^x = 2^6 \Rightarrow 2x = 6 \Rightarrow$ $x = \mathbf{3}$.

(c) $2^{^-x} = 64 \Rightarrow \frac{1}{2^x} = \frac{1}{2^{^-6}} \Rightarrow x = {^-}\mathbf{6}$, (if the numerators are equal the denominators must be equal).

(d) $\sqrt{\frac{3x}{2}} = 36 \Rightarrow \left(\sqrt{\frac{3x}{2}}\right)^2 = 36^2 \Rightarrow \frac{3x}{2} = (6^2)^2 \Rightarrow 3x = 2 \cdot 6^4 \Rightarrow x = \frac{2592}{3} = \mathbf{864}$.

(e) $\frac{3}{\sqrt{3}} = \sqrt{x} \Rightarrow \left(\frac{3}{\sqrt{3}}\right)^2 = \sqrt{x}^2 \Rightarrow \frac{9}{3} = x \Rightarrow \mathbf{3} = x$.

21. $a \geq 2 \Rightarrow \sqrt{a} \geq \sqrt{2} \Rightarrow \frac{\sqrt{a}}{\sqrt{2}} \geq 1 \Rightarrow \frac{1}{\sqrt{2}} \geq \frac{1}{\sqrt{a}}$. Thus $0 < \frac{1}{\sqrt{a}} \leq \frac{1}{\sqrt{2}}$ on a number line.

Assessment 8-2A: Variables

1. (a) Use the format for an arithmetic sequence: $a_n = a_1 + (n-1)d$, if $a_1 = 10$. Then $a_3 = 10 + (3-1)d = \mathbf{10 + 2d}$.

(b) If n is the number, then twice the number is $2n$ and 10 less than that is $\mathbf{2n - 10}$.

(c) If n is the number, then its square is n^2 and 10 times that is $\mathbf{10n^2}$.

(d) If n is the number, its square is n^2 and twice the number is $2n$. Their difference is $\mathbf{n^2 - 2n}$.

3. (a) There are 4, 6, 8, and 10 shaded tiles, respectively, in the four figures. Assume and arithmetic sequence with $a_1 = 4$ and $d = 2$. Thus $a_n = 4 + (n-1)2 = \mathbf{2n + 2}$ or $\mathbf{2(n+1)}$.

(b) There are a total of $(n+2)^2$ squares in each figure. Assume the pattern continues; then the number of white squares is $(n+2)^2$ – the number of shaded squares, or $(n+2)^2 - (2n+2) = n^2 + 4n + 4 - 2n - 2 = \mathbf{n^2 + 2n + 2}$.

5. If the number of students is 20 times the number of professors, then $\mathbf{S = 20P}$.

7. Let m be the number of matchsticks Ryan uses. Then $m_1 = 10, m_2 = 16, m_3 = 22, \ldots$ Assume an arithmetic sequence with $m_1 = 10$ and $d = 6$. Thus $m_n = 10 + (n-1) \cdot 6 = \mathbf{6n + 4}$ matchsticks.

9. Let r be total revenue. Then $r = 5x + 13(100) = \mathbf{\$(5x + 1300)}$.

11. (a) (*i*) The pattern appears to be subtracting 4 from each term to obtain its successor. At the 100th term $^-4$ will be subtracted 99 times. So the 100th term is $-3 - 4(99) = \mathbf{-399}$.

(*ii*) The nth term can be found using the expression $\mathbf{-3 - 4(n-1)} = -3 - 4n + 4 = \mathbf{-4n + 1}$.

(b) (*i*) The pattern appears to be adding $\sqrt{2}$ to each term to obtain its successor. At the 100th term $\sqrt{2}$ will be added 99 times. The 100th term is $1 + 99\sqrt{2}$.

(*ii*) The nth term can be found using the expression $\mathbf{1 + (n-1)\sqrt{2}} = 1 + \sqrt{2}\,n - \sqrt{2} = \mathbf{\sqrt{2}n + 1 - \sqrt{2}}$.

(c) (*i*) The pattern appears to be adding 2 to each term to obtain its successor. At the 100th term 2 will be added 99 times. The 100th term is $\pi + 0.5 + 2(99) = \mathbf{\pi + 198.5}$.

(*ii*) $\mathbf{\pi + 0.5 + (n-1)2 = 2n + \pi - 1.5}$

13. By "difference" the authors mean the difference between successive terms, e.g., $a_{n+1} - a_n$. Thus, the 1st term is $\mathbf{14 - 9\sqrt{2}}$.

15. (a)

n	F_n
1	1
2	1
3	$1 + 1 = 2$
4	$1 + 2 = 3$
5	$2 + 3 = 5$
6	$3 + 5 = 8$
7	$5 + 8 = 13$

(b) $1 + 1 + 2 = \mathbf{4}$

(c) $1 + 1 + 2 + 3 = \mathbf{7}$

(d) $1 + 1 + 2 + 3 + 5 = 7 + 5 = \mathbf{12}$

(e) $12 + 8 = \mathbf{20}$

(f) $20 + 13 = \mathbf{33}$.

(g) Answers may vary. The sum of the first n Fibonacci numbers is 1 less than the Fibonacci number two terms later in the sequence.

(h) $F_1 + F_2 + F_3 + \cdots + F_n = F_{n+2} - 1$.

Assessment 8-2B

1. **(a)** If n is the number, then 10 more than the number is $\mathbf{n + 10}$.
   (b) If n is the number, then 10 less than the number is $\mathbf{n - 10}$.
   (c) If n is the number, then 10 times it is $\mathbf{10n}$.
   (d) If n is the number, then the sum of it and 10 is $\mathbf{n + 10}$.
   (e) If n is the number, its square is n^2 and the difference between its square and itself is $\mathbf{n^2 - n}$.

3. **(a)** There are 10,13, and 16 shaded tiles, respectively, in the three figures. Assume an arithmetic sequence with $a_1 = 10$ and $d = 3$. Thus $a_n = 10 + (n - 1)3 = \mathbf{3n + 7}$.
   (b) There are a total of $(n + 3)^2$ squares in each figure. The number of white squares is then $(n + 3)^2$ – the number of shaded squares, or $(n + 3)^2 - (3n + 7) = n^2 + 6n + 9 - 3n - 7 = \mathbf{n^2 + 3n + 2}$.

5. If there are 100 more men than women $\Rightarrow m = \mathbf{w + 100}$.

7. Let m be the number of matchsticks Ryan uses. Then $m_1 = 4, m_2 = 7, m_3 = 10, \ldots, m_{22} = 67$. Assume an arithmetic series with $a_1 = 4$ and $d = 3$.
   (*i*) $a_n = 4 + (n - 1) \cdot 3 \Rightarrow \mathbf{3n + 1}$ matchsticks.
   (*ii*) $a_{n-1} = 3(n - 1) + 1 = \mathbf{3n - 2}$ matchsticks.

9. Let n be the student's odd number. Then $\dfrac{4n + 16}{2} - 7 = \dfrac{4n + 16 - 14}{2} = 2n + 1$ is the new number. If the teacher then subtracted 1 from each answer and divided the difference by 2 she would come back to the original number.

11. **(a)** **(*i*)** The pattern appears to subtract 4 from each term to obtain its successor. The 100th term is given by $3 - 4(99) = \mathbf{{}^{-}393}$.
    (*ii*) The nth term is given by $3 - 4(n - 1) = 3 - 4n + 4 = \mathbf{{}^{-}4n + 7}$.
    (b) **(*i*)** The pattern appears to subtract $\sqrt{2}$ from each term to obtain its successor. The 100th term is given by $1 - \sqrt{2}(99) = \mathbf{1 - 99\sqrt{2}}$
    (*ii*) The nth term is given by $1 - \sqrt{2}(n - 1) = 1 - \sqrt{2}n + \sqrt{2} = \mathbf{{}^{-}\sqrt{2}n + 1 + \sqrt{2}}$.
    (c) **(*i*)** The pattern appears to add π to each term to obtain its successor. The 100th term is given by $0.5 + \pi + \pi(99) = \mathbf{0.5 + 100\pi}$.
    (*ii*) The nth term is given by $0.5 + \pi + \pi(n - 1) = \mathbf{0.5 + \pi n}$.

13. By "difference" the author means the difference in successive terms, e.g., $a_{n+1} - a_n$. So the first term is given by $\mathbf{16 - 9\sqrt{2}}$.

15. **(a)**

n	F_n
1	1
2	2
3	$1 + 2 = 3$
4	$2 + 3 = 5$
5	$3 + 5 = 8$
6	$5 + 8 = 13$
7	$8 + 13 = 21$

   (b) $1 + 2 + 3 = \mathbf{6}$.
   (c) $1 + 2 + 3 + 5 = \mathbf{11}$.
   (d) $11 + 8 = \mathbf{19}$.
   (e) $19 + 13 = \mathbf{32}$.
   (f) $32 + 21 = \mathbf{53}$.
   (g) Answers may vary. The sum of the first n terms in this Fibonacci-type sequence is 2 less than the Fibonacci number two terms later in the sequence.
   (h) $F_1 + F_2 + F_3 + \cdots + F_n = F_{n+2} - 2$

8-2 Review Problems

11. **Integer**, since $\sqrt[3]{729} = \sqrt[3]{9^3} = 9$.

13. $\sqrt{0.\overline{9}} = \sqrt{1} = 1$.

Assessment 8.3A: Equations

1. The balance scales imply addition, so:

(*i*) $\Delta + \square = 12.$

(*ii*) $O + O + \Delta + \square = 18;$

$2O + 12 = 18;$ so

$O = 3.$

(*iii*) $2O + \square = 10;$

$6 + \square = 10;$ so

$\square = 4.$

(*iv*) $\Delta + \square = 12;$

$\Delta + 4 = 12;$ so

$\Delta = 8.$

3. Let m be the number of matchsticks Ryan uses. Then $m_1 = 4, m_2 = 7, m_3 = 10, \ldots, m_n = 67.$ Assume an arithmetic series with $m_1 = 4, d = 3,$ and $m_n = 67.$

Thus $67 = 4 + (n - 1) \cdot 3 \Rightarrow 67 = 4 + 3n - 3 \Rightarrow 66 = 3n \Rightarrow n =$ **22 squares.**

5. Let e be the amount the eldest receives,
m be the amount the middle sibling receives,
y be the amount the youngest receives.
Then $e = 3y$ and $m = y + 14{,}000$

So
$$e + m + y = 486{,}000$$
$$3y + (y + 14{,}000) + y = 486{,}000$$
$$5y + 14{,}000 = 486{,}000$$
$$5y = 472{,}000$$

Thus $\mathbf{y = \$94{,}400}$

$\mathbf{m = \$108{,}400}$

$\mathbf{e = \$283{,}200}$

7. Let d be the number of dimes; then $67 - d$ is the number of nickels. $0.10d$ is the amount of money in dimes and $0.05(67 - d)$ is the amount of money in nickels.

Thus $0.10d + 0.05(67 - d) = 4.20$, or
$$10d + 5(67 - d) = 420$$
$$10d + 335 - 5d = 420$$
$5d = 85 \Rightarrow \quad d =$ **17 dimes**

$67 - d =$ **50 nickels.**

9. Let g be the number of graduate students; $15g$ is the number of undergraduates. Then
$$g + 15g = 10{,}000$$
$16g = 10{,}000 \Rightarrow$ **625 graduate students.**

11. In the given sequence $d = 3$, thus $n + (n - 3) = 299 \Rightarrow 2n - 3 = 299 \Rightarrow \mathbf{n = 151}$ and $\mathbf{n - 3 = 148}$.

Assessment 8-3B

1. If $\square = 2\Delta$, then $\Delta < \square$.
If $\Delta = 2O$, then $O < \Delta$.
So $O < \Delta < \square$.

(a) $\square$ **weighs the most.**

(b) O weighs the least.

3. Let m be the number of matchsticks Ryan uses. Then $m_1 = 4, m_2 = 7, m_3 = 10, \ldots$. Assume an arithmetic sequence with $d = 3$. Let m_n be the number of matchsticks in the last figure; then $m_n + (m_n - 3) = 599 \Rightarrow m_n = 301$ matchsticks. Thus the last two figures used **298** and **301** matchsticks, respectively.

5. Let e be the amount left to the eldest, m be the amount left to each of two middle siblings, and y be the amount left to the youngest.
It is given that $e = 2y$ and $m = y + 16{,}000$.
Thus if $e + m + m + y = 1{,}000{,}000$ then
$2y + 2(y + 16{,}000) + y = 1{,}000{,}000 \Rightarrow$
$5y + 32{,}000 = 1{,}000{,}000 \Rightarrow y = 193{,}600.$

So $e = 2y = \mathbf{\$387{,}200}$

$m = y + 16{,}000 = \mathbf{\$209{,}600}$

$\mathbf{y = \$193{,}600.}$

7. Let d be the number of Dave's stickers; Matt then has $2d$ stickers.

So $d + 2d = 2(120) \Rightarrow d = 80$ and $2d = 160$. If **Matt gives Dave 40 stickers**, each will then have 120.

9. Let s be the number of students; then the number of professors is $\frac{1}{13}s$.

Thus $s + \frac{1}{13}s = 28{,}000 \Rightarrow 1\frac{1}{13} \cdot s = 28{,}000$
$\Rightarrow s =$ **26,000 students.**

11. In a geometric sequence, $a_n = a_1r^{n-1}$ and the first two terms are thus a_1 and a_1r. Then $a_1 + a_1r =$

$100a_1 \Rightarrow a_1(1 + r) = 100a_1 \Rightarrow 1 + r = 100$
$\Rightarrow \mathbf{r = 99}$.

Review Problems

11. If the middle even number is n, then the next two are $n + 2$ and $n + 4$; the previous two are $n - 4$ and $n - 2$. Their sum is $(n - 4) + (n - 2) + n + (n + 2) + (n + 4) = 5n + 6 - 6 = \mathbf{5n}$.

13. **(a)** The sum is $30 + 35 + \ldots + [30 + 5(t - 1)]$, which sums to $\frac{\mathbf{(60+5(t-1))t}}{\mathbf{2}}$

(b) $\$(d + 2d + 4d + 8d) = \mathbf{\$15d}$.

15. The sixth term is given by $\sqrt{5}(0.5)^5 = \sqrt{5} \cdot \left(\frac{1}{2}\right)^5$
$= \sqrt{5}\left(\frac{1}{32}\right) = \frac{\mathbf{\sqrt{5}}}{\mathbf{32}}$.

Assessment 8-4A
Functions

1. Where x is the first element in each ordered pair and $f(x)$ is the second:

(a) **Double** the input number; i.e., $f(x) = 2x$.

(b) **Add 6** to the input number; i.e., $f(x) = x + 6$.

3. **(a)** Answers may vary; e.g.:

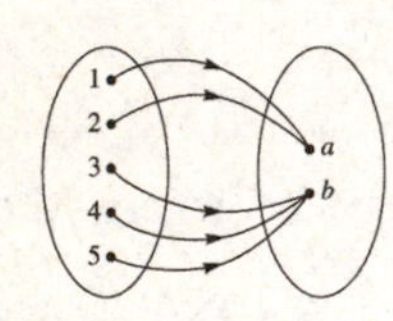

(b) Each of the five elements in the domain have two choices for a pairing $\Rightarrow 2^5 = 32$ possible functions.

5. **(a)** **Function from $\mathbb{R}$ to {2}**, a subset of R.

$f(x) = 2$ is called a constant function and its only output is 2.

(b) **Function from $\mathbb{R}$ to $\{x \mid x \geq 0\}$**, a subset of R. $f(x) = \sqrt{x}$ has output of nonnegative numbers only because $\sqrt{x}$ is the principal square root.

7. This is an example of **step function**; note in the following graph that a child weighing exactly 30, 32, 34,... pounds uses the lower dosage at the break point:

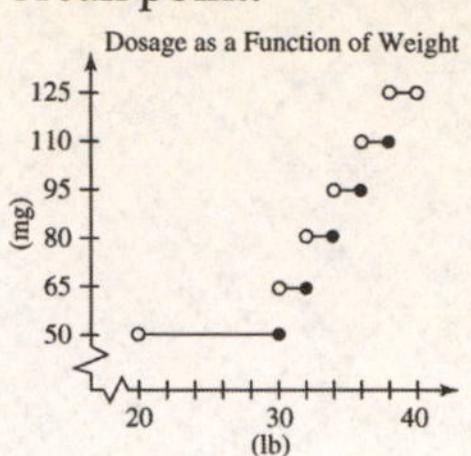

9. **(a)**

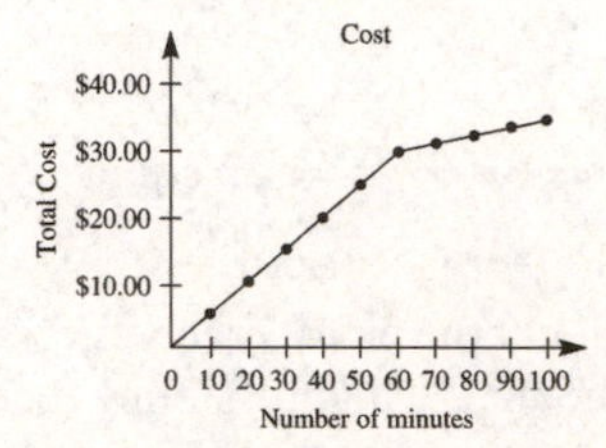

(b) It must be assumed that the company charges only for the **exact fraction of minutes used.**

(c) The two segments represent the **two different rates** per minute; the steeper one comes from the 50¢ per minute charge for the first 60 minutes.

(d) If $c(t)$ in dollars is cost as a function of time t in minutes, then

$$\mathbf{c(t)} = \begin{cases} \mathbf{0.50t} & \textbf{if } \mathbf{t \leq 60} \\ \mathbf{30 + 0.10(t - 60)} & \textbf{if } \mathbf{t > 60} \end{cases}$$

Where \$30 is the cost of the first 60 minutes.

11. **(a)** $(g \circ f)(5) = g[f(5)] = f(5) - 5 = 7(5) - 5 = \mathbf{30}$

(b) $(g \circ f)(10) = g[f(10)] = f(10) - 5 = 7(10) - 5 = \mathbf{65}$.

(c) $(g \circ f)(\sqrt{7}) = g[f(\sqrt{7})] = f(\sqrt{7}) - 5 = \mathbf{7(\sqrt{7}) - 5}$.

(b) $(g \circ f)(0) = g[f(0)] = f(0) - 5 = \mathbf{{}^{-}5}$.

13. **(a)** **(*i*)** $(1, 7) \Rightarrow 2 \cdot 1 + 2 \cdot 7 = \mathbf{16}$.

(*ii*) $(2,6) \Rightarrow 2 \cdot 2 + 2 \cdot 6 = \mathbf{16}$.

(*iii*) $(6,2) \Rightarrow 2 \cdot 6 + 2 \cdot 2 = \mathbf{16}$.

(*iv*) $(\sqrt{5},\sqrt{5}) \Rightarrow 2 \cdot \sqrt{5} + 2 \cdot \sqrt{5} = \mathbf{4\sqrt{5}}$.

(b) If output (or perimeter) = 20 then

$2\ell + 2w = 20 \Rightarrow 2(\ell + w) = 20$

$\Rightarrow \ell + w = 10$.

The possibilities are **any ordered pair $(l, 10-l)$ with $0 < l < 10$.**

(c) Domain:

$\{(\ell, w) \mid \ell$ and w are any positive real numbers$\}$.

The range is any positive real number.

15. (a) (*i*) $H(2) = 128(2) - 16(2)^2 = \mathbf{192}$ **feet.**

(*ii*) $H(6) = 128(6) - 16(6)^2 = \mathbf{192}$ **feet.**

(*iii*) $H(3) = 128(3) - 16(3)^2 = \mathbf{240}$ **feet.**

(*iv*) $H(5) = 128(5) - 16(5)^2 = \mathbf{240}$ **feet.**

Some of the heights correspond to the height of the ball as it rises; some to its height as it falls.

(b) Plot:

t	$H(t) = 128t - 16t^2$
0	0
1	112
2	192
3	240
4	256
5	240
6	192
7	112
8	0

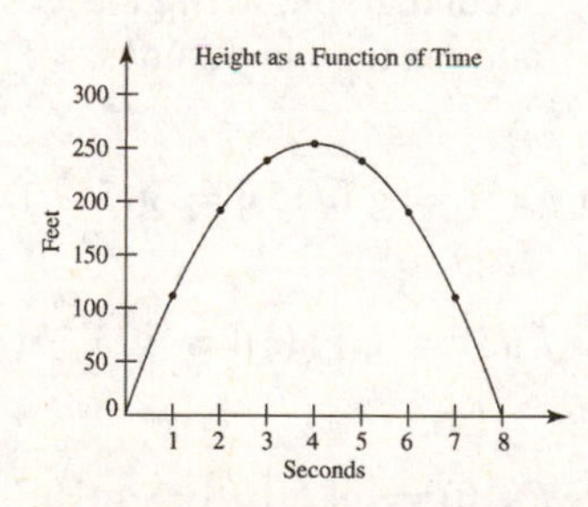

The ball will reach its highest point of **256 feet** at t = **4 seconds.**

(c) $H(t) = 0$ at $t = 8$; i.e., the ball will hit the ground in **8 seconds**. Note that $H(t) = 0$ also at $t = 0$.

(d) Domain: $\{t \mid 0 \leq t \leq 8\}$ **seconds.**

(e) Range: $\{H(t) \mid 0 \leq H(t) \leq 256\}$ **feet.**

17. (a) If each figure adds one row and one column to the preceding figure, there is one more column in each then the number of rows. Thus $S(n) = \frac{n(n+1)}{2}$.

(b) If the number of squares corresponds to powers of four; i.e., $4^0, 4^1, 4^2, 4^3, \ldots$. Then $S(n) = 4^{n-1}$.

19. (a) A function. $x + y = 2 \Rightarrow y = 2 - x$; for any input x, y is unique.

(b) Not a function. $x - y < 2 \Rightarrow y > x - 2$; for any input x, y may be any value greater than $x - 2$.

(c) A function. y is unique for any input x.

(d) A function. $xy = 2 \Rightarrow y = \frac{2}{x}$; for any input x (except $x = 0$, for which y is undefined), y is unique.

21. (a) B and H must be boys since there are no "is the sister of arrows" from either. The remainder; A, C, D, G, I, J, E, and F; must be girls.

(b) $\{(A, B), (A, C), (A, D), (C, A), (C, B), (C, D), (D, A), (D, B), (D, C), (E, H), (F, G), (G, F), (I, J), (J, I)\}$.

(c) No. A "is the sister of" three different people $\Rightarrow$ the relation is not a function.

23. (a) Not an equivalence relation:

(*i*) Not reflexive. A person cannot be a parent to him/herself.

(*ii*) Not symmetric. John can be a parent to Jane, but Jane cannot be a parent to John.

(*iii*) Not transitive. If John is the parent of James and James is the parent of Joseph, John is not the parent of Joseph.

(b) An equivalence relation (reflexive, symmetric, and transitive):

(*i*) Reflexive. Juan is the same age as Juan.

(*ii*) Symmetric. If Juan is the same age as Juanita, then Juanita is the same age as Juan.

(*iii*) Transitive. If Juan is the same age as Jose and Jose is the same age as Victor, then Juan is the same age as Victor.

(c) An equivalence relation:

(*i*) Reflexive. Jo Ann has the same last name as herself.

(*ii*) Symmetric. If Jo Ann has the same last name as Cheryl, then Cheryl has the same last name as Jo Ann.

(*iii*) Transitive. If Jo Ann has the same last name as Cheryl and Cheryl has the same last name as Penelope, then Jo Ann has the same last name as Penelope.

(d) An equivalence relation:

(*i*) Reflexive. Vicky is the same height as herself.

(*ii*) Symmetric. If Barbara is the same height as Margarita, then Margarita is the same height as Barbara.

(*iii*) Transitive. If Willy is the same height as Billy and Billy is the same height as Don, then Willy is the same height as Don.

(e) Not an equivalence relation:

(*i*) Not reflexive. Cindy cannot be married to herself.

(*ii*) Symmetric. If Arnold is married to Pam, then Pam is married to Arnold.

(*iii*) Not transitive. If John is married to Clara and Clara is married to John, then John is not married to John.

(f) Not an equivalence relation:

(*i*) Reflexive. Peter lives within 10 miles of himself.

(*ii*) Symmetric. If Jon lives within 10 miles of Evangeline, then Evangeline lives within 10 miles of Jon.

(*iii*) Not transitive. If Fred lives within 10 miles of Jim and Jim lives within 10 miles of Herb, then Fred does not necessarily live within 10 miles of Herb.

(g) Not an equivalence relation:

(*i*) Not reflexive. Juan cannot be older than himself.

(*ii*) Not symmetric. If Jose is older than Mireya then Mireya cannot be older then Jose.

(*iii*) Transitive. If Jean is older than Mike and Mike is older than Cybil, then Jean is older than Cybil.

Assessment 8-4B

1. **(a) Subtract 2** from the input number; i.e., $f(x) = x - 2$.

   (b) Square the input number and **add 1**; i.e., $f(x) = x^2 + 1$.

3. **(a)** Answers may vary; for example:

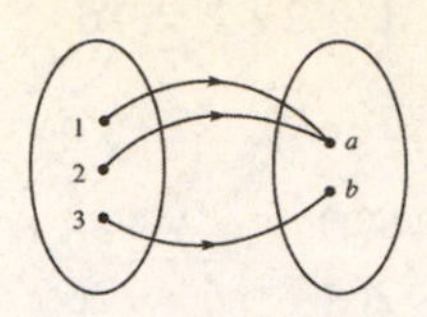

   (b) Each of the three elements in the domain have two choices for a pairing $\Rightarrow$ $\mathbf{2^3 = 8}$ possible functions.

5. **(a) A function**. The function is {(0, 0), (1, 0), (2, 0), (3, 0), (x, 3)} where x is any real number other than 0,1,2 or 3.

   (b) Not a function. Some elements of the domain correspond with more than one element of the range; e.g., (3, 0) and (3, 1), (4, 0) and (4, 1),....

   (c) A function. {(0, 0), (1, 1), (2, 2),..., (9, 9), (10, 0), (11, 1), (12, 2),...}.

7. **(a)** $T = 70°$ F. Then $C = 70 - 40 = 30$ chirps per 15 seconds $=$ **2 chirps** each second.

   (b) 40 chirps per minute $=$ 40 chirps per 60 seconds $=$ 10 chirps per 15 seconds. $10 = T - 40 \Rightarrow T = 10 + 40 =$ **50°F**.

9. **(a)** The "Cost per Minute" graph shows 45¢ for the **cost of the sixth minute**; the "Total Cost for Calls" graph shows the **cost for a six-minute call** to be $2.70.

   (b) It must be assumed that the cost per minute is always the same regardless of the length of the call, and that the company charges 45¢ per minute for each fraction of a minute.

   (c) (*i*) $\mathbf{c = 45}$**¢** per minute in the first graph.

   (*ii*) At 45¢ per minute, $\mathbf{c = 45t}$ in the second graph, where c is cost in cents and t is time in minutes.

11. **(a)** $(g \circ f)(5) = g[f(5)] = g(5 - 1) = 7 \cdot 4 = \mathbf{28}$.

    (b) $(g \circ f)(3) = g[f(3)] = g(3 - 1) = 7 \cdot 2 = \mathbf{14}$.

    (c) $(g \circ f)(10) = g[f(10)] = g(10 - 1) = 7 \cdot 9 = \mathbf{63}$.

    (d) $(g \circ f)(a) = g[f(a)] = g(a - 1) = \mathbf{7(a - 1)}$.

13. **(a)** (*i*) $(1, 4) \Rightarrow 2 \cdot 1 + 2 \cdot 4 = \mathbf{10}$.

    (*ii*) $(2, 1) \Rightarrow 2 \cdot 2 + 2 \cdot 1 = \mathbf{6}$

(*iii*) $(1, 2) \Rightarrow 2 \cdot 1 + 2 \cdot 2 = $ **6**.

(*iv*) $(\sqrt{3}, \sqrt{3}) \Rightarrow 2 \cdot \sqrt{3} + 2 \cdot \sqrt{3} = \mathbf{4\sqrt{3}}$.

(*v*) $(x, y) \Rightarrow \mathbf{2 \cdot x + 2 \cdot y}$.

(b) If output (or perimeter) = 20 then $2\ell + 2w = 20 \Rightarrow 2(\ell + w) = 20 \Rightarrow \ell + w = 10$. The possibilities are **{(1, 9), (2, 8), (3, 7), (4, 6), (5, 5), (6, 4), (7, 3), (8, 2), (9, 1)}**.

(c) **No**. Outputs are single numbers, not ordered pairs.

15. **(a)** (*i*) $H(2) = 128(2) - 16(2)^2 = $ **192 feet**.

(*ii*) $H(6) = 128(6) - 16(6)^2 = $ **192 feet**.

(*iii*) $H(3) = 128(3) - 16(3)^2 = $ **240 feet**.

(*iv*) $H(5) = 128(5) - 16(5)^2 = $ **240 feet**.

Some of the heights correspond to the height of the ball as it rises; some to its height as it falls. The graph is a parabola with vertex (4, 256), so **256** is the highest point.

(b) Plot:

t	$H(t) = 128t - 16t^2$
0	0
1	**112**
2	192
3	240
4	256
5	240
6	192
7	**112**
8	0

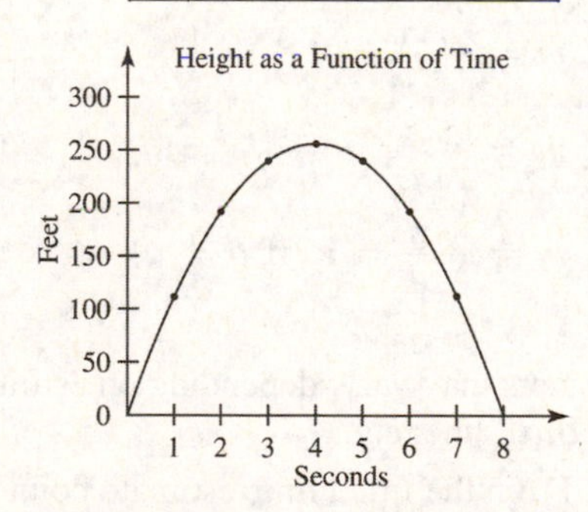

The ball will reach its highest point of **256 feet** at $\mathbf{t = 4}$ **seconds**.

(c) $H(t) = 0$ at $t = 8$; i.e., the ball will hit the ground in **8 seconds**. Note that $H(t) = 0$ also at $t = 0$.

(d) Domain: $\{\mathbf{t \mid 0 \le t \le 8}\}$ **seconds**.

(e) Range: $\{\mathbf{H(t) \mid 0 \le H(t) \le 256}\}$ **feet**.

17. **(a)** $S(n) = 1 + 2 + 3 + \cdots + n = \frac{n(n+1)}{2}$.

(b) $S(n) = n + 2 + 2n = \mathbf{3n + 2}$.

19. **(a)** **A function**. $x - y = 2 \Rightarrow y = x - 2$; for any input x, y is unique.

(b) **Not a function**. $x + y < 20 \Rightarrow y < 20 - x$; for any input x, y may be any value less than $20 - x$. Thus, it is easy to find two outputs y for each input x.

(c) **A function**. For any input x, y is unique.

(d) **A function**. For any input x, y is unique.

21. **(a)** (*i*) **Symmetric**. If $x = a$ and $y = b$ satisfies the relation, then $x = b$ and $y = a$ also satisfies it (by the associative property).

(*ii*) **Not symmetric**. $a - b \neq b - a$.

(*iii*) **Symmetric**. If $x = a$ and $y = b$ satisfies the relation, then $x = b$ and $y = a$ also satisfies it.

(*iv*) **Symmetric**. If $y = x$, then $x = y$.

(*v*) **Not symmetric**. $(2, 4) \neq (4, 2)$.

(b) **All** are functions of y in terms of x. In each, for any input x there is a unique output y.

23. **(a)** **A function**. Each element of the domain is paired with one and only one element of the range.

(b) A relation but **not a function**. New York from the domain is paired with two elements of the range.

(c) A relation but **not a function**. "mother" from the domain could be paired with more than one element of the range.

(d) **A function**. Each element of the domain is paired with one and only one element of the range.

(e) A relation but **not a function**. The element 1 from the domain could be paired with any odd number from the range to produce an even number.

Review Problems

17. Since distance = rate × time, the fast car will travel $70t$ miles in the same time as the slow car travels $60t$ miles. Then $70t - 40 = 60t \Rightarrow 10t = 40 \Rightarrow \mathbf{t = 4}$ **hours**.

19. Answers may vary. For example, $\sqrt{2} + {}^-\sqrt{2} = 0$, a rational number.

Assessment 8-5A

Lines in a Cartesian Coordinate System

1. The graph of $y = mx + 3$ contains the point $(0, 3)$ and is parallel to the line $y = mx$. Similarly, the graph of $y = mx - 3$ contains the point $(0, {}^-3)$ and is parallel to $y = mx$ (see below).

(a) Parallel line; y-intercept $= 3$.

(b) Parallel line; y-intercept $= {}^-3$.

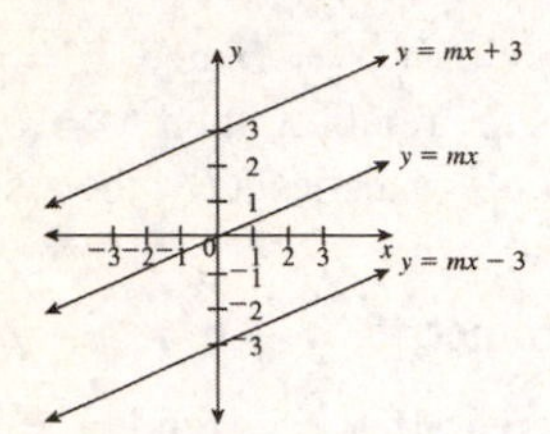

3. When the equations in Problem 2 are in the form $y = mx + b$, b is the y-intercept and is located at $(0, b)$. The x-intercept may be found by setting y equal to 0 and solving for x. If x is some value a, the x-intercept is located at $(a, 0)$.

(a) $y = \frac{{}^-3}{4}x + 3$:

(*i*) $b = 3$. y-intercept is at **(0, 3)**.

(*ii*) If $y = 0 = \frac{{}^-3}{4}x + 3 \Rightarrow x = 4$. x-intercept is at **(4, 0)**.

(b) $y = {}^-3$:

(*i*) $b = {}^-30$. y-intercept is at **(0, ¯3)**.

(*ii*) There is **no x-intercept**; i.e., $y = {}^-3$ is a horizontal line parallel to the x-axis.

(c) $y = 15x - 30$.

(*i*) $b = {}^-30$. y-intercept is at **(0, ¯30)**.

(*ii*) If $y = 0 = 15x - 30 \Rightarrow x = 2$. x-intercept is at **(2, 0)**.

5. When the equations are in slope-intercept form, $y = mx + b$, m is the slope of the line and b is the y-intercept, located at $(0, b)$.

(a) (*i*) $3y - x = 0 \Rightarrow 3y = x \Rightarrow \mathbf{y = \frac{1}{3}x}$.

(*ii*) $m = \frac{1}{3}$; y-intercept is at $(0, 0)$.

(b) (*i*) $x + y = 3 \Rightarrow \mathbf{y = {}^-x + 3}$.

(*ii*) $m = {}^-1$; y-intercept is at $(0, 3)$.

(c) (*i*) $x = 3y \Rightarrow \mathbf{y = \frac{1}{3}x}$.

(*ii*) $m = \frac{1}{3}$; y-intercept is at $(0, 0)$.

7. Answers may vary.

(a) Both points include the coordinate $y = 2$, thus the equation of the line is $y = 2$. Other points could be $({}^-3, 2), (5, 2), \ldots$.

(b) Both points include the coordinate $x = 0$, thus the line is the y-axis. Other points could be $(0, 1), (0, {}^-6), \ldots$.

9. The rectangle is shown below:

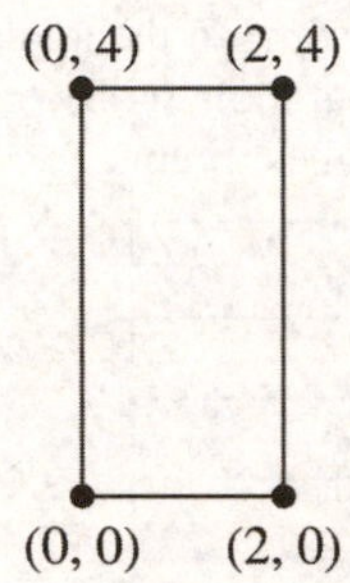

It has dimensions of $2 - 0 = 2$ and $4 - 0 = 4$.

(*i*) Area $= 2 \cdot 4 =$ **8 square units.**

(*ii*) Perimeter $= 2 \cdot 2 + 2 \cdot 4 =$ **12 units.**

11. In each case, slope $(m) = \frac{\text{rise}}{\text{run}} = \frac{y_2 - y_1}{x_2 - x_1}$.

(a) $m = \frac{0-3}{{}^-5-4} = \frac{{}^-3}{-9} = \mathbf{\frac{1}{3}}$.

(b) $m = \frac{2-2}{1-\sqrt{5}} = \frac{0}{1-\sqrt{5}} = \mathbf{0}$.

(c) $m = \frac{b-a}{b-a} = \mathbf{1}$ (if $b \neq a$).

13. Answers may vary, depending on estimates from the fitted line; e.g.:

(a) From the fitted line, estimate point coordinates of $(50, 10)$ and $(80, 40)$. $m = \frac{40-10}{80-50} = \frac{30}{30} = 1$.

Use the point $(50, 10)$ and substitute $T = 50$ and $C = 10$ into $C = 1T + b$ (i.e., an equation of the form $y = mx + b$). $(10) = 1(50) + b \Rightarrow b = {}^-40$. The equation is then $\mathbf{C = T - 40}$.

(b) If $T = 90°$, then $C = (90) - 40 =$ **50 chirps** per 15-second interval.

(c) $N = 4C$; i.e., there are four 15-second intervals in one minute. $N = 4(T - 40)$, or $\mathbf{N = 4T - 160}$.

15. A horizontal line containing $(3, 4)$ has equation $\mathbf{y = 4}$.

17. **(a)** (*i*) The plotted points are shown below:

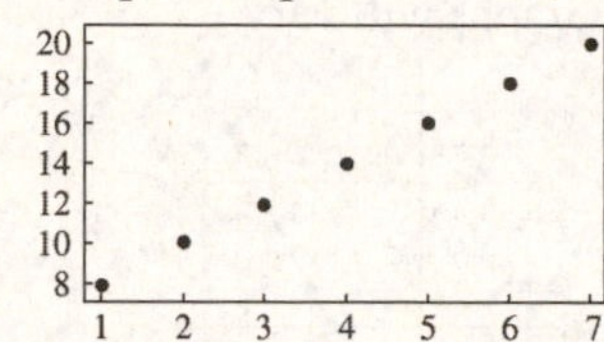

(*ii*) Answers may vary, depending on the points selected from which to derive an equation. Using the points $(3,12)$ and $(6,18)$, $m = \frac{18-12}{6-3} = \frac{6}{3} = 2$.

Using $(3, 12)$ and substituting into the form $y = mx + b, (12) = 2(3) + b$ $\Rightarrow b = 6$. So an equation of the best-fitting line might be $\mathbf{y = 2x + 6}$.

(b) When $x = 10, y = 2(10) + 6 =$ **26**.

19. Let g be the number of gallons of gasoline and let k be the number of gallons of kerosene on the truck.

Then: $g + k = 5000 \Rightarrow g = 5000 - k$. $\$0.13g + \$0.12k = \$640 \Rightarrow 13g + 12k = 64000$.

Substitute $g = 5000 - k$ into $13g + 12k = 64000 \Rightarrow 13(5000 - k) + 12k = 64000 \Rightarrow$ $k =$ **1000 gallons of kerosene.**

Substitute $k = 1000$ into $g = 5000 - k \Rightarrow$ $g =$ **4000 gallons of gasoline.**

Assessment 8-5B

1. The graph of $y = mx + 5$ contains the point $(0, 5)$ and is parallel to the line $y = mx$. Similarly, the graph of $y = mx - 5$ is contains the point $(0, -5)$ and is parallel to the line $y = mx$.

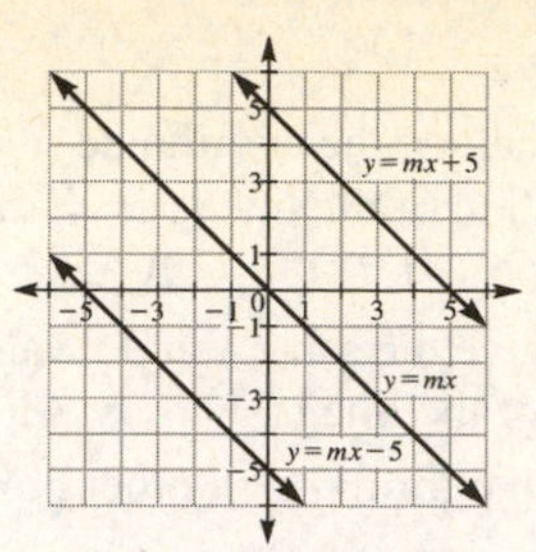

3. When the equations in problem lare in the form $y = mx + b, b$ is the y-intercept and is located at $(0, b)$. The x-intercept may be found by setting y equal to 0 and solving for x. If x is some value a, the x-intercept is located at $(a, 0)$.

(a) $x = {}^-2$:

(*i*) There is **no *y*-intercept**; i.e., $x = {}^-2$ is a vertical line.

(*ii*) The x-intercept is at $x = {}^-2$, or $\mathbf{({}^-2, 0)}$.

(b) $y = 3x - 1$:

(*i*) $b = {}^-1$. y-intercept is at $\mathbf{(0, {}^-1)}$.

(*ii*) if $y = 0 = 3x - 1 \Rightarrow x = \frac{1}{3}$. x-intercept is at $\mathbf{\left(\frac{1}{3}, 0\right)}$.

(c) $y = \frac{1}{20}x$:

(*i*) $b = 0$. y-intercept is at $\mathbf{(0, 0)}$.

(*ii*) If $y = 0 = \frac{1}{20}x \Rightarrow x = 0$. x-intercept is at $\mathbf{(0, 0)}$.

5. When the equations are in slope-intercept form, $y = mx + b, m$ is the slope of the line and b is the y-intercept, located at $(0, b)$.

(a) (*i*) $\frac{x}{3} + \frac{y}{4} = 1 \Rightarrow 12\left(\frac{x}{3} + \frac{y}{4}\right) = 12(1) \Rightarrow$ $4x + 3y = 12 \Rightarrow 3y = {}^-4x + 12$. $\mathbf{y = \frac{{}^-4}{3}x + 4}$.

(*ii*) $m = \frac{{}^-4}{3}$; y-intercept is at $(0, 4)$.

(b) (*i*) $3x - 4y + 7 = 0 \Rightarrow {}^-4y = {}^-3x - 7 \Rightarrow 4y = 3x + 7$. $\mathbf{y = \frac{3}{4}x + \frac{7}{4}}$.

(*ii*) $m = \frac{3}{4}$; y-intercept is at $\left(0, \frac{7}{4}\right)$.

(c) (*i*) $x - y = 4(x - y) \Rightarrow x - y = 4x - 4y \Rightarrow 0 = 3x - 3y$. $\mathbf{y = x}$.

(*ii*) $m = 1$; y-intercept is at $(0, 0)$.

7. Answers may vary.

(a) Both points include the coordinate $x = {}^{-}1$, thus the equation of the line is $x = {}^{-}1$. Other points could be $({}^{-}1, 7), ({}^{-}1, {}^{-}5), \ldots$.

(b) Both points have the same x and y value; i.e., the equation of the line is $y = x$. All other points are of the form (a, a); other points could be $(3, 3), ({}^{-}1, {}^{-}1), \ldots$.

9. **(a)** Sketches may vary.

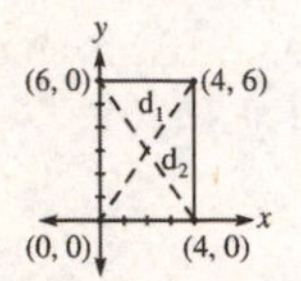

(b) Coordinates of the other vertices are (0, 0), (0, 6), and (4, 0).

(c) There are two diagonals, one with positive slope and one with negative slope.

(*i*) Positive slope: $m = \frac{6-0}{4-0} = \frac{3}{2}$. The y-intercept is at (0, 0). Thus $\mathbf{y = \frac{3}{2}x}$.

(*ii*) Negative slope: $m = \frac{0-6}{4-0} = \frac{{}^{-}3}{2}$. The y-intercept is at (0, 6). Thus $\mathbf{y = \frac{{}^{-}3}{2}x + 6}$.

(d) The point of intersection is at the intersection of the two diagonals, or where $\frac{3}{2}x = \frac{{}^{-}3}{2}x + 6 \Rightarrow 3x = 6 \Rightarrow x = 2$. If $x = 2$, $y = \frac{3}{2}(2) = 3$. The coordinates of the intersection are then **(2, 3)**.

11. **(a)** $m = \frac{2-1}{5-{}^{-}4} = \mathbf{\frac{1}{9}}$.

(b) $m = \frac{198-81}{{}^{-}3-{}^{-}3} = \frac{117}{0} \Rightarrow$ **undefined** (or no slope).

(c) $m = \frac{10-12}{1-1.0001} = \frac{{}^{-}2}{{}^{-}0.0001} = \mathbf{20{,}000}$.

13. **(a)** In eight months, \$2180 – \$2100 = \$80 simple interest was earned, or \$10 per month. (10 months)·(\$10 per month) = \$100 interest earned in the first ten months. The original balance was \$2100 – \$100 = **\$2000**.

(b) Simple interest (I) = principal (p). annual rate $(r)\cdot$ time (t) in years.

$r = \frac{I}{pt} = \frac{\$100}{\$2000 \cdot \frac{10}{12}} = 0.06$, or **6%**.

15. A vertical line through $({}^{-}7, {}^{-}8)$ contains all points of the form $({}^{-}7, b)$ where b is any real number. Thus, the equation is $\mathbf{x = {}^{-}7}$.

17. **(a)** Graphs will vary.

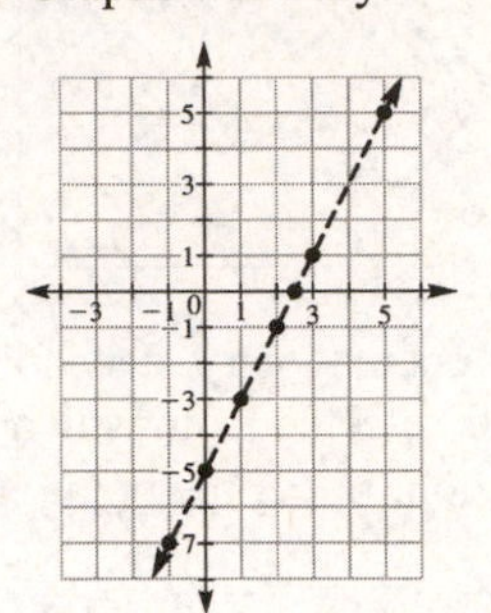

To find an equation, one could choose two points on or very near the line drawn above. For example, choose $({}^{-}1, {}^{-}7)$ and $(3, 1)$. Then the *a* possible equation would be

$y - 1 = \frac{1-({}^{-}7)}{3-({}^{-}1)}(x - 3) \Rightarrow \mathbf{y = 2x + {}^{-}5}$.

(b) $y = 2(7) + {}^{-}5 = 14 - 5 = \mathbf{9}$.

19. The equation of the segment connecting (5,0) and (6,8) is $\mathbf{y = 8x - 40}$ (determined by finding m; then substituting x and y from one of the points into the equation $y = mx + b$ to find b).

The equation of the segment connecting (10,0) and (3,4) is $\mathbf{y = \frac{-4}{7}x + \frac{40}{7}}$.

The equation of the segment connecting (0,0) and (8,4) is $\mathbf{y = \frac{1}{2}x}$.

Equating these three values of y (i.e., $8x - 40 = \frac{{}^{-}4}{7}x + \frac{40}{7} = \frac{1}{2}x$) yields $x = \frac{16}{3}$.

Substituting $x = \frac{16}{3}$ into any of the three equations yields $y = \frac{8}{3}$. The coordinates of the common intersection are $\mathbf{\left(\frac{16}{3}, \frac{8}{3}\right)}$.

8-5 Review Problems

11. $\sqrt{6} \approx \mathbf{2.45}$.

13. **(a)** $f(3) = 3\sqrt{7} - \sqrt{7} = \mathbf{2\sqrt{7}}$.

(b) $f(\sqrt{7}) = \sqrt{7} \cdot \sqrt{7} - \sqrt{7} = \mathbf{7 - \sqrt{7}}$.

(c) $f(^-4) = {}^-4\sqrt{7} - \sqrt{7} = {}^-5\sqrt{7}$.

15. **(a)** For $f(x) = \sqrt{x+1}$ to have real-valued range values, $x + 1$ must not be negative. So $x + 1 \geq 0 \Rightarrow \mathbf{x \geq {}^-1}$. Write as $\{x \mid x \in \mathbb{R}$ and $x \geq {}^-1\}$.

(b) For $f(x) = \sqrt{^-x}$ to have real-valued range values, $-x$ must not be negative. So $-x \geq 0 \Rightarrow \mathbf{x \leq 0}$. Write as $\mathbf{\{x | x \in R \text{ and } x \leq 0\}}$.

Chapter 8 Review

1. **(a)** **Irrational**. The pattern never repeats in blocks of the same length.

(b) **Irrational**. Any non-zero rational number divided by any irrational number is irrational.

(c) **Rational**. The ratio of two integers.

(d) **Rational**. The pattern repeats in blocks of 0011.

(e) **Irrational**. The pattern does not repeat.

3. **(a)** **No**. ${}^-\sqrt{2} + \sqrt{2} = 0$ is rational.

(b) **No**. $\sqrt{2} - \sqrt{2} = 0$ is rational.

(c) **No**. $\sqrt{2} \cdot \sqrt{2} = 2$ is rational.

(d) **No**. $\frac{\sqrt{2}}{\sqrt{2}} = 1$ is rational.

5. $1 < \sqrt[3]{2} < 2$

$1.2^3 = 1.728$ and $1.3^3 = 2.197$.

$1 \cdot 2 < \sqrt[3]{2} < 1.3$.

$1.25^3 = 1.953125$ and $1.26^3 = 2.000376$.

Thus, $1.25 < \sqrt[3]{2} < 1.26$.

2 is closer to 2.000376 that it is to 1.953125, so $\sqrt[3]{2} \approx \mathbf{1.26}$.

7. $\sqrt{7}(^-1)^{10} = \sqrt{7}$.

9. $\mathbf{S = 13P}$.

11. $\mathbf{f = 3y}$.

13. Let n be the whole number. Then $12\left(\frac{n}{13}\right) - 20 + 89 = 93 \Rightarrow 12\left(\frac{n}{13}\right) = 24 \Rightarrow \frac{n}{13} = 2 \Rightarrow \mathbf{n = 26}$.

15. **(a)** $4x - 2 = 3x + \sqrt{10} \Rightarrow \mathbf{x = 2 + \sqrt{10}}$.

(b) $4(x - 12) = 2x + 10 \Rightarrow 4x - 48 = 2x + 10 \Rightarrow 2x = 58 \Rightarrow \mathbf{x = 29}$.

(c) $4(7x - 21) = 14(7x - 21) \Rightarrow 28x - 84 = 98x - 294 \Rightarrow 210 = 70x \Rightarrow \mathbf{x = 3}$.

(d) $2(3x + 5) = 6x + 11 \Rightarrow 6x + 10 = 6x + 11 \Rightarrow$ **no solution**.

(e) $3(x + 1) + 1 = 3x + 4 \Rightarrow 3x + 3 + 1 = 3x + 4 \Rightarrow$ **all whole numbers** are solutions.

17. Let s be the number of science book overdue days and c be the number of children's book overdue days. It is given that $c = s - 14$.

Then $8(0.20)(s - 14) + 2(0.20)s = 11.60 \Rightarrow 1.6s - 22.4 + 0.4s = 11.60 \Rightarrow 2s = 34 \Rightarrow s = 17$.

Each science book was overdue by **17 days**

Each children's book was overdue by $17 - 14 =$ **3 days**.

19. **(a)** **A function**. Each component of the domain corresponds to a unique component of the range.

(b) **Not a function**. a and b both correspond to two components of the range.

(c) **A function**. a and b correspond to unique components of the range.

21. **(a)** **Not a function**. A student may have more than one major.

(b) **A function**. The range is the subset of the natural numbers that includes the number of pages in each book in the library.

(c) **A function**. The range is $\{x | x \geq 6$ and x is even$\}$.

(d) **A function**. The range is $\{0, 1\}$.

(e) **A function**. The range is the set of all natural numbers N.

23. If $4x - 5 = 15$, then $4x = 20$ and $\mathbf{x = 5}$ is the input.

25. Assume Jilly starts with unpainted blocks each time:

(a)

Number of cubes	Number of squares to paint
1	6
2	10
3	14
4	18
5	22
6	26

(b)

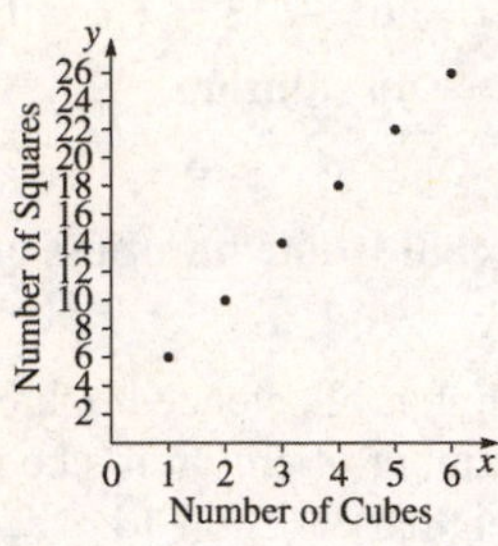

(c) This is a function which is an arithmetic sequence with $a_1 = 6, d = 4,$ and $a_6 = 26 \Rightarrow a_n = 6 + (n - 1)4 \Rightarrow 4n + 2.$ The function is $y = \mathbf{4x + 2}$ for $x = 1, 2, ..., 6.$

(d) **No**. The graph does represent a function, but not a straight line because values of x cannot assume anything but natural numbers.

27. Graph the system using spreadsheets:

$y = {}^{-}x + 6$

A	B
3	3
4	2
$4.\overline{3}$	$1.\overline{6}$
$4.\overline{6}$	$1.\overline{3}$
5	1

$y = 2x - 7$

A	B
3	$^{-}1$
4	1
$4.\overline{3}$	$1.\overline{6}$
$4.\overline{6}$	$2.\overline{3}$
5	3

The graphs intersect at approximately $\left(\mathbf{4\tfrac{1}{3}, 1\tfrac{2}{3}}\right)$ **or** $\mathbf{(4.\overline{3}, 1.\overline{6})}$.

29. If $a_1 = \pi$ and $r = \frac{1}{\pi}$, then $a_3 = \pi\left(\frac{1}{\pi}\right)^{3-1} = \frac{\pi}{\pi^2} = \frac{1}{\pi}.$

Chapter 9

PROBABILITY

Assessment 9-1A: How Probabilities Are Determined

1. (a) **No.** There are fewer face cards than there are not face cards.

   (b) **Yes.** Each suit has the same number of cards.

   (c) **Yes.** There are equal numbers of black and red cards.

   (d) **No.** There are 4 each kings, queens, jacks, and aces; 20 even-numbered cards; and 16 odd-numbered cards in a standard deck.

3. (a) $P(1,5 \text{ or } 7) = \frac{1}{8} + \frac{1}{8} + \frac{1}{8} = \mathbf{\frac{3}{8}}$. Factors of 35 are 1, 5, or 7 and their probabilities are mutually exclusive.

   (b) $P(3 \text{ or } 6) = \frac{1}{8} + \frac{1}{8} = \frac{2}{8} = \mathbf{\frac{1}{4}}$. Multiples of 3 are 3 or 6 and their probabilities are mutually exclusive.

   (c) $P(2,4,6 \text{ or } 8) = \frac{1}{8} + \frac{1}{8} + \frac{1}{8} + \frac{1}{8} = \frac{4}{8} = \mathbf{\frac{1}{2}}$.

   (d) $P(6 \text{ or } 2) = \frac{1}{8} + \frac{1}{8} = \frac{2}{8} = \mathbf{\frac{1}{4}}$.

   (e) $P(11) = \mathbf{0}$. The probability of an impossible event is 0.

   (f) $P(4, 6 \text{ or } 8) = \frac{1}{8} + \frac{1}{8} + \frac{1}{8} = \mathbf{\frac{3}{8}}$. 4, 6, and 8 are the only composite numbers on the spinner.

   (g) $P(1) = \mathbf{\frac{1}{8}}$. 1 is the only number on the spinner which is neither prime nor composite.

5. Assume the socks are randomly mixed in the drawer.

   (a) $P(brown) = \frac{n(brown)}{n(S)} = \frac{4}{12} = \mathbf{\frac{1}{3}}$. There are four brown socks out of the twelve in the drawer.

   (b) $P(black \text{ or } green) = \frac{n(black)}{n(S)} + \frac{n(green)}{n(S)} = \frac{6}{12} + \frac{2}{12} = \frac{8}{12} = \mathbf{\frac{2}{3}}$. The two events are mutually exclusive.

   (c) $P(red) = \mathbf{0}$. There are no red socks in the drawer, so this is an impossible event.

   (d) $P(not\ black) = 1 - P(black) = 1 - \frac{n(black)}{n(S)} = 1 - \frac{6}{12} = \frac{6}{12} = \mathbf{\frac{1}{2}}$.

   or

   $P(not\ black) = \frac{n(broun)}{n(S)} + \frac{n(green)}{n(S)} = \frac{4}{12} + \frac{2}{12} = \frac{6}{12} = \mathbf{\frac{1}{2}}$.

   (e) $P(pair\ of\ same\ color) = \mathbf{1}$. Since there are only three colors, if four socks are pulled out at least two of them must be of the same color.

7. $P(no\ rain) = 1 - P(rain) = 1 - 0.30 = 0.70$, or **70%**.

9. $P(0 \text{ or } 00) = \frac{n(0 \text{ or } 00)}{n(S)} = \frac{2}{38} = \frac{1}{19}$. In 190 spins one would expect the ball to land in one of these slots $\frac{1}{19} \cdot 190 =$ **10 times**.

11. Because A and B are mutually exclusive, $P(A \cup B) = P(A) + P(B) = 0.3 + 0.4 = \mathbf{0.7}$.

13. (a) If r red balls are added then $P(red\ ball) = \frac{2+r}{10+r} = \frac{3}{4} \Rightarrow 4(2 + r) = 3(10 + r)$. $r =$ **22 red balls.**

    (b) There are originally $\frac{5 \text{ white}}{10 \text{ total}}$ balls. b black balls are added so that $P(white) = \frac{5}{10+b} = \frac{1}{4} \Rightarrow 20 = 10 + b$. $b =$ **10 black balls** added.

15. A fair coin will always exhibit equal probabilities of a head or a tail when flipped, regardless of past history. Thus the probability of a head on toss 16 is $\mathbf{\frac{1}{2}}$.

Assessment 9-1B

1. (a) $P(point\ up) = \frac{n(up)}{n(s)} = \frac{56}{56+24} = \frac{56}{80} = \mathbf{\frac{7}{10}}$.

(b) $P(\textit{point down}) = \frac{n(\textit{down})}{n(s)} = \frac{24}{80} = \frac{\mathbf{3}}{\mathbf{10}}$.

Another method of solving this problem is to recognize that *point up* and *point down* are mutually exclusive events; thus $P(\textit{point down}) = 1 - P(\textit{point up}) = 1 - \frac{7}{10} = \frac{3}{10}$.

(c) **Probably not**. When running the experiment again with only a relatively small number of tries the outcome will probably not be identical. Experimental probability is based only on the number of times the experiment is repeated, not what will happen in the long run.

(d) **Yes**. The thumbtack would be unchanged, so the results would be determined by the same dynamics. Experimental probability, though, does not guarantee exact results.

3. $n(S) = 28$:

(a) $P(\textit{diet soda}) = \frac{n(\textit{diet sodas})}{n(S)} = \frac{16}{28} = \frac{\mathbf{4}}{\mathbf{7}}$.

(b) $P(\textit{regular soda}) = \frac{n(\textit{regular sodas})}{n(S)} = \frac{8}{28} = \frac{\mathbf{2}}{\mathbf{7}}$.

(c) $P(\textit{water}) = \frac{n(\textit{waters})}{n(S)} = \frac{4}{28} = \frac{\mathbf{1}}{\mathbf{7}}$.

Part (*iii*) could also be solved by realizing that the probability of drawing some bottle $= 1 \Rightarrow P(\textit{water}) = 1 - \left(\frac{4}{7} + \frac{2}{7}\right) = \frac{\mathbf{1}}{\mathbf{7}}$.

5. **(a)** **No**. $P(\textit{I win or you lose}) = P(H) + P(T) = \frac{1}{2} + \frac{1}{2} = 1$. Each player's probability of winning is not equal.

(b) **Yes**. $P(\textit{heads I win}) = P(H) = \frac{1}{2}$.

$P(\textit{otherwise}) = P(T) = \frac{1}{2}$.

(c) **Yes**. $P(1; \textit{I win}) = \frac{1}{6}$. $P(6; \textit{you win}) = \frac{1}{6}$.

(d) **Yes**. $P(\textit{even}; \textit{I win}) = \frac{n(\textit{even})}{n(S)} = \frac{3}{6}$.

$P(\textit{odd}; \textit{you win}) = \frac{n(\textit{odd})}{n(S)} = \frac{3}{6}$.

(e) **No**. $P(\geq 3; \textit{I win}) = \frac{n(\geq 3)}{n(S)} = \frac{4}{6}$.

$P(< 3; \textit{you win}) = \frac{n(<3)}{n(S)} = \frac{2}{6}$.

(f) **Yes**. $P(1 \textit{ on each}; \textit{I win}) = \frac{1}{36}$; $P(6 \textit{ on each}; \textit{you win}) = \frac{1}{36}$.

(g) **No**. $P(3\,; \textit{I win}) = \frac{n(3)}{n(S)} = \frac{2}{36}$.

$P(2; \textit{you win}) = \frac{n(2)}{n(S)} = \frac{1}{36}$.

7. Assume drawing without replacement; i.e., there will be 50 cards remaining in the deck after the first two have been dealt.

(a) There are five cards between 5 and jack in each of the four suits $\Rightarrow P(\textit{between 5 and jack}) = \frac{5 \cdot 4}{50} = \frac{20}{50} = \frac{\mathbf{2}}{\mathbf{5}}$.

(b) There are ten cards between 2 and king in each of the four suits. $P(\textit{between 2 and king}) = \frac{10 \cdot 4}{50} = \frac{40}{50} = \frac{\mathbf{4}}{\mathbf{5}}$.

(c) $P(\textit{between 5 and 6}) = \mathbf{0}$. This is an impossible event.

9. **(a)** $P(A \cup C) =$ the probability of students taking Algebra **or** Chemistry, or both.

(b) $P(A \cap C) =$ the probability of students taking both Algebra **and** Chemistry.

(c) $1 - P(C) =$ the complement of the probability of a student taking Chemistry, or the probability of a student **not** taking chemistry.

11. $P(\textit{not study}) = \frac{n(\textit{not studied})}{n(S)} = \frac{\mathbf{2}}{\mathbf{3}}$.

13. **(a)** **No**. Suppose x white, x black, and x red balls are added to the box. Then $P(\textit{black ball}) = \frac{3+x}{(5+x)+(3+x)+(2+x)} = \frac{3+x}{10+3x}$.

If $P(\textit{black ball}) = \frac{1}{3}$ then $\frac{3+x}{10+3x} = \frac{1}{3} \Rightarrow 3(3 + x) = 10 + 3x \Rightarrow 9 = 10$. There is no solution to the problem.

(b) **Yes**. If x white, x black, and x red balls are added to the box then $P(\textit{black ball}) = \frac{3+x}{10+3x}$.

If P (*black ball*) $= 0.32$ then $\frac{3+x}{10+3x} = 0.32 \Rightarrow 3 + x = 0.32(10 + 3x) \Rightarrow 3 + x = 3.2 + 0.96x \Rightarrow 0.04x = 0.2$.

$x = 5$; i.e., add five balls of each color.

15. **(a)** Since there are currently fewer women than men on the U.S. Supreme Court this event is **likely**.

(b) **Unlikely**, there are fewer African–Americans than non-African–Americans in the current senate.

(c) **Likely**, there are more people with Asian roots than people with non-Asian roots.

Assessment 9-2A Multistage Experiments with Tree Diagrams and Geometric Probabilities

1. **(a)** **(*i*)** A tree diagram showing paths for each possible outcome is shown below. Paths are defined by finding the probability of each event in draw one followed by the probability of each event in draw two; for example, the probability of drawing a white ball first is $\frac{3}{5}$, and the probability of drawing another is then $\frac{2}{4}$, since there will be two white balls of a total of four:

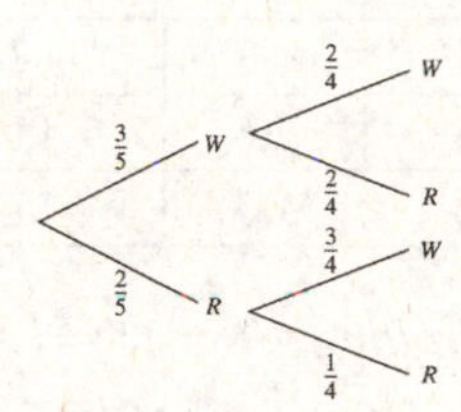

(*ii*) The possible outcomes to obtain different colors are white and red or red and white. $P(\text{different colors}) = \frac{3}{5}\cdot\frac{2}{4}+\frac{2}{5}\cdot\frac{3}{4} = \frac{6}{20}+\frac{6}{20}=\frac{12}{20}=\mathbf{\frac{3}{5}}$.

(b) **(*i*)** A tree diagram is shown below. Probabilities in each path differ from (a) above in that balls are replaced; thus the probability of drawing a white or red ball is the same in each path:

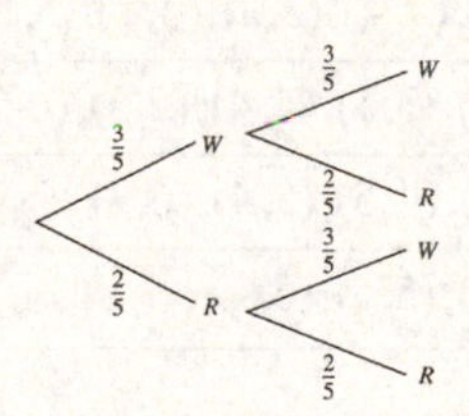

(*ii*) $P(\text{different colors}) = \frac{3}{5}\cdot\frac{2}{5}+\frac{2}{5}\cdot\frac{3}{5} = \mathbf{\frac{12}{25}}$.

3. This is a multistage event without replacement. $P(\text{three women}) = \frac{4}{10}\cdot\frac{3}{9}\cdot\frac{2}{8}=\frac{24}{720}=\mathbf{\frac{1}{30}}$.

5. $P(\text{all boys}) = P\ (\text{first child boy})\cdot P(\text{second child boy})\cdot P(\text{third child boy})\cdot P(\text{fourth child boy}) = \frac{1}{2}\cdot\frac{1}{2}\cdot\frac{1}{2}\cdot\frac{1}{2}=\mathbf{\frac{1}{16}}$.

7. Each case is a multistage, independent event.

(a) $P(\text{three plums}) = \frac{5}{20}\cdot\frac{1}{20}\cdot\frac{5}{20}=\frac{25}{8000}=\mathbf{\frac{1}{320}}$.

(b) $P(\text{three oranges}) = \frac{3}{20}\cdot\frac{6}{20}\cdot\frac{7}{20}=\frac{126}{8000}=\mathbf{\frac{63}{4000}}$.

(c) $P(\text{three lemons}) = \frac{3}{20}\cdot\frac{0}{20}\cdot\frac{4}{20}=\mathbf{0}$.

(d) $P(\text{no plums}) = \frac{15}{20}\cdot\frac{19}{20}\cdot\frac{15}{20}=\frac{4275}{8000}=\mathbf{\frac{171}{320}}$.

9. Each question has a probability of $\frac{1}{2}$ of being right, and the result of each question has no effect on subsequent questions. $P(100\%) = \frac{1}{2}\cdot\frac{1}{2}\cdot\frac{1}{2}\cdot\frac{1}{2}\cdot\frac{1}{2}=\mathbf{\frac{1}{32}}$.

11. The total area of the dart board $= 5x$ by $5x = 25x^2$.

(a) The area of $A = x$ by $x = x^2$. $P(\text{section } A) = \frac{x^2}{25x^2}=\mathbf{\frac{1}{25}}$.

(b) The area of $B = (3x)^2 - x^2 = 8x^2$.

$P(\text{section } B) = \frac{8x^2}{25x^2}=\mathbf{\frac{8}{25}}$.

(c) The area of $C = (5x)^2 - (3x)^2 = 16x^2$.

$P(\text{section } C) = \frac{16x^2}{25x^2}=\mathbf{\frac{16}{25}}$.

or

$P(\text{section } C) = 1 - [P(\text{section } A + P(\text{section } B)] = 1-\left(\frac{1}{25}+\frac{8}{25}\right)=1-\frac{9}{25}=\mathbf{\frac{16}{25}}$.

13. Of the ten secretaries, two are male. $P(\text{secretary given male}) = \frac{n(\text{male and secretary})}{n(\text{male})}=\frac{2}{28}=\mathbf{\frac{1}{14}}$.

15. **(a)** "At least as many heads as tails" means:

0 tails and 4 heads, which can occur in 1 way;
1 tail and 3 heads, which can occur in 4 ways;
2 tails and 2 heads, which can occur in 6 ways.

The total number of possible outcomes is $2^4 = 16$.

$P(\textit{as many heads as tails}) = \frac{1+4+6}{16} = \mathbf{\frac{11}{16}}.$

(b) If the quarter is fair (i.e., heads and tails occur the same number of times) then this probability is the same as in (a), $\mathbf{\frac{11}{16}}$.

17.

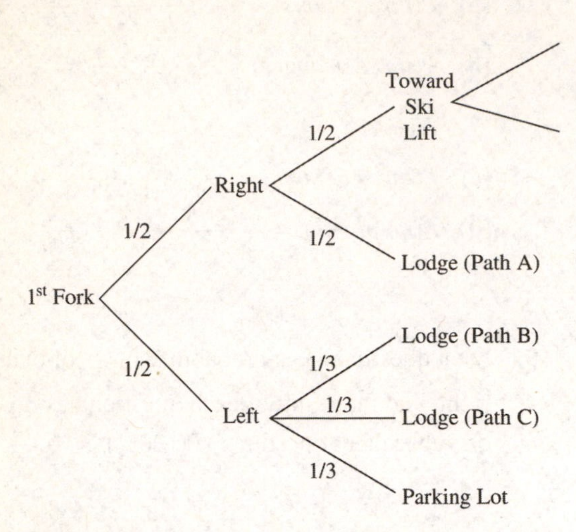

There are three paths to the lodge. These paths represent mutually exclusive events since a skiers cannot be on more than one path at a time.
$P(A \cup B \cup C) = P(A) + P(B) + P(C) =$
$\frac{1}{2} \cdot \frac{1}{2} + \frac{1}{2} \cdot \frac{1}{3} + \frac{1}{2} \cdot \frac{1}{3} = \frac{1}{4} + \frac{1}{6} + \frac{1}{6} =$
$\mathbf{\frac{7}{12}} \approx \mathbf{0.58}.$

Assessment 9-2B

1. **(a)** $S = \{(1,1), (1,2), (1,3), (2,1), (2,2), (2,3)\}$.

(b) $A = \{(2,2)\}$.

(c) $B = \{(1,2), (2,1), (2,2), (2,3)\}$.

(d) $C = \{(1,2), (2,1), (2,3)\}$.

3. The denomination of the coins is irrelevant; "at least three heads" means either three heads or four heads. There are $2^4 = 16$ possible outcomes if four coins are tossed. There is only one way four heads can be obtained, with probability $\frac{1}{2} \cdot \frac{1}{2} \cdot \frac{1}{2} \cdot \frac{1}{2} = \frac{1}{16}$. There are four ways three heads can be obtained (*HHHT, HHTH, HTHH, THHH*), each with probability $\frac{1}{2} \cdot \frac{1}{2} \cdot \frac{1}{2} \cdot \frac{1}{2} = \frac{1}{16}$.

$P(\textit{at least three heads}) = \frac{1}{16} + \frac{1}{16} + \frac{1}{16} + \frac{1}{16} + \frac{1}{16} = \mathbf{\frac{5}{16}}.$

5. A tree diagram would show three event paths:

(i) First inspector will catch the defect, with a probability of 0.95; or

(ii) First inspector will miss the defect and the second will catch it, with a probability of $(0.05)(0.99) = 0.0495$; or

(iii) First inspector will miss the defect and the second will miss it, with a probability of $(0.05)(0.01) = 0.0005$.

The probability of a defect passing both inspectors (i.e., both inspectors missing) $= \mathbf{0.0005}$.

7. **(a)** If we think of S, the sample space, as ordered pairs (1st die, 2nd die), then there are $6 \times 6 = 36$ possible outcomes. The number of ways to roll a 7 or 11 can be seen in the table below.

1st die	2st die	Total
1	6	7
2	5	7
3	4	7
4	3	7
5	2,6	7,11
6	1,5	7,11

Thus, $P(\textit{rolling a } 7 \text{ or } 11) = \frac{8}{36} = \mathbf{\frac{2}{9}} \approx \mathbf{.22}$

(b) There is one way to roll a $2: (1,1)$. There are two ways to roll a $3: (2,1)$ and $(1,2)$. There is one way to roll a $12: (6,6)$. Thus, P(rolling a 2, 3, or 12) $= \frac{4}{36} = \mathbf{\frac{1}{9}} \approx \mathbf{.11}$.

(c)

Total	Ways of rolling the total	# of ways
4	(1,3),(2,2),(3,1)	3
5	(1,4),(2,3),(3,2),(4,1)	4
6	(1,5),(2,4),(3,3),(4,2),(5,1)	5
8	(2,6),(3,5),(4,4),(5,3),(6,2)	5
9	(3,6),(4,5),(5,4),(6,3)	4
10	(4,6),(5,5),(6,4)	3

$P(\textit{rolling}$ a 4, 5, 6, 8, 9, or 10$) =$
$\frac{3+4+5+5+4+3}{36} = \frac{24}{36} = \mathbf{\frac{2}{3}} \approx \mathbf{.66}.$

(d) Either **6 or 8**, since there are more ways to roll this sum.

(e) **0**, the smallest possible sum is 2.

(f) **1**, all sums are less than 13.

(g) There are six ways to roll a 7. So the probability of the event of rolling a 7 when two dice are rolled is $\frac{6}{36} = \frac{1}{6}$. In 60 rolls of two dice, we expect to see **10** sevens.

9. There are five different ways of ascending four steps: $S = \{(1, 1, 1, 1), (1, 2, 1), (1, 1, 2), (2, 1, 1), (2, 2)\}$. Their probabilities are, respectively:

(*i*) $\frac{1}{2} \cdot \frac{1}{2} \cdot \frac{1}{2} \cdot 1$ (i.e., once she ascends the first three steps the probability is 1 that she will take one step to ascend the fourth) $= \frac{1}{8}$;

(*ii*) $\frac{1}{2} \cdot \frac{1}{2} \cdot 1 = \frac{1}{4}$;

(*iii*) $\frac{1}{2} \cdot \frac{1}{2} \cdot \frac{1}{2} = \frac{1}{8}$;

(*iv*) $\frac{1}{2} \cdot \frac{1}{2} \cdot 1 = \frac{1}{4}$;

(*v*) $\frac{1}{2} \cdot \frac{1}{2} = \frac{1}{4}$.

(a) $P(\textit{two strides}) = \mathbf{\frac{1}{4}}$.

(b) $P(\textit{three strides}) = \frac{1}{4} + \frac{1}{8} + \frac{1}{4} = \mathbf{\frac{5}{8}}$.

(c) $P(\textit{four strides}) = \mathbf{\frac{1}{8}}$.

11. (a) 10 units by 10 units = **100 square units**.

(b) (*i*) $P(\textit{region A}) = \frac{4}{100} = \mathbf{\frac{1}{25}}$.

(*ii*) $P(\textit{region B}) = \frac{12}{100} = \mathbf{\frac{3}{25}}$.

(*iii*) $P(\textit{region C}) = \frac{20}{100} = \mathbf{\frac{1}{5}}$.

(*iv*) $P(\textit{region D}) = \frac{28}{100} = \mathbf{\frac{7}{25}}$.

(*v*) $P(\textit{region E}) = \frac{36}{100} = \mathbf{\frac{9}{25}}$.

(c) 20 points can be scored if and only it both darts land in region A.
$P(20\ pts) = \frac{1}{25} \cdot \frac{1}{25} = \mathbf{\frac{1}{625}}$.

(d) $P(A \text{ or}, B \text{ or } C) = \frac{4}{100} + \frac{12}{100} + \frac{20}{100} = \frac{36}{100} = \mathbf{\frac{9}{25}}$.

13. $P(MISSISSIPPI) = P(M) \cdot P(I) \cdot P(S) \cdot P(S) \cdot P(I) \cdot P(S) \cdot P(S) \cdot P(I) \cdot P(P) \cdot P(P) \cdot P(I) = \frac{1}{11} \cdot \frac{4}{10} \cdot \frac{4}{9} \cdot \frac{3}{8} \cdot \frac{3}{7} \cdot \frac{2}{6} \cdot \frac{1}{5} \cdot \frac{2}{4} \cdot \frac{2}{3} \cdot \frac{1}{2} \cdot 1$ (i.e., without replacement) $= \frac{1152}{39{,}916{,}800} = \mathbf{\frac{1}{34{,}650}}$.

15. "At least one" means one, two, or three children. There are seven ways of having at least one child inherit the disease (where y indicates yes and n indicates no): $S = \{(ynn), \{nyn), (nny), (yyn), (yny), (nyy), (yyy)\}$. Three of these have probabilities of $0.1 \cdot 0.9 \cdot 0.9 = 0.081$; three have probabilities of $0.1 \cdot 0.1 \cdot 0.9 = 0.009$; and one has probability of $(0.1)^3 = 0.001$. $P(\textit{at least one}) = 3 \cdot 0.081 + 3 \cdot 0.009 + 0.001 = \mathbf{0.271}$.

or

$P(\textit{at least one}) = 1 - P(\textit{none}) = 1 - (0.9)^3 = \mathbf{0.271}$.

17. Each room has the same probability of being chosen. An equivalent tree diagram is shown below:

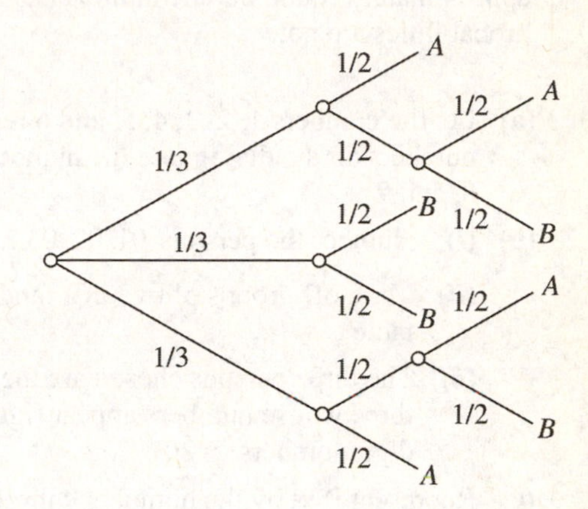

$P(A) = \frac{1}{3} \cdot \frac{1}{2} + \frac{1}{3} \cdot \frac{1}{2} \cdot \frac{1}{2} + \frac{1}{3} \cdot \frac{1}{2} \cdot \frac{1}{2} + \frac{1}{3} \cdot \frac{1}{2} = \frac{1}{6} + \frac{1}{12} + \frac{1}{12} + \frac{1}{6} = \frac{1}{2}$.

$P(B) = 1 - P(A) = 1 - \frac{1}{2} = \frac{1}{2}$.

Thus it makes **no difference** where the car is placed.

19. It is known that $P(A \text{ given } B) = \frac{P(A \cap B)}{P(B)} \Rightarrow P(A \cap B) = P(B) \cdot P(A \text{ given } B)$. (See the discussion of conditional probabilities in Section 9-4 of the text.) Thus:

$P(\textit{eaten} \cap \textit{sickly}) = P(\textit{sickly}) \cdot P(\textit{eaten given sickly}) = \frac{1}{20} \cdot \frac{1}{3} = \frac{1}{60}$.

$P(\textit{eaten} \cap \textit{not sickly}) = P(\textit{not sickly}) \cdot P(\textit{eaten given not sickly}) = \frac{19}{20} \cdot \frac{1}{150} = \frac{19}{3000}$.

$P(\textit{eaten}) = \frac{1}{60} + \frac{19}{3000} = \frac{69}{3000} = \mathbf{\frac{23}{1000}}$.

Review Problems

17. (a) $P(\textit{April 7}) = \mathbf{\frac{1}{30}}$.

(b) $P(April\,31) = \mathbf{0}$.

(c) $P(before\ April\,20) = \mathbf{\frac{19}{30}}$.

Assessment 9-3A Using Simulations in Probability

In discussing methods of using simulations, all the following answers could vary.

1. If the use of the thumbtack was to involve dropping it and then observing whether it landed point up or point down, it would not be possible. The probabilities for birth of boys versus girls are approximately equal, but the thumbtack probabilities are not.

3. **(a)** Let the numbers 1, 2, 3, 4, 5, and 6 represent numbers on the die; ignore the numbers 0, 7, 8, and 9

(b) *(i)* Number the persons 01, 02, 03,...,20

(ii) Mark off groups of two in a random digit table

(iii) The three persons chosen are the first three whose numbers appear (ignore two-digit numbers > 20)

(c) Represent Red by the numbers 0 through 4; Green by the numbers 5, 6, and 7; Yellow by the number 8; and White by the number 9; then pick a number from a random digit table

5. To simulate Monday, let the digits 1 through 8 represent rain and 0 and 9 represent no rain. If rain occurred on Monday, repeat the same process for Tuesday. If it did not rain on Monday, let the digits 1 through 7 represent rain and 0, 8, and 9 represent dry. Repeat a similar process for the rest of the week.

For example, 60304 13976 was chosen from the random digits table

Day of the week	digit	weather
Mon	6	rain
Tues	0	no rain
Wed	3	rain
Thurs	0	no rain
Fri	4	rain
Sat	1	rain

7. $P(< 30) = \frac{n(numbers < 30)}{n(possible\ two-digit\ numbers)} = \frac{30}{100} = \mathbf{\frac{3}{10}}$.

9. **(a)** **7 games.** It is possible for the losing team to win three games while the winning team wins four.

(b) Answers may vary. E.g., because the teams are evenly matched use a table of random digits and let a number between 0 and 4 represent a win by Team *A*; let a number between 5 and 9 represent a win by Team *B*.

Pick a starting spot and count the number of digits it takes before a Team *A* or Team *B* series win is recorded. Repeat the experiment many times and then base your answers on:

(i)

$$P(four\text{-}game\ series) = \frac{n(four\text{-}game\ series)}{n(total\ series)}.$$

(ii)

$$P(seven\text{-}game\ series) = \frac{n(seven\text{-}game\ series)}{n(total\ series)}.$$

Given evenly matched teams, the probability of a four-game series would be expected to be low.

Assessment 9-3B

In discussing methods of using simulations, all the following answers could vary.

1. The red half of the spinner might represent the birth of a boy and the blue half of the spinner might represent the birth of a girl. Spin the arrow and record the results.

3. One way would be to consider the people to be numbered as 00, 01, 02,..., 09, 10, 11,..., 24. Then from the random-digit generator or random-number table select a starting point; look at sets of numbers from the starting point in pairs. If, for example, the numbers 405809664176912311... are seen, the first pair that represents a person is 09. The next pair would be 23. Continue until the desired four people are selected.

5. *(i)* Use a random-number table or random-digit generator and look at pairs of digits, where every odd digit could represent a tail and every even digit could represent a head. For example, 83 could represent a head and a tail, while 71 could represent two tails. Then look at 25 such pairs to determine the experimental probability.

(ii) Theoretically, the probability of obtaining two tails from two coin losses is $\frac{1}{4}$. The expected number of such results from 25 trials would be $25 \cdot \frac{1}{4}$, or about six times.

7. The probability of a single student choosing the square with the dot is $\frac{1}{100}$. In 100 trials (i.e., 100 students making a choice) the expected number of successes would be $100 \cdot \frac{1}{100} = \mathbf{1}$.

9. For the simulation you choose two-digit numbers to represent the 12 zodiac signs. For example, we could chose 00, 01, 02,...,11 to represent these signs. Close your eyes and put your finger on a number in Table 9-4, and collect strings of five two-digit numbers of the above form. For example, my finger landed on 89523. Working to the right, row by row, I created the string

$$\overbrace{04, 04, 04, 03, 02}^{\text{Pair}},\ \overbrace{08, 09, 07, 05, 01}^{\text{No pair}}$$

$$\overbrace{06, 02, 07, 02, 06}^{\text{Pair}},\ \overbrace{11, 00, 09, 07, 03}^{\text{No pair}}$$

$$\overbrace{03, 02, 01, 04, 04}^{\text{Pair}}$$

this simulation of at least one pair in zodiac signs is $\frac{\mathbf{3}}{\mathbf{5}} = \mathbf{.6}$.

Review Problems

13. (a) $P(club) = \frac{13}{52} = \frac{\mathbf{1}}{\mathbf{4}}$. There are 13 cards of each suit in a deck of 52 cards.

(b) $P(queen\ and\ spade) = \frac{\mathbf{1}}{\mathbf{52}}$. There is only 1 Queen of Spades in a deck.

(c) $P(not\ a\ queen) = 1 - P(queen) = 1 - \frac{4}{52} = 1 - \frac{1}{13} = \frac{\mathbf{12}}{\mathbf{13}}$. There are four queens in a deck of 52 cards.

(d) $P(not\ a\ heart) = 1 - P(heart) = 1 - \frac{13}{52} = 1 - \frac{1}{4} = \frac{\mathbf{3}}{\mathbf{4}}$.

(e) $P(spade\ or\ heart) = P(spade) + P(heart) = \frac{13}{52} + \frac{13}{52} = \frac{26}{52} = \frac{\mathbf{1}}{\mathbf{2}}$. These are mutually exclusive events.

(f) $P(6\ of\ diamonds) = \frac{\mathbf{1}}{\mathbf{52}}$. There is only one 6 of Diamonds.

(g) $P(queen\ or\ spade) = P(queen) + P(spade) - P(queen\ and\ spade) = \frac{4}{52} + \frac{13}{52} - \frac{4}{52} \cdot \frac{13}{52} = \frac{1}{13} + \frac{1}{4} - \frac{1}{13} \cdot \frac{1}{4} = \frac{\mathbf{4}}{\mathbf{13}}$. These are non-mutually exclusive events.

(h) $P(either\ red\ or\ black) = \frac{1}{2} + \frac{1}{2} = \mathbf{1}$. This is a certain event.

15. If the probability of making a free throw is $\frac{1}{3}$, the probability of missing is $\frac{2}{3}$. The probability of missing three in a row, assuming the player's skill level is unchanged from shot to shot, would be $\frac{2}{3} \cdot \frac{2}{3} \cdot \frac{2}{3} = \frac{\mathbf{8}}{\mathbf{27}}$.

Assessment 9-4A
Odds, Conditional Probability, and Expected Value

1. (a) $P(drawing\ a\ face\ card) = \frac{12}{52} = \frac{3}{13}$. Odds in favor $= \frac{P(face\ card)}{1-P(face\ card)} = \frac{\left(\frac{3}{13}\right)}{\left(\frac{10}{13}\right)} = \frac{3}{10}$ or **3:10**.

(b) If odds in favor are 3:10, then odds against are **10:3**.

3. If $P(boy) = \frac{1}{2}$ then $P(four\ boys) = \left(\frac{1}{2}\right)^4 = \frac{1}{16}$.

Odds against $= \frac{1-P(four\ boys)}{P(four\ boys)} = \frac{\left(\frac{15}{16}\right)}{\left(\frac{1}{16}\right)} = \frac{15}{1}$ or **15:1**.

5. Given $P(cat) = 0.27$, then the odds against $= \frac{1-P(cat)}{P(cat)} = \frac{0.73}{0.27} = \mathbf{73:27}$.

7. Odds against $= \frac{1-P(red)}{P(red)} = \frac{\left(\frac{3}{8}\right)}{\left(\frac{5}{8}\right)} = \frac{3}{5}$ or **3:5**.

9. The prime numbers of the six on a die are 2, 3, and 5, so $P(prime) = \frac{3}{6} = \frac{1}{2}$. Then odds of a prime are $\frac{P(prime)}{1-P(prime)} = \frac{\left(\frac{1}{2}\right)}{\left(\frac{1}{2}\right)} = \frac{1}{1}$, or **1:1**.

11. If the probability of rain is 90% or 0.9, then the probability of no rain is 10% or 0.1. The odds are 0.9 : 0.1, or equivalently, **9:1**.

13. (a) The first ball was white, so two white and two red remain. $P(red) = \frac{2}{4} = \frac{\mathbf{1}}{\mathbf{2}}$.

(b) The first ball was white, so three white and one red remain. $P(red) = \frac{\mathbf{1}}{\mathbf{4}}$.

15. Odds in favor of winning $= 5{:}2 \Rightarrow \frac{P(win)}{1-P(win)} = \frac{5}{2} \Rightarrow P(win) = \frac{5}{7}$. $E = \$14{,}000 \cdot \frac{5}{7} = \mathbf{\$10{,}000}$.

17.

$$P(A\cap B) = P(B\cap A) = P(B\,|\,A)\cdot P(A) = \frac{3}{8}\cdot\frac{3}{4} = \frac{9}{32}.$$

$$P(A\,|\,B) = \frac{P(A\cap B)}{P(B)} = \frac{\frac{9}{32}}{\frac{1}{3}} = \mathbf{\frac{27}{32}}$$

Assessment 9-4B

1. **(a)** **No.**

(b) $P(black\ card) = \frac{26}{52} = \frac{1}{2}$. Odds in favor are $\frac{P(black\ card)}{1-P(black\ card)} = \frac{\left(\frac{1}{2}\right)}{\left(\frac{1}{2}\right)} = \frac{1}{1} = \mathbf{1{:}1}$.

3. The probabilities of a boy or a girl are essentially $\frac{1}{2}$.

(a) Absent other genetic factors, $P(boy) = \mathbf{\frac{1}{2}}$.

(b) There are four elements in the sample space for a two-child family: $\{b,b\}$, $\{b,g\}$, $\{g,b\}$, and $\{g,g\}$. Thus $P(girl \cap boy) = \frac{1}{4}$ and $P(boy\,|\,girl) = \frac{P(girl\cap boy)}{P(girl)} = \frac{\left(\frac{1}{4}\right)}{\left(\frac{1}{2}\right)} = \frac{2}{4} = \mathbf{\frac{1}{2}}$. **Thus, the odds are 1:1.**

5. $S = \{(hhh), (hht), (hth), (htt), (thh), (tht), (tth), (ttt)\}$. At least two heads appear in four of the eight outcomes, thus $P(at\ least\ two\ heads) = \frac{1}{2}$. Odds in favor $= \frac{P(at\ least\ two)}{1-P(at\ least\ two)} = \frac{\left(\frac{1}{2}\right)}{\left(\frac{1}{2}\right)} = \frac{1}{1}$ **or 1:1.**

7. There are 7 unfavorable outcomes, i.e., letters not S, and 4 S_s. The odds are **7:4**.

9. Given $P(event) = 0.\overline{3}$ then the odds against $= \frac{1-P(event)}{P(event)} = \frac{0.\overline{6}}{0.\overline{3}} = \mathbf{2{:}1}$.

11. Odds against $= \frac{1-P(rain)}{P(rain)} = \frac{0.4}{0.6} = \frac{2}{3}$ or **2:3**.

13. **(a)** $P(17) = \mathbf{\frac{1}{38}}$.

(b) Odds against $17 = \frac{1-P(17)}{P(17)} = \frac{\left(\frac{37}{38}\right)}{\left(\frac{1}{38}\right)} = \frac{37}{1}$ or **37:1**.

(c) $E = \$35\cdot\frac{1}{38} - \$1\cdot\frac{37}{38} = {}^{-}\$\frac{2}{38} = \mathbf{{}^{-}\$\frac{1}{19}}$, or about a 5¢ loss. Another viewpoint is that only $\$\frac{18}{19}$, or about 95¢, will be gained for each \$1 bet.

15. In order to make a fair game, the cost should be equal to the expected value. Each of the numbers 1 through 6 will show on the top of an unbiased die with equal probability $\frac{1}{6}$. Thus:

$E = \$(6+5+4+3+2+1)\cdot\frac{1}{6}$

$= \$\frac{21}{6} = \mathbf{\$3.50}$.

17. $P(A\cap B) = P(A)\cdot P(B|A) = \frac{2}{3}\cdot\frac{1}{3} = \frac{2}{9}$.

$P(A|B) = \frac{P(A\cap B)}{P(B)} = \frac{\left(\frac{2}{9}\right)}{\left(\frac{1}{2}\right)} = \frac{4}{9}$.

Review Problems

11. $P(blue, blue) = P(blue)\cdot P(blue.) \Rightarrow P(blue) = \sqrt{P(blue, blue)} = \sqrt{\frac{25}{36}} = \frac{5}{6}$. Thus the **blue** section must have $\frac{5}{6}\cdot 360° = \mathbf{300°}$ and the **red** section must have $360° - 300° = \mathbf{60°}$.

Assessment 9-5A
Using Permutations and Combinations in Probability

1. The event "girls" can occur in 16 ways; the event "boys" can occur in 14 ways. Thus the event "girls *and* boys" can occur as $16\cdot 14 =$ **224 unique pairings**.

3. There are $3\cdot 15\cdot 4 =$ **180** possible different three-course meals.

5. **(a)** There are 8 unlike letters in SCRAMBLE. $\mathbf{8! = 40{,}320}$.

(b) There are 9 unlike letters in PERMUTATION and 2 like letters. $\mathbf{\frac{11!}{2!} = 19{,}958{,}400}$.

7. (a) Order is distinct, implying a permutation of 30 persons taken three at a time. ${}_{30}P_3 = \frac{30!}{(30-3)!} = 30 \cdot 29 \cdot 28 = \mathbf{24{,}360}$ **ways** in which to select the three officers.
   (b) Order is not distinct, implying a combination of 30 persons taken three at a time. ${}_{30}C_3 = \frac{30!}{(30-3)!3!} = 29 \cdot 14 \cdot 10 = \mathbf{4060}$ **ways** in which to choose three-person committees.

9. The problem is to choose ten points two at a time. Since a line may be drawn either way order is not distinct. ${}_{10}C_2 = \frac{10!}{(10-2)!2!} = \mathbf{45}$ **lines**.

11. If there were n people at the party there were n combinations of people two at a time shaking hands. ${}_nC_2 = \frac{n!}{(n-2)!2!} = \frac{n(n-1)}{2!} = 28 \Rightarrow$ $n(n-1) = 56$. This is the product of two consecutive whole numbers; look for factors of 56 which yield $8 \cdot 7$. Thus $n(n-1) = 8 \cdot 7$, and $n = \mathbf{8}$ **people** at the party.

13. Order is not important, implying a combination of 54 numbers taken six at a time. ${}_{54}C_6 = \frac{54!}{(54-6)!6!} = \frac{54 \cdot 53 \cdot 52 \cdot 51 \cdot 50 \cdot 49}{6 \cdot 5 \cdot 4 \cdot 3 \cdot 2 \cdot 1} = 25{,}827{,}165$ possible combinations of numbers, only one of which will win. Then $P(win) = \mathbf{\frac{1}{25{,}827{,}165}}$.

15. (a) $P(\text{two Britons, four Italians, two Danes}) = \frac{{}_{20}C_2 \cdot {}_{21}C_4 \cdot {}_4C_2}{{}_{45}C_8} = \frac{190 \cdot 5985 \cdot 6}{215{,}553{,}195}$, or about **0.032**.
   (b) $P(\text{no Britons}) = \frac{{}_{20}C_0 \cdot {}_{25}C_8}{{}_{45}C_8}$, or about **0.005**.
   (c) $P(\text{at least one Briton}) = 1 - P(\text{no Britons}) = 1 - \frac{{}_{20}C_0 \cdot {}_{25}C_8}{{}_{45}C_8} \approx \mathbf{0.995}$.
   (d) $P(\text{all Britons}) \frac{{}_{20}C_8 \cdot {}_{25}C_0}{{}_{45}C_8}$, or about $\mathbf{5.84 \cdot 10^{-4}}$.

17. (a) Since numbers can be repeated, there would be $10^4 = \mathbf{10{,}000}$ possibilities.
   (b) If numbers cannot be repeated, there would be $10 \cdot 9 \cdot 8 \cdot 7 = \mathbf{5040}$ possibilities.
   (c) There are five choices for each digit. So there are $5^4 = 625$ ways to create four digit numbers where the digits are all even.

   $P(\text{all even}) = \frac{625}{10{,}000} = \mathbf{\frac{1}{16}}$.

19. Order is not important, so the number of choices is ${}_8C_3 = \frac{8!}{3! \cdot 5!} = \frac{40{,}320}{6 \cdot 120} = \mathbf{56}$ **choices** for the exercise.

Assessment 9-5B

1. Each coin toss will result in two possible outcomes (head or tail). Five tosses will result in $2^5 = \mathbf{32}$ **different combinations** of heads and tails.

3. If we only consider gender, then we are arranging 7 objects, 3 of which are alike and 4 of which are alike. The number of ways the people can be arranged is $\frac{7!}{3!4!} = \frac{7 \cdot 6 \cdot 5}{3 \cdot 2 \cdot 1} = \mathbf{35}$.

5. (a) There are two O's repeated in OH1O, so there are $\frac{4 \cdot 3 \cdot 2 \cdot 1}{2!} = \frac{4!}{2!} = \mathbf{12}$ **possible arrangements.**
   (b) There are four A's repeated in ALABAMA, so there are $\frac{7 \cdot 6 \cdot 5 \cdot 4 \cdot 3 \cdot 2 \cdot 1}{4 \cdot 3 \cdot 2 \cdot 1} = \frac{7!}{4!} = 7 \cdot 6 \cdot 5 =$ **210 possible arrangements.**
   (c) There are three I's and two L's repeated in ILLINOIS, so there are $\frac{8 \cdot 7 \cdot 6 \cdot 5 \cdot 4 \cdot 3 \cdot 2 \cdot 1}{(3 \cdot 2 \cdot 1)(2 \cdot 1)} = \frac{8 \cdot 7 \cdot 6 \cdot 5 \cdot 4}{2 \cdot 1} = 8 \cdot 7 \cdot 6 \cdot 5 \cdot 2 = \mathbf{3360}$ **possible arrangements.**
   (d) There are four I's, four S's, and two P's in MISSISSIPPI, so there are $\frac{11 \cdot 10 \cdot 9 \cdot 8 \cdot 7 \cdot 6 \cdot 5 \cdot 4 \cdot 3 \cdot 2 \cdot 1}{(4 \cdot 3 \cdot 2 \cdot 1)(4 \cdot 3 \cdot 2 \cdot 1)(2 \cdot 1)} = \frac{11!}{4!4!2!} = 11 \cdot 10 \cdot 9 \cdot 7 \cdot 5 = \mathbf{34{,}650}$ **possible arangements.**
   (e) There are four E's, two N's, and two S's in TENNESSEE, so there are $\frac{9 \cdot 8 \cdot 7 \cdot 6 \cdot 5 \cdot 4 \cdot 3 \cdot 2 \cdot 1}{(4 \cdot 3 \cdot 2 \cdot 1)(2 \cdot 1)(2 \cdot 1)} = \frac{9!}{4!2!2!} = 9 \cdot 7 \cdot 5 \cdot 4 \cdot 3 = \mathbf{3780}$ **possible arrangements.**

7. Player's positions are not distinct, implying a combination of 12 persons taken five at a time ${}_{12}C_5 = \frac{12!}{(12-5)!5!} = \mathbf{792}$ **possible** team combinations.

9. (a) From the starting point there are 3 possible paths, each of which can get to the next point in the shortest distance. From the ends of each of these paths there are 2 shortest, and from

the ends of each of these there is only 1 path which will end at the other point. Thus there are $3 \cdot 2 \cdot 1 = 3! = \mathbf{6}$ **shortest paths** on the cubes.

(b) Using the fundamental principal of counting, there are $6 \cdot 6 = \mathbf{36}$ **shortest paths**.

11. Order is not important, thus a combination of 24 people taken 12 at a time. ${}_{24}C_{12} = \mathbf{2,704,156}$ different juries.

13. **(a)** $P(1 \text{ on each roll}) = \left(\frac{1}{6}\right)^8$.

(b) If $P(6) = \frac{1}{6}$ then $P(\text{not } 6) = 1 - \frac{1}{6} = \frac{5}{6}$.

Thus $P(\text{two 6's and six others}) = \left(\frac{1}{6}\right)^2 \cdot \left(\frac{5}{6}\right)^6$.

There are ${}_8C_2 = \frac{8!}{6! \cdot 2!} = 28$ ways of rolling two sixes out of eight tries (order is not distinct).

$P(6 \text{ exactly twice}) = {}_8C_2 \cdot \left(\frac{1}{6}\right)^2 \cdot \left(\frac{5}{6}\right)^6 = \frac{28 \cdot 5^6}{6^8} = \frac{437,500}{1,679,616} \approx \mathbf{0.260}$.

(c) $P(\text{at least one } 6) = 1 - P(\text{zero 6's}) = 1 - \left(\frac{5}{6}\right)^8 \approx \mathbf{0.767}$.

15. Order is not important in the choosing of the executive committee, implying the combination ${}_{32}C_5 = \frac{32!}{5!(32-5)!} = 201,376$ choices. Once the committee members are selected there are then five possibilities for president, so the total number of choices is $5({}_{32}C_5) = \mathbf{1,006,880}$.

17. $P(\text{royal flush})$
$= \frac{4}{{}_{52}C_5} = \frac{4}{2,598,960} = \frac{1}{649,740}$, where the 4 in the numerator is from choosing the suit.
$P(\text{no royal flush}) = 1 - P(\text{royal flush}) = 1 - \frac{1}{649,740} = \frac{649,739}{649,740}$.

Odds against royal flush $= \frac{P(\text{no royal flush})}{P(\text{royal flush})} = \frac{\left(\frac{649,739}{649,740}\right)}{\left(\frac{1}{649,740}\right)} = \frac{649,739}{1} = \mathbf{649,739:1}$.

19. **(a)** $P(7)$ on one roll of two dice $= \frac{6}{36} = \frac{1}{6}$.
$P(7 \text{ on each roll}) = \frac{1}{6} \cdot \frac{1}{6} \cdot \frac{1}{6} \cdot \frac{1}{6} \cdot \frac{1}{6} = \left(\frac{1}{6}\right)^5 \approx \mathbf{0.00013}$.

(b) $P(7)$ on one roll of two dice $= \frac{1}{6}$; thus $P(\text{not } 7)$ on one roll of two dice $= 1 - P(7) = 1 - \frac{1}{6} = \frac{5}{6}$. $P(\text{7 twice and not-7 three times})$ on five rolls of two dice $= \left(\frac{1}{6}\right)^2 \cdot \left(\frac{5}{6}\right)^3$.

There are ${}_5C_2 = 10$ ways of obtaining 7 twice and not-7 three times in five rolls of two dice.

$P(\text{7 exactly twice}) = 10 \cdot \left(\frac{1}{6}\right)^2 \cdot \left(\frac{5}{6}\right)^3 = \frac{1250}{7776}$, or **about 0.161**.

Review Problems

11. Since $P(11)$ and $P(12)$ are mutually exclusive events, $P(>10) = P(11) + P(12) = \frac{2}{36} + \frac{1}{36} = \frac{3}{36} = \mathbf{\frac{1}{12}}$.

13. **(a)** **No.** $E = \$36 \cdot \frac{1}{38} = \$\frac{36}{38}$, or about 95¢. It is not a fair game because expected gain and cost are not the same.

(b) Each time you play you either win \$36 or lose \$1.Then $\$36\left(\frac{1}{38}\right) - \$1\left(\frac{37}{38}\right) = \$\left(\frac{-1}{38}\right)$, or an expected loss of about $2\frac{1}{2}$¢.

Chapter 9 Review

1. **(a)** $S=$ {**(hhh), (hht), (hth), (htt), (ttt), (tth), (thh), (tht)**}. There are 2^3 = eight elements in the sample space.

(b) "At least two" means two or three heads, so $S =$ {**(hhh), (hht), (hth), (thh)**}.

(c) $P(\text{at least two heads}) = \frac{n(\text{at least two heads})}{n(S)} = \frac{4}{8} = \mathbf{\frac{1}{2}}$.

3. Answers may vary; e.g.,

(i) $\frac{4}{5} \cdot 1000 = 800$, so there are 800 blue beans;

(ii) $\frac{1}{8} \cdot 1000 = 125$, so there are 125 red beans;

(*iii*) $\frac{4}{5} + \frac{1}{8} < 1$, so there are jelly beans in the jar that are neither red nor blue; or

(*iv*) There are $1000 - 800 - 125 = 75$ jelly beans that are neither red nor blue.

5. **(a)** $P(black) = \frac{n(black)}{n(S)} = \mathbf{\frac{5}{12}}$.

(b) $P(black\ or\ white) = P(black) + P(white) = \frac{5}{12} + \frac{4}{12} = \frac{9}{12} = \mathbf{\frac{3}{4}}$. These are mutually exclusive events.

(c) $P(neither\ red\ nor\ white) = P(black) = \mathbf{\frac{5}{12}}$.

(d) $P(red\ not\ drawn) = P(black\ or\ white) = \mathbf{\frac{3}{4}}$.

(e) $P(black\ and\ white) = \mathbf{0}$. An impossible event.

(f) $P(black\ or\ white\ or\ red) = \mathbf{1}$. A certain event.

7. **(a)** $P(all\ white) = \frac{4}{9} \cdot \frac{4}{9} \cdot \frac{4}{9} = \mathbf{\frac{64}{729}}$. With replacement the probabilities of each draw are the same.

(b) $P(all\ white) = \frac{4}{9} \cdot \frac{3}{8} \cdot \frac{2}{7} = \frac{24}{504} = \mathbf{\frac{1}{21}}$. Without replacement the probabilities are changed for each draw.

9. $P(A\ from\ any\ box) = P(choosing\ that\ box) \cdot P(A\ from\ chosen\ box)$. The probabilities are added because the probabilities of A from each of the boxes are mutually exclusive events.
$P(A) = \frac{1}{4} \cdot \frac{0}{2} + \frac{1}{4} \cdot \frac{1}{4} + \frac{1}{4} \cdot \frac{1}{4} + \frac{1}{4} \cdot \frac{1}{5} = 0 + \frac{1}{16} + \frac{1}{16} + \frac{1}{20} = \mathbf{\frac{7}{40}}$.

11. $P(jack) = \frac{4}{52} = \frac{1}{13}$. Odds in favor $= \frac{P(jack)}{1-P(jack)} = \frac{\left(\frac{1}{13}\right)}{\left(\frac{12}{13}\right)} = \frac{1}{12}$ or **1:12**.

13. $\frac{P(event)}{1-P(event)} = \frac{3}{5} \Rightarrow 5.P(event) = 3 \cdot [1 - P(event)] \Rightarrow 5 \cdot P(event) = 3 - 3 \cdot P(event) \Rightarrow 8 \cdot P(event) = 3 \Rightarrow P(event) = \mathbf{\frac{3}{8}}$.

Note that by Theorem 9-8 if the odds in favor of an event are m to n, the probability of that event is $\frac{m}{m+n}$.

15. $P(win) = \frac{1}{3000}$. $E = \$1000 \cdot \frac{1}{3000} = \$\frac{1}{3} = 33\frac{1}{3}¢$. There is **no actual value,** to the nearest cent, that would produce a fair game.

17. There are $9 \cdot 10 \cdot 10 \cdot 1 = \mathbf{900}$ possible different numbers.

19. Order is distinct, implying a permutation of 10 flags taken 4 at a time. ${}_{10}P_4 = \frac{10!}{(10-4)!} = 10 \cdot 9 \cdot 8 \cdot 7 = \mathbf{5040}$ possible different ways.

21. **(a)** Order is distinct, implying a permutation of 5 finishes taken 3 at a time. ${}_5P_3 = \frac{5!}{(5-3)!} = 5 \cdot 4 \cdot 3 = \mathbf{60}$ possible different finishes.

(b) There are ${}_5P_2 = 20$ possible ways for a first/second place finish; there is only one way for a Deadbeat/Bandy finish. $P(Deadbeat/Bandy\ finish) = \mathbf{\frac{1}{20}}$.

(c) There are ${}_5P_3 = 60$ possible first-, second-, and third-place finishes; there is only one way for a Deadbeat/Egglegs/Cash finish. $P(Deadbeat/Egglegs/Cash\ finish) = \mathbf{\frac{1}{60}}$.

23. Order is not distinct, implying a combination of 5 questions taken 3 at a time. There are ${}_5C_3$ ways of selecting three questions out of five; since the number 1 is not to be selected, there are four options left and three of them will be selected. Thus $P(1\ not\ chosen) = \frac{{}_4C_3}{{}_5C_3} = \frac{4}{10} = \mathbf{\frac{2}{5}}$.

25. $P(all\ green) = 0.3 \cdot 0.3 \cdot 0.3 = \mathbf{0.027}$.

27. Answers may vary. E.g.,

(a) Randomly select digits 1 through 6 from a random digit table; discard digits 0 and 7 through 9.

(b) Block off twelve two-digit blocks in a random digit table (discarding any other than 01 through 12); randomly select three.

(c) Let random digits 0 through 2 represent red; digits 3 through 5 represent white; digits 6 through 8 represent blue; discard any nines.

29. Divide the figure into 16 equal triangles:

(a) A would represent two of the triangles. $P(A) = \frac{2}{16} = \mathbf{\frac{1}{8}}$.

(b) B would represent four of the triangles.

$P(B) = \frac{4}{16} = \frac{1}{4}$.

(c) C would represent one of the triangles.

$P(C) = \frac{1}{16}$.

31. *P(at least one face card)=1-P(no face card)* $= 1 - \frac{_{40}C_3}{_{52}C_3} = 1 - \frac{\frac{40!}{3!37!}}{\frac{52!}{3!49!}} = \frac{47}{85} \approx 0.553$

Chapter 10

DESIGNING EXPERIMENTS/COLLECTING DATA

Assessment 10-1A: Designing Experiments/Collecting Data

This entire assessment is subject to varying answers. Each of the following are representative possibilities.

1. Among the questions the class must determine are:

(i) Do you count the houses all around the block or only on the side of your house?

(ii) Are the houses across the street from you on your block, and do you count them?

(iii) What happens if you are in a new part of town where the blocks are not developed yet?

(iv) Do you count any businesses that might be on your block?

Data to be collected will be be determined by the questions asked, but in the second grade will likely be a frequency count.

The frequency count could be shown in a histogram or a bar graph. Any interpretations would be made about the graph.

3. Among the questions the class must determine are:

(i) What is the definition of "better?" Taste? Chemical content? Water temperature?

(ii) How will "better" be measured? By whom?

(iii) Does local government do water testing already?

(iv) Is there a water baseline?

5. The question in **(a) is fair** because it does not make any assumptions about favorite subjects. The question in (b) is bias because it uses words such as "calm" and "soothing" to suggest a person should like the ocean.

7. Among the questions the class must determine are:

(i) What is the definition of "visiting?" An airline layover? Overnight? An extended vacation?

(ii) What is the definition of "a country?" Would Monte Carlo qualify?

(iii) If a country is now separated from another, such as with the former Soviet Union, are visits before the breakup to now-independent states counted?

(iv) If a country was a protectorate of another state when a visit occurred, does it count?

9. How are "positive" and "negative" comments determined? A better and fairer way would be to list a representative sample of comments on the website.

11. It must be determined how accurate the second-grade observations really are. Second graders may not actually see what shoes are worn, but instead use personal knowledge of each other; e.g., "Alphie always wears tennis."

It might be observed that on Tuesday the most popular are tennis shoes and crocs, but an equally valid interpretation might be that the students just prefer soft-soled shoes.

Assessment 10-1B

1. *(i)* What is the definition of "families?" Parents and siblings? Extended family, such as cousins? Half-or-step brothers or sisters? Those who might be living away from home, such as college students?

(ii) Once the answers to those or other questions are determined, then the decision can be made as to who will collect the data, from whom it will be collected, and who will analyze it.

(iii) The data might be displayed in a bar graph and/or a frequency table. Interpretation might include measures of central tendency, and a discussion of what measures might be appropriate.

3. Among the questions the class must determine are:

(i) What is "affluent?" Would all those questioned have the same definition?

(ii) Would affluency be defined by the government or by a yardstick developed in the class?

(iii) Who should ask the questions? A neutral observer or someone from the class?

(iv) How could the question of affluency be asked without embarrassing some in the population?

5. Question (b) is fair because it makes no assumptions about whether the person being asked has a clothing preference. Question (a) is bias because it suggest that all exploration is a United States tradition.

7. Among the questions the class must determine (similar to those in question 5 of 10-4A) are:

(*i*) What is a "visit?" Driving through? An airline layover? Overnight? A vacation?

(*ii*) Would it count if Hawaii and/or Alaska were visited before they became states? What would be the status of the commonwealth of Puerto Rico?

(*iii*) If Hawaii is visited, must each island be part of it?

9. Each choice is "good" or better, not a likely reflection of the class. In any group, there would be some negative factors.

11. It is reasonable for students to talk about the most popular types of shoes worn and the least popular. They likely will begin to identify different people's shoes, but that is added information that they might already know that has nothing to do with the graph. It is important that they recognize what the graph says and what they are adding to the conversation from outside knowledge.

Assessment 10-2A: Displaying Data: Part I

1. (a) **Tuesday**.

(b) There were approximately 4500 pieces of mail processed on Tuesday and 3000 on Monday. $4500 - 3000 =$ **1500**.

(c) **Three**.

3.

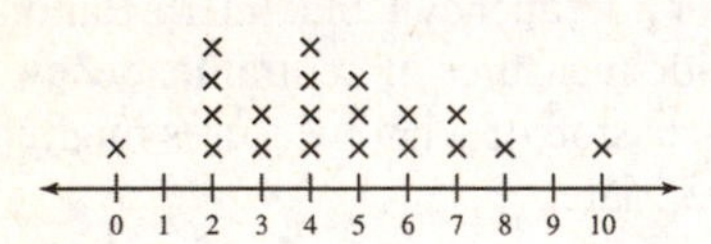

5. The histogram below groups weights into 10 pound classes:

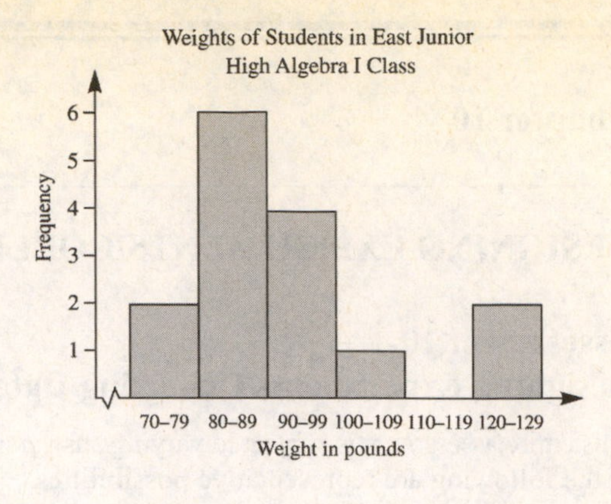

7. (a) **November** (with the highest bar) had the most rain-fall; approximately **30 cm**.

(b) 15 cm (October) + 25 cm (December) + 10 cm (January) = **50 cm**.

9. The bar graph below reflects that two or three heads showed up most frequently. Zero or five heads showed up the least number of times.

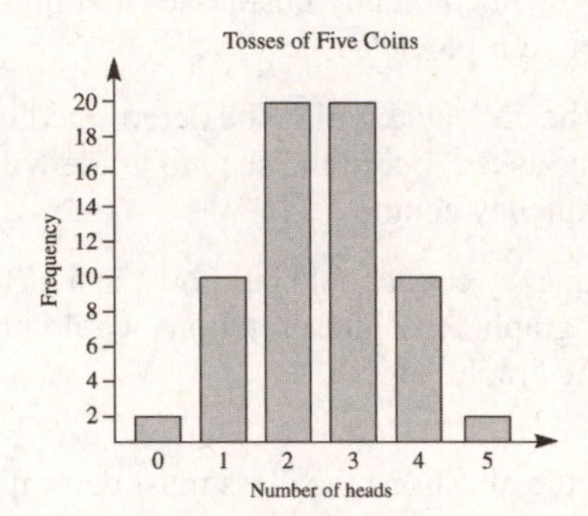

11.

13. Numbers are rounded to the nearest whole:

19% of 21 people is about **four**.

14% of 21 people is about **three**.

52% of 21 people is about **eleven**.

Each of the 5% sectors is about **one**.

15. (a) Ordered stem-and-leaf plot showing fall textbook costs:

Stem	Leaf
1	6
2	3 3
3	0 3 5 7 7 9 9
4	0 1 2 2 5 8 9
5	0 0 1 3 8
6	0 2 2

2 | 3 represents $23

(b) Grouped frequency table showing fall textbook costs:

Classes	Tally	Frequency
$15–19	|	1
$20–24	||	2
$25–29		0
$30–34	||	2
$35–39	~~||||~~	5
$40–44	||||	4
$45–49	|||	3
$50–54	||||	4
$55–59	|	1
$60–64	|||	3
		25

(c) The bar graph below shows the number of students on its vertical axis and the dollar amount paid for their textbooks on the horizontal axis:

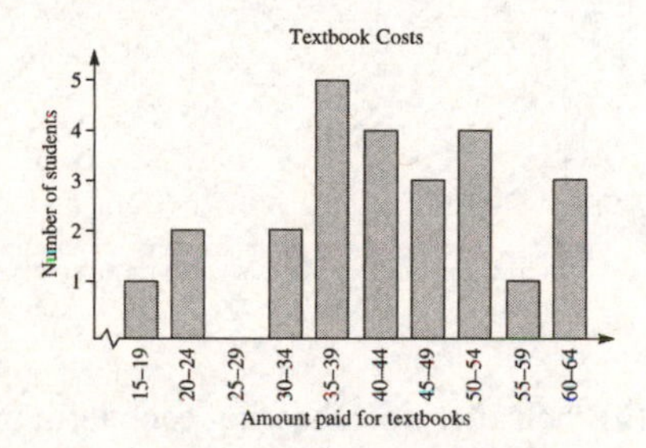

17. Measures are rounded to the nearest tenth of a centimeter.

Savings: 10% of 8 cm = **0.8 cm.**

Rent: 30% of 8 cm = **2.4 cm.**

Food: 12% of 8 cm = 0.96 cm ≈ **1.0 cm.**

Auto payment: 27% of 8 cm = 2.16 cm ≈ **2.2 cm.**

Tuition: 100% − (10 + 30 + 12 + 27)% = 21% of 8 cm = 1.68 cm ≈ **1.7 cm.**

Assessment 10-2B

1. Boys soccer, 4500; Girls soccer, 2250; Boys basketball, 4750; Girls basketball, 4000.

3. The line plot below shows that ages 10, 13, and 14 are the most common, while age 8 occurs the least frequently:

Student Ages at Washington School

5.

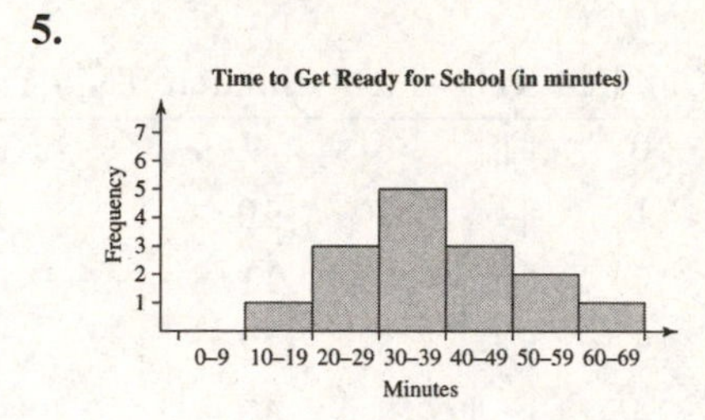

7. (a) The length of the Mississippi bar is about $\frac{8}{10}$ of the way between 3000 and 4000, so the Mississippi is **about 3800 km** long.

(b) The length of the Columbia bar is about $\frac{9}{10}$ of the way between 1000 and 2000, so the Columbia is about 1900 km long.

9.

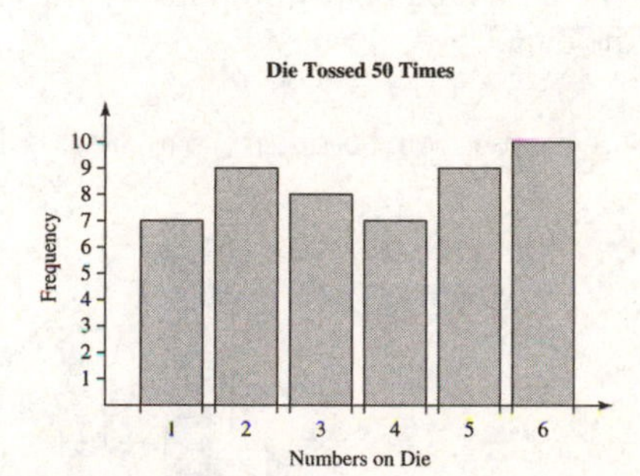

11.

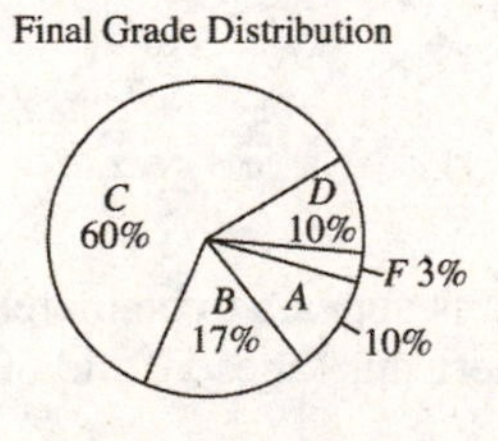

13. $\frac{32°}{360°} = 0.0\overline{8}$, or **about 9%.**

15. **(a)** A dot plot of the number of children of U.S. presidents is shown below:

Children of U.S. Presidents

0 1 2 3 4 5 6 7 8 9 10 11 12 13 14 15

(b) A frequency table is shown below:

<table>
<tr><th>Nr. of Children</th><th>Tally</th><th>Freq</th><th>Nr. of Children</th><th>Tally</th><th>Freq</th></tr>
<tr><td>0</td><td>||||| |</td><td>6</td><td>9</td><td></td><td>0</td></tr>
<tr><td>1</td><td>||</td><td>2</td><td>10</td><td>|</td><td>1</td></tr>
<tr><td>2</td><td>||||| |||||</td><td>10</td><td>11</td><td></td><td>0</td></tr>
<tr><td>3</td><td>|||||</td><td>5</td><td>12</td><td></td><td>0</td></tr>
<tr><td>4</td><td>||||| |||</td><td>8</td><td>13</td><td></td><td>0</td></tr>
<tr><td>5</td><td>||||</td><td>4</td><td>15</td><td>|</td><td>1</td></tr>
<tr><td>6</td><td>||||</td><td>4</td><td></td><td></td><td></td></tr>
<tr><td>7</td><td>|</td><td>1</td><td>Total</td><td></td><td>42</td></tr>
<tr><td>8</td><td>|</td><td>1</td><td></td><td></td><td></td></tr>
</table>

(c) Ten presidents had **two children**.

17. Answers may vary.

(a) The bar graph below graphically illustrates the data.

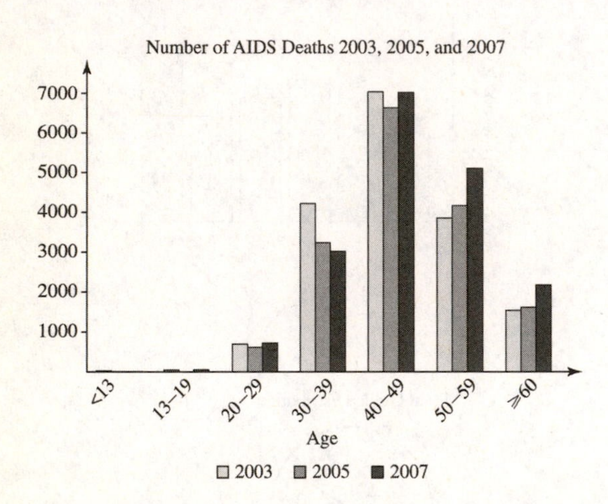

(b) AIDs deaths appears to be increasing among older Americans (ages 50 and 60).

Review Problems

15. Both have some bias. Method (i) is biased in that you might only get the opinions of the very satisfied or unsatisfied. Method (ii) is biased in that you are only asking the person paying the bill.

Assessment 10-3A
Displaying Data: Part II

1. **(a)** 70% of \$12,000, or **about \$8400**.

(b) (100 − 30)% of \$20,000 = 70% of \$20,000 in depreciation, or **about \$14,000**.

(c) 35% of \$20,000, **about \$7000**.

(d) Right after **two years**. Average trade-in value is about 55% at two years.

3. **(a)** **January** corresponds to the largest vertical values on both graphs.

(b) **March**.

(c) Although other factors might influence sales, in 4 of the five months illustrated, more snow shovels were sold in **2011**.

5.

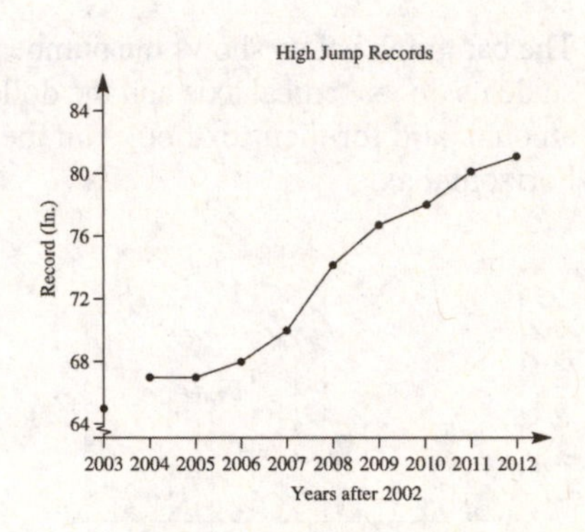

7. **(a)** Scatterplot, showing each term on the vertical axis (y) with the number of the term on the horizontal axis (x):

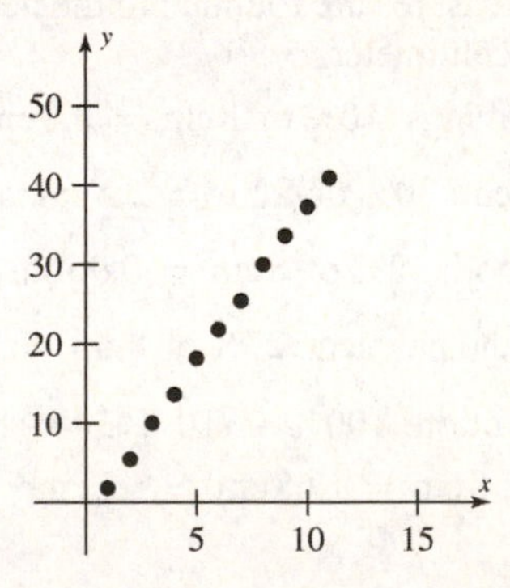

(b) Trend line, depicting the scatterplot points as a set of continuous data:

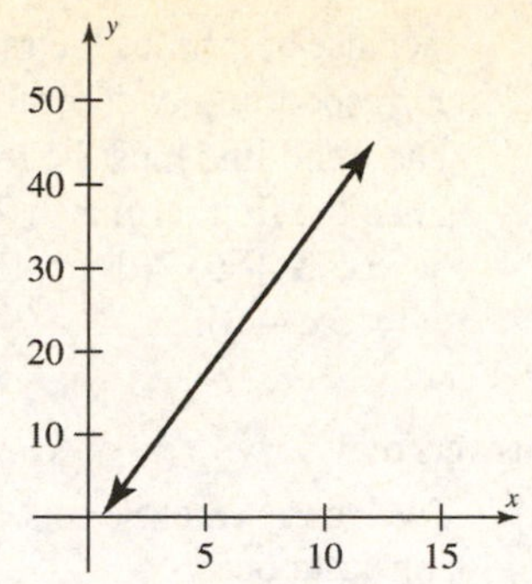

(c) Use a spreadsheet approach, where values of x are in column A and values of y are in column B:

A	B
1	2
2	6
3	10
⋮	⋮

Each time the value of the x-coordinate increases by 1 the value of the y-coordinate increases by 4 $\Rightarrow$ the y-coordinates increase at 4 times the rate of increase of the corresponding x-coordinates; i.e., $y = 4x$. The trend line must be lowered by 2, though, since the first term is 2. Thus the equation is $\boldsymbol{y = {}^-2 + 4x}$ (or in slope-intercept form, $y = 4x - 2$).

9. Answers may vary.

(a) Two representative sets of coordinates appear to be

x	y
$^-2$	$^-1$
3	1

When the x-coordinates of these points go from $^-2$ to 3, an increase of 5, the y-coordinates go from $^-1$ to 1, an increase of 2 $\Rightarrow y = \frac{2}{5}x = 0.4x$. The trend line must be lowered by about 0.2, as nearly as can be determined from the graph, so the equation would be $\boldsymbol{y = {}^-0.2 + 0.4x}$ (or in slope-intercept form, $y = 0.4x - 0.2$).

(b) Two representative sets of coordinates appear to be

x	y
$^-1$	1
1	2

When the x-coordinates of these points go from $^-1$ to 1, an increase of 2, the y-coordinates go from 1 to 2, in increase of 1 $\Rightarrow y = \frac{1}{2}x = 0.5x$. The trend line must be raised by about 1.5, as nearly as can be determined from the graph, so the equation would be $\boldsymbol{y = 1.5 + 0.5x}$ (or in slope-intercept form, $y = 0.5x + 1.5$).

(c) Two representative sets of coordinates appear to be

x	y
$^-3$	$^-2$
1	1

When the x-coordinates of these points go from $^-3$ to 1, an increase of 4, the y-coordinates go from $^-2$ to 1, an increase of 3 $\Rightarrow y = \frac{3}{4}x = 0.75x$. The trend line must be raised by about 0.25, as nearly as can be determined from the graph, so the equation would be $\boldsymbol{y = 0.25 + 0.75x}$ (or in slope-intercept form, $y = 0.75x + 0.25$).

11. If the equation is $y = 3x + 5$:

(a) $y = 3(^-2) + 5 = \boldsymbol{^-1}$.

(b) $y = 3(14) + 5 = \mathbf{47}$.

(c) $y = 3(0) + 5 = \mathbf{5}$.

(d) $y = 3(5) + 5 = \mathbf{20}$.

13. The data is **constant**. Regardless of the value of x, y will always equal 12.

Assessment 10-3B

1. (a) ***(i)*** **22**

(ii) **20.4**

(iii) **25.4**

(b) Between **1940 and 1950**, the slope of the line connecting the ordered pairs corresponding to these years is steepest (in the negative direction) than between any other two ordered pairs.

(c) **1980–1990** for similar reasons as in 1 (b).

3. **(a)** About **75,000**.

(b) The slopes of the lines connecting the ordered pairs that correspond to 2000–01 to 2000–02 are both negative, indicating that the participation dropped.

(c) The trend is positive and participation among women appears to be increasing more rapidly than among men.

5. ***(i)***

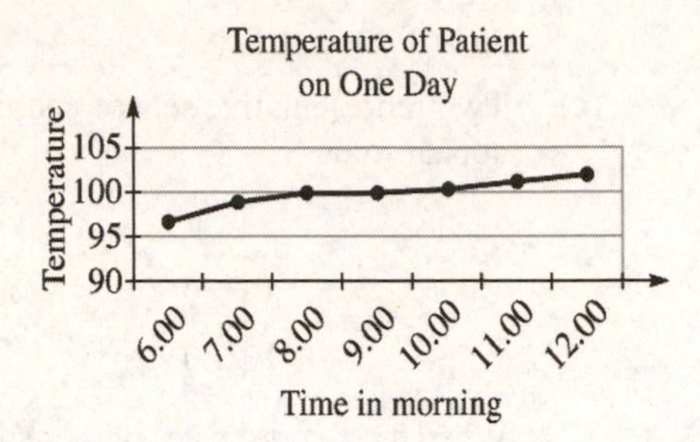

(ii) The patient's temperature is rising steadily throughout the morning. It might be inferred that this is not normal.

7. **(a)** Scatterplot, showing each term on the vertical axis (y) with the number of the term on the horizontal axis (x), and:

(b) Trend line, depicting the scatterplot points as a set of continuous data:

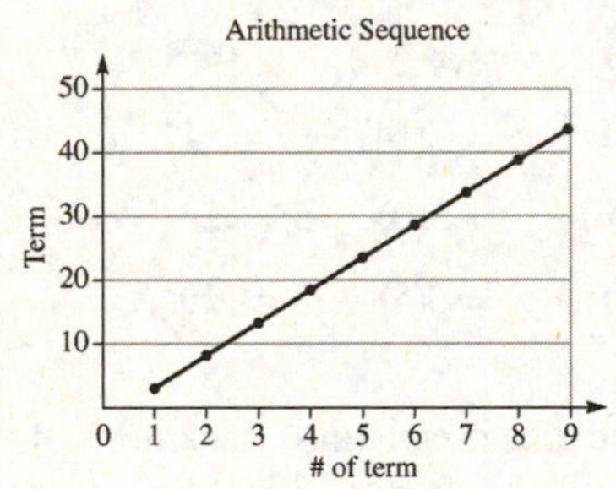

(c) Use a spreadsheet approach, where values of x are in column A and values of y are in column B:

A	B
1	3
2	8
3	13
⋮	⋮

Each time the value of the x-coordinate increases by 1 the value of the y-coordinate increases by 5 $\Rightarrow$ the y-coordinates increase at 5 times the rate of increase of the corresponding x-coordinates; i.e., $y = 5x$. The trend line must be lowered by 3, though, since the first term is 3. Thus the equation is $\mathbf{y = {}^-3 + 5x}$ (or in slope-intercept form, $y = 5x - 3$).

9. Answers may vary.

(a) Two representative sets of coordinates appear to be:

x	y
$^-3$	$^-1.8$
2	1

When the x-coordinates of these points go from $^-3$ to 2, an increase of 5, the y-coordinates go from $^-1.8$ to 1, an increase of 2.8 $\Rightarrow y = \frac{2.8}{5}x = 0.56x$. The trend line must be lowered by about 0.1, as nearly as can be determined from the graph, so the equation would be $\mathbf{y = {}^-0.1 + 0.56x}$.

(b) Two representative sets of coordinates appear to be:

x	y
$^-2$	1
1	0

When the x-coordinates of these points go from $^-2$ to 1, an increase of 3, the y-coordinates go from 1 to 0, a decrease of 1 $\Rightarrow y = \frac{^-1}{3}x$. The trend line must be raised by about $\frac{1}{3}$, as nearly as can be determined from the graph, so the equation would by $\mathbf{y = \frac{^-1}{3}x + \frac{1}{3}}$.

11. If $y = 2x - 7$:

(a) $y = 2(^-2) - 7 = \mathbf{^-11}$.

(b) $y = 2(14) - 7 = \mathbf{21}$.

(c) $y = 2(0) - 7 = \mathbf{^-7}$.

(d) $y = 2(5) - 7 = 3$.

13. The data is **constant**. Regardless of the value of x, y will always equal 15.

Review Problems

9. Reducing the miscellaneous category to 10% produces the percentage graph below:

Smith Expenses

Misc	**10%**
Gas	**13%**
Utilities	**5%**
Food	**20%**
Rent	**32%**
Taxes	**20%**

Assessment 10-4A
Measures of Central Tendency and Variation

1. **(a)** Ordering data: 2, 5, 5, 7, 8, 8, 8, 10.

   (i) Mean $= \frac{2+5+5+7+8+8+8+10}{8} = \frac{53}{8} = \mathbf{6.625}$.

   (ii) Median $=$ **7.5**; i.e., between 7 and 8, or $\frac{7+8}{2}$.

   (iii) Mode $=$ **8**, the value which occurs most frequently.

   (b) Ordering data: 10, 11, 12, 12, 12, 14, 14, 16, 20.

   (i) Mean $= \frac{10+11+12+12+12+14+14+16+20}{9} = \frac{121}{9} = \mathbf{13.\bar{4}}$.

   (ii) Median$=$ **12**; i.e., the midpoint value.

   (iii) Mode$=$ **12**, the value which occurs most frequently.

   (c) Ordering data: 12, 17, 18, 18, 22, 22, 30.

   (i) Mean $= \frac{12+17+18+18+22+22+30}{7} = \frac{139}{7} = \mathbf{19\frac{6}{7}}$.

   (ii) Median $=$ **18**; i.e., the midpoint value.

   (iii) Mode $=$ **18** and **22**. These are dual modes; each occurs twice.

3. **(a)** ***(i)*** $\frac{3(60)+3(80)}{6} = \mathbf{70}$.

   (ii) There is an even number of data points. Average the two middle values: $\frac{60+80}{2} = \mathbf{70}$.

   (iii) Bi-modal: **60 and 80**.

   (b) Mean $= \frac{\text{sum of test scores}}{\text{number of test scores}}$. $75 = \frac{\text{sum}}{20} \Rightarrow$ sum $= 75 \cdot 20 = \mathbf{1500}$.

5. Mean $= \frac{3 \cdot 100 + 9 \cdot 50}{12} = \mathbf{62.5}$.

7. The number of points for each course is the product of the number of credits and the point value of each. 15 points for math, 12 for english, 10 for physics, 3 for German, and 4 for handball gives a total of 44 points. GPA $= \frac{44 \text{ points}}{17 \text{ credits}}$, or **about 2.59**.

9. **(a)** The total salary for the 40 dancers is shown below:

Salary ($)	Nr. Dancers	Total ($)
18,000	2	36,000
22,000	4	88,000
26,000	4	104,000
35,000	3	105,000
38,000	12	456,000
44,000	8	352,000
50,000	4	200,000
80,000	2	160,000
150,000	1	150,000
Total	40	1,651,000

   Mean annual salary $= \frac{\$1{,}651{,}000}{40} = \mathbf{\$41{,}275}$.

   (b) Ordering the salaries: $18,000, 18,000, 22,000, ..., 150,000; the median is between the 20th and 21st and all salaries between the 13th and 25th are $38,000.
   Median $=$ **$38,000**.

   (c) **$38,000**, or the most frequent salary.

11. **(a)** Answers may vary. One possible viewpoint might be that the mean of $41,275 with mean

average deviation of $12,143.75 would give a realistic picture.

(b) While the data does not form a perfect bell curve (normal distribution), the majority at data points are near $41,275 and symmetrically spread around this value.

13. $24 + 34 =$ **58 years old.**

15. (a) Theater **A: $25**; Theater **B: $50**. The median in a box plot is the middle line through the box; i.e., 25 in the Theater A box and 50 in the Theater B box.

(b) Theater **B**. Range is the difference between upper and lower extremes. Theater A range $= 40 - 15 = 25$; Theater B range $= 80 - 15 = 65$.

(c) **$80** at Theater B, its upper extreme. Theater A's upper extreme is $40.

(d) Answers may vary. There is significantly more variation and generally higher prices at Theater B.

17. (a) *L. A.*: $Q_1 = 599$; $Q_2 = 662$; $Q_3 = 742.5$.
S. L.: $Q_1 = 395$; $Q_2 = 427$; $Q_3 = 564$.

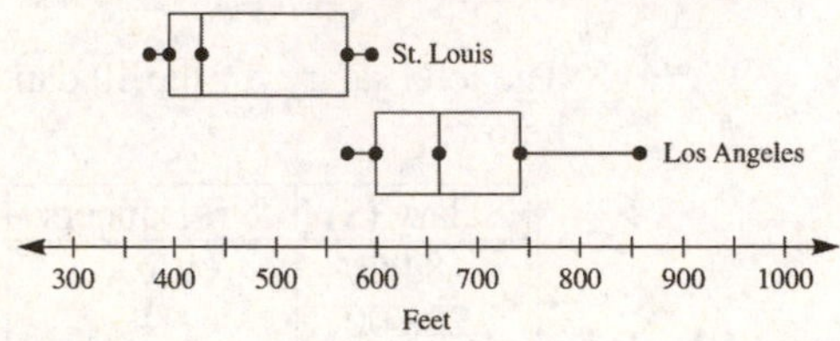

(b) More than 50% at the buildings in *LA* are taller than those in St. Louis.

19. (a) Sum of grades $= 96 + 71 + 43 + 77 + 75 + 76 + 61 + 83 + 71 + 58 + 97 + 76 + 74 + 91 + 74 + 71 + 77 + 83 + 87 + 93 + 79 = 1613$. $\bar{x} = \frac{1613}{21} \approx$ **76.8.**

(b) Ordering the scores: 43, 58, 61, 71, 71, 71, 74, 74, 75, 76, 76, 77, 77, 79, 83, 83, 87, 91, 93, 96, 97. Median = **76** (the 11th score).

(c) The most frequent score is 71. Mode = **71.**

(d) Upper quartile $= \frac{83+87}{2} = 85$; lower quartile $= \frac{71+71}{2} = 71$. IQR $= 85 - 71 =$ **14.**

(e) $(x - \bar{x})^2$ is tabularized below:

x	$x - \bar{x}$	$(x - \bar{x})^2$
96	19.2	368.64
71	⁻5.8	33.64
43	⁻33.8	1142.44
77	0.2	0.04
75	⁻1.8	3.24
76	⁻0.8	0.64
61	⁻15.8	249.64
83	6.2	38.44
71	⁻5.8	33.64
58	⁻18.8	353.44
97	20.2	408.04
76	⁻0.8	0.64
74	⁻2.8	7.84
91	14.2	201.64
74	⁻2.8	7.84
71	⁻5.8	33.64
77	0.2	0.04
83	6.2	38.44
87	10.2	104.04
93	16.2	262.44
79	2.2	4.84
		Total: 3293.24

$v = \frac{3293.24}{21} \approx$ **156.8.**

(f) $s = \sqrt{v} = \sqrt{156.8} \approx$ **12.5.**

(g) The mean, $\bar{x}$, is 76.8.

x	$\lvert x - \bar{x} \rvert$
96	19.2
71	5.8
43	33.8
77	0.2
75	1.8
76	0.8
61	15.8
83	6.2
71	5.8
58	18.8
97	20.2

x	$\lvert x - \bar{x} \rvert$
76	0.8
74	2.8
91	14.2
74	2.8
71	5.8
77	0.2
83	6.2
87	10.2
93	16.2
79	2.2
	Total: 189.8

MAD $= \frac{189.8}{21} \approx$ **9.04.**

21. 1 minutes is three standard deviation below the mean. 99.8% of the area under the normal curve lies below 3 standard deviation above and below the mean. $\frac{100\%-99.8\%}{2}$ =**1%** of the calls should last less than 1 minutes.

23. \$4.50 is one standard deviation below the mean; \$5.50 is one standard deviation above the mean. 68% of the area under the normal curve lies within ± 1 standard deviation from the mean. Therefore the probability of having a wage between \$4.50 and \$5.50 $=$ **0.68**.

25. About 68% of the Values 1 is within are standard deviation of the mean an a normal distribution plot. Al's score is exactly equal to the mean plus one standard deviation. Thus, $68\% + \frac{32\%}{2} = 68\% + 16\% = 84\%$ of the students scored below Al, or $.84 \cdot 10{,}000 =$ **8400 students**.

27. Jack's percentile ranking is $\frac{200-70}{200} = \frac{130}{200} = .65$, meaning the 65th percentile. Thus, **Jill's** standing is higher than Jack's.

Assessment 10-4B

1. (a) Ordering data: 63, 75, 80, 80, 80, 80, 82, 90, 92, 92.

(i) Mean $= \frac{63+75+80+80+80+80+82+90+92+92}{10} = \frac{814}{10} = \mathbf{81.4}$.

(ii) Median $=$ **80**, the midpoint between the fifth and sixth 80's.

(iii) Mode $=$ **80**, the value which occurs most frequently.

(b) *(i)* Mean $= \frac{5+5+5+5+5+10}{6} = \frac{35}{6} = \mathbf{5.8\overline{3}}$.

(ii) Median $=$ **5**, the midpoint between the third and fourth 5's.

(iii) Mode $=$ **5**, the value which occurs most frequently.

3. (a) If all scores $= 80$:

(i) Mean $=$ **80**; i.e., $\frac{80+80+80+80+80+80}{6} = 80$.

(ii) Median $=$ **80**, since the midpoint is between the third and fourth 80's.

(iii) Mode $=$ **80**, since all values are 80.

(b) Answers may vary. One set might be {70, 80, 80, 80, 80, 90}.

5. (a) $\frac{40+36+8+6+2}{5} = \frac{92}{5} =$ **18.4 years**.

(b) In five years the family's ages will be 45, 41, 13, 11, and 7. $= \frac{45+41+13+11+7}{5} = \frac{117}{5} =$ **23.4 years**.

(c) In ten years the family's ages will be 50, 46, 18, 16, and 12. Mean $= \frac{50+46+18+16+12}{5} = \frac{142}{5} =$ **28.4 years**.

(d) Means will increase by the same amount as the increase in each individual data point (if those increases are the same for each data point).

7. (a) If a total of 100 points could be awarded $(60\% + 25\% + 15\%)$, then the following number could be gained in each category:

Term paper; $0.60 \cdot 85 = 51$,

Homework; $0.25 \cdot 78 = 19.5$, and

Final; $0.15 \cdot 90 = 13.5$.

The grade would then be $51 + 19.5 + 13.5 =$ **84**.

(b) Let x be the term paper percentage, y be the homework percentage, and z be the final exam percentage. Let t be the term paper score, h be the homework score, and f be the final exam score. Then the overall grade would be $\mathbf{(xt + yh + zf)\%}$. Note that $x + y + z$ must equal 100%.

9. (a) *(i)* Balance beam - **Olga**. Mean $= \frac{9.5+9.6+9.6+9.6}{4} = 9.575$.

(ii) Uneven bars - **Lisa**. Mean $= \frac{9.8+9.8+9.9+.9.9}{4} = 9.85$.

(iii) Floor exercises - **Lisa**. Mean $= \frac{9.8+9.9+10+10}{4} = 9.925$.

(b) **Lisa**, with 29.20 points.

11. Mean of five numbers $= 6$ implies that the sum of the numbers $= 5 \cdot 6 = 30$. Removing a number; $\frac{30 - \text{number}}{5-1} = 7 \Rightarrow 30 - \text{number} = 28$. Number $=$ **2**.

13. (a) $66 - 21 =$ **45**.

(b) **Not necessarily**, the range can be effected by extreme values.

15. (a) About **23.7**.

(b) About **21**.

(c) The ranges are **about the same**, although the range of women's ages is slightly greater.

(d) In the sample, women tend to marry about 2 years younger than men. The spread of the data is about the same.

17. (a) Minneapolis:

Lower extreme = 366;

upper extreme = 950;

lower quartile = 416;

upper quartile = 668;

and median = $\frac{447+561}{2}$ = 504.

IQR = 668 − 416 = 252.

Lower quartile − 1.5 · **IQR** = 38,

so there is no lower outlier.

Upper quartile + 1.5 · **IQR** = 1046,

so there is no upper outlier.

Los Angeles:

Lower extreme = 516;

upper extreme = 858;

lower quartile = 571;

upper quartile = 735;

and median = $\frac{620+625}{2}$ = 622.5.

IQR = 735 − 571 = 164.

Lower quartile − 1.5 · **IQR** = 325,

so there is no lower outlier.

Upper quartile + 1.5 · **IQR** = 981,

so there is no upper outlier.

Giving the following box plot:

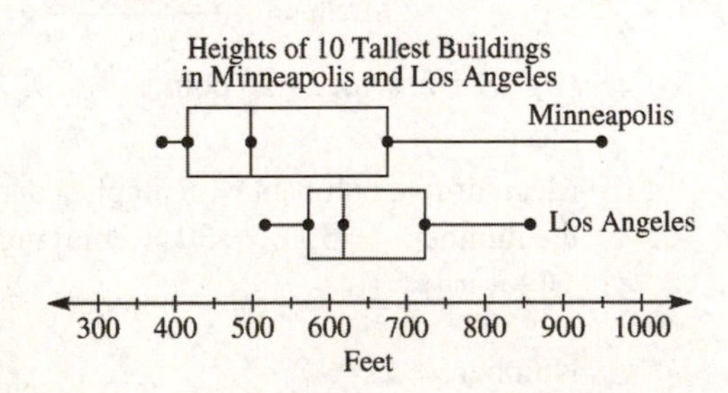

(b) **No**. In neither case are there values below the lower quartile minus 1.5 times the interquartile range (IQR) nor above the upper quartile plus 1.5 times the IQR.

(c) Answers may vary. Minneapolis has more spread in their ten tallest buildings and has at least one building higher than any in Los Angles. All the tallest buildings in Los Angeles are higher than over 50% of the tallest buildings in Minneapolis. All the tallest buildings in Los Angeles are between 516 and 858 feet; the highest building in either group is in Minneapolis and is 950 feet high.

19. (a) (*i*) New mean = $\frac{\text{Total of old salaries} + 1000 \cdot nr \text{ of teachers}}{\text{Nr of teachers}}$ = Old mean + \$1000. The mean **increases by \$1000**.

(*ii*) The median **increases by \$1000**.

(*iii*) The extremes **increase by \$1000**.

(*iv*) The quartiles **increase by \$1000**.

(*v*) The standard deviation is **unchanged**.

(*vi*) The IQR is **unchanged**.

(b) (*i*) The mean **increases by 5%**.

(*ii*) Variations about the mean would be unchanged, so the standard deviation **increases by 5%**.

21. 16 ounces is two standard deviations below the mean; 13.5% + 34% + 50% = 97.5% of the area under the normal curve is above 2 standard deviations below the mean. **97.5%** of the boxes should contain more than 16 ounces.

23. 125 ounces is one standard deviation above the mean. 50% + 34% = 84% of the area under the normal curve lies below 1 standard deviation above the mean. Thus the probability a weight less than 125 oz = **0.84**.

25. A score of 440 is one standard deviation below the mean. 0.1% + 2.4% + 13.5% = 16% of the area under the normal curve lies below 1 standard deviation below the mean. Thus 16% of 10,000 is **1600 students**.

27. Jill has a percentile rank of $\frac{200-80}{200}$ = 60. Thus **Nathan** with a percentile rank of 80 has a higher standing.

Review Problems

21. (a) **Mount Everest** is the highest mountain, at about **8500** meters.

(b) Mounts **Aconcagus, Everest, and McKinley**.

Assessment 10-5A
Abuses of Statistics

This entire assessment is subject to varying answers. Each the following are representative possibilities.

1. **(a)** The claim cannot be substantiated without knowing more about the noise characteristics of the car and glider in question. Many gliders are quite noisy in flight, or the car may not be running (in which case it would be exceedingly quiet).

(b) Without knowing sales figures for the last 15 years this claim cannot be substantiated. It may be that 95% of the manufacturer's cycles sold in the United States were sold in the last year.

(c) 10% more than 10% (which is 11%) is not much of an increase.

(d) Fresher than what? 40% of what?

3. **No**. It could very well be that most of the pickups sold in the past ten years were actually sold during the previous two years. In such case most of the pickups would have been on the road for only two years and therefore the given information would not imply that the average life of a pickup is around ten years.

5. The three-dimensional drawing distorts the graph. The result of doubling the radius and the height of the can is to increase the volume by a factor of 8, whereas sales had only doubled.

7. The scores below the median are further from it than the scores above the median.

9. **(a)** This bar graph would have perhaps 20 accidents as the baseline and some larger number (e.g., 100) as its maximum. Then 24 in 2008 would be 4 units above the baseline while 38 in 2012 would appear as 3.5 times or 350% taller than 24 of 2001, when in fact it is just 58% higher.

(b) This bar graph would have zero accidents as its baseline and some larger number (e.g., 100) as its maximum value. The effect would be to show values from 24 to 38 as lying a line with a small positive slope.

11. The mode might be representative if enough randomly-selected spots were reported—although it is possible that a mode might not exist. A median or mean might be misleading, depending upon the number of data points selected. A report of mean, median, mode, and standard deviation would probably be the most helpful.

13. The student is incorrectly mixing percentages of "effectiveness" with percentages of "times taking the drug." Further, "up to" 92% effective could mean something much lower as the norm. Finally, an 8% increase on 92% does not yield 100%; it would result in 99.36%.

15. The data is simply an indicator of passenger complaints and not a overall airline rating. Larger airlines might also have more total complaints but a lower percentage per passenger-mile.

17. Only part (b) might not be helpful, even though it could have some bearing on insurance cost. All the others have some bearing on the true cost of operating an automobile.

Assessment 10-5B

1. **(a)** "Up to" is quite indefinite (6 mpg is "up to" 30 mpg). The conditions under which 30 mpg may have been realized are not stated (the car may have been rolling downhill).

(b) "Brighter" than what?

(c) How many dentists responded? Were they paid? What brand of sugarless gum did they recommend?

(d) Most accidents occur in the home because a majority of time is spent in the home. This is an example of carrying an argument to a deceptive extreme.

(e) Is there another airline flying to the city?

3. We don't know who was surveyed, how many were surveyed, or how the question was phrased. The question might have been biased. For example, the survey might have asked "Don't you agree that 2 hours of homework per night in needed?"

5. The three-dimensional graph distorts the data. It makes it appear as though more cars were sold in 2011 then it 2009, when the two-dimensional graph demonstrater that the quantities were the same. Worsening the distortion is the fact that no scale is shown on the vertical axis in the three-dimensional graph.

7. She could have taken a different number of quizzes during the first part of the quarter than during the second part.

9. **(a)** The survey shows that a majority of teachers rate their textbooks as good to excellent. The national "experts", on the other hand, seem not to agree.

(b) The teachers and the "experts", who are not identified, obviously disagree. It may be that the teachers are actually in the classroom and thus have a more intimate knowledge of student reaction to the texts.

11. **(a)** In grades K-4 homework of less than one hour might be justified because the children are younger and 75% of the teachers surveyed fall within this range.

In grades 5-8 the spread is wider; it may be that older students are assigned more homework than those who are younger. 71% of the teachers surveyed assign between 31 and 120 minutes.

(b) A potential misuse in grades 5-8 might note that of the teachers who assign at least 91 minutes of homework, about half assign at least two hours.

Based on the survey data, it would be difficult to justify at least two hours at the K-4 level.

13. Any sample should have sufficient size to be statistically significant, and should be representative of the demographics and the population sampled.

15. **(a)** 647 to 649, or $\frac{647}{649} \approx$ **99.7%**.

(b) 622 to 649, or $\frac{622}{649} \approx$ **95.8%**.

(c) **(*i*)** *P(lung cancer and smoker|smoker)* = $\frac{647}{1269} \approx$ **51%**.

(*ii*) *P(lung cancer and non-smoker|non-smoker)* = $\frac{2}{29} \approx$ **7%**.

(d) There is a high association between smoking and lung cancer because of the disparate probabilities. Insufficient data to arrive at this conclusion.

17. Answer vary. In the next 200 years, western Texas or Eastern New Mexico would be reasonable guesses. In the next 400 years western, Arizona is a reasonable guess.

Review Problems

1. **(a)** $\overline{x} = \frac{43+91+73+65+56+77+84+91+82+65+98+65}{12} = \frac{890}{12} \approx$ **74.17**.

(b) Ordering the data: 43, 56, 65, 65, 65, 73, 77, 82, 84, 91, 91, 98. Median $= \frac{73+77}{2} =$ **75.**

(c) Mode = **65**, the largest number of scores.

(d) The sum of the $(x - \overline{x})^2$ terms = 2855.68. $v = \frac{2855.68}{12} \approx$ **237.97**.

(e) $s = \sqrt{v} = \sqrt{237.97} \approx$ **15.43**.

(f) The sum of the $|x_n - \overline{x}|$ terms = 156. MAD $= \frac{156}{12} =$ **13**.

3. First ten-paper mean = 70 so $10 \cdot 70 = 700$ points. Second 20-paper mean = 80 so $20 \cdot 80 = 1600$ points. Combined mean $= \frac{700+1600}{10+20} = \frac{2300}{30} =$ **76.$\overline{6}$**.

5. Freestyle: Lower extreme = 53.12;
Lower quartile = 54.17;
Median = 54.93;
Upper quartile = 59.045; and
Upper extreme = 61.20.

Butterfly: Lower extreme = 56.61;
Lower quartile = 58.17;
Median = 59.26;
Upper quartile = 64.02; and
Upper extreme = 69.50.

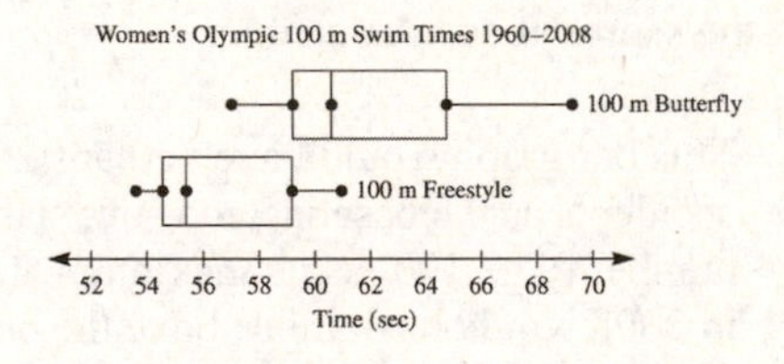

The times for the 100-m butterfly are (relatively) much greater than those for the 100-m freestyle.

Chapter 10 Review

1. **(*i*)** If the average is reported as 2.41 children, the mean is probably being used. The mode would not have a decimal number, and the median would be either a whole number or one ending in .5.

(*ii*) If the average is 2.5, then either the mean or the median might have been used.

3. **(a)** *(i)* Mean $= \frac{10+50+30+40+10+60+10}{7} = \mathbf{30}$.

(ii) Ordering the data: 10, 10, 10, 30, 40, 50, 60. Median = **30**, the middle data point.

(iii) Mode = **10**, the most frequent data point.

(b) *(i)* Mean $= \frac{5+8+6+3+5+4+3+6+1+9}{10} = \mathbf{5}$.

(ii) Ordering the data: 1, 3, 3, 4, 5, 5, 6, 6, 8, 9. Median $= \frac{5+5}{2} = \mathbf{5}$.

(iii) Modes = **3**, **5**, and **6**.

5. **(a)** Dot plot of Miss Rider's class (masses in kilograms):

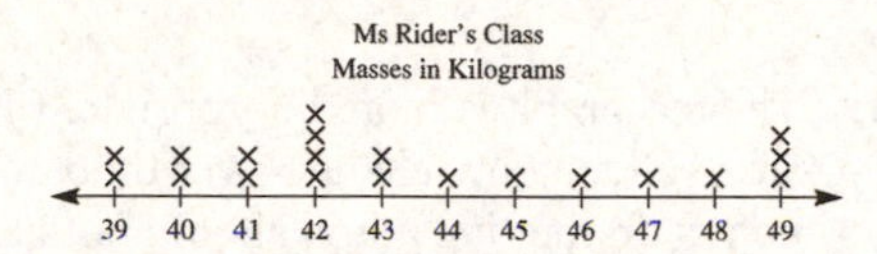

(b) Ordered stem-and-leaf plot of Miss Rider's class (masses in kilograms):

3	99
4	00112222333345678999

4 | 0 represents 40 kg

(c) Frequency table of Miss Rider's class (masses in kilograms):

Mass	Tally	Frequency
39	||	2
40	||	2
41	||	2
42	||||	4
43	||	2
44	|	1
45	|	1
46	|	1
47	|	1
48	|	1
49	|||	3
		20

(d) Bar graph of Miss Rider's class (masses in kilograms):

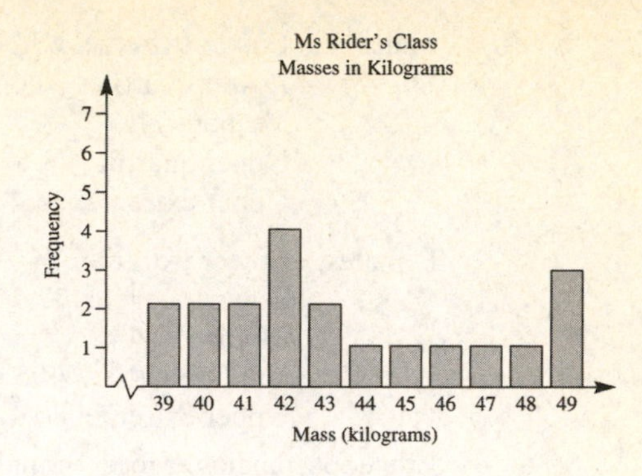

7. Wegetum expenditures, where:

Bribes $= \frac{\$600{,}000}{\$2{,}000{,}000} = 0.30$; **30%** of 360° = 108°.

Legal fees $= \frac{\$400.000}{\$2{,}000{,}000} = 0.20$; **20%** of 360° = 72°.

Public relations $= \frac{\$400{,}000}{\$2{,}000{,}000} = 0.20$; **20%** of 360° = 72°.

Bail money $= \frac{\$300{,}000}{\$2{,}000{,}000} = 0.15$; **15%** of 360° = 54°.

Contracts $= \frac{\$300{,}000}{\$2{,}000{,}000} = 0.15$; **15%** of 360° = 54°.

9. Total salary $= 24 \cdot 9000 = \$216{,}000$. New mean $= \frac{\$216{,}000+\$80{,}000}{24+1} = \frac{296{,}000}{25} = \$11{,}840$. The mean was increased by $\$11{,}840 - \$9000 = \mathbf{\$2840}$.

11. **(a)** Life expectancies at birth of males and females (from 1970 through 2002)

Females		Males
	67	1446
	68	28
	69	156
	70	0049
	71	02235578
	72	01145
	73	1689
7	74	123599
9310	75	14
86	76	
88532	77	
9999854332211	78	
99655443210	79	
42	80	
7|74| represents 74.7 years old		|67| 1 represents 67.1 years old

(b) Males: Lower extreme: 67.1
Lower quartile: 70.0
Median: 71.6
Upper quartile: 73.8
Upper extreme: 75.4

Females: Lower extreme: 74.7
Lower quartile: 77.5
Median: 78.45
Upper quartile: 79.5
Upper extreme: 80.4

Life expectancies at birth of males and females (from 1970 through 2002):

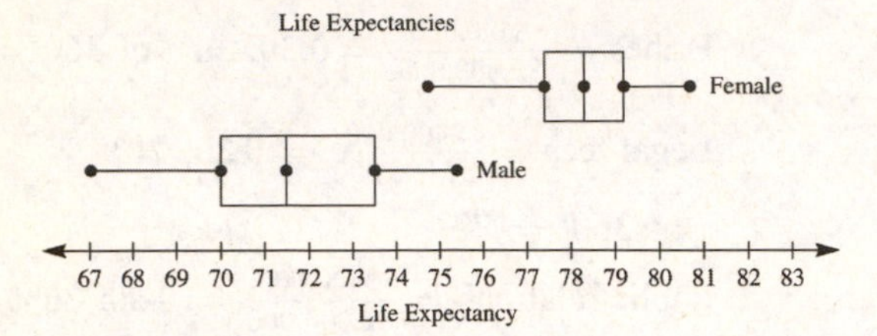

(c) Because the IQR (as well as all other data) is significantly above that of men, women are expected to live longer. Note that the upper extreme for men is only 0.7 years above the lower extreme for women.

13. (a) Ordering the data: 160, 180, 330, 350, 360, 380, 450, 460, 480. Median $=$ **360** yards, the middle data point.

(b) There is **no mode**, since no two or more lengths are the same.

(c) $\overline{x} = \frac{160+180+330+350+360+380+450+460+480}{9} =$ **350** yards.

(d) The sum of the $(x - \overline{x})^2$ terms $= 105{,}400$.

$s = \sqrt{\frac{105{,}400}{9}} \approx \mathbf{108.2}$ yards.

(e) Range $= 480 - 160 =$ **320** yards.

(f) Lower quartile $= \frac{180+330}{2} = 255$;

upper quartile $= \frac{450+460}{2} = 455$.

IQR $= 455 - 255 =$ **200** yards.

(g) Variance $=$ the sum of the $(x - \overline{x})^2$ terms divided by $9 = \frac{105{,}400}{9} = \mathbf{11711.\overline{1}\ yd^2}$.

(h) The sum of the $|x_n - \overline{x}|$ terms $= 760$.

MAD $= \frac{760}{9} = \mathbf{84.\overline{4}}$ yards.

15. (a) **Positive** association. Weights tend to increase as heights increase.

(b) **170 pounds**; i.e., the data point corresponding to 72 inches on the horizontal axis.

(c) **67 inches**; i.e., the data point corresponding to 145 pounds on the vertical axis.

(d) **64 inches**. There are four girls 64 inches tall; more than for any other height.

(e) 170 pounds $-$120 pounds $=$ **50 pounds**.

17. 2.5% of the area under the standard normal curve lies more than two standard deviations above the mean. Thus the probability that a student would score more than 2 standard deviations above the mean $=$ **0.025**.

19. The length of the columns in the bar graph should be approximately the same.

21. The bar graph is more appropriate. Line graphs are used to show change over time. Moreover, the points on the line graph should not be connected since there are no values "between" colors.

23. Answers may vary. One method might be to pull boxes at random, weigh them, and then determine whether some predetermined percentage of the boxes fall within a predetermined weight range.

25. Answers may vary. The three-dimensional graph distorts the data, as does the fact that the vertical axis begins at 110. The graph makes it look as though Tom is twice as heavy as Dick, when in reality Tom is only about 7 lbs heavier than Dick.

27. Answers may vary. Graphs may show area or volume instead of relative size. Another way is to select a horizontal or vertical baseline that will support the point trying to be made.

29. (a) 68 inches is two standard deviations above the mean, and 2.5% of the area under the normal curve lies above 2 standard deviations above the mean. 2.5% of 1000 is **25 girls**.

(b) 60 inches is two standard deviations below the mean. 13.5% + 34% = 47.5% of the area under the normal curve lies between the mean and 2 standard deviations below the mean. 47.5% of 1000 is **475 girls**.

(c) 66 inches is one standard deviation above the mean. 50% of the data lies below the mean and 34% lies between the mean and one standard deviation above the mean. This is 50% + 34% = 84% of the data, so

$100\% - 84\% =$ **16%** is the probability that a girl will be over 66 inches tall.

31. (a) P_{16} represents the bottom 16% of all scores, which in the normal distribution falls one standard deviation below the mean. Thus $P_{16} = 600 - 75 =$ **525**.

(b) D_5 is the decile ranking below which is 50% of the data; i.e., the mean, or **600**.

(c) P_{84} represents the value below which are 84% of the scores, which in the normal distribution falls one standard deviation above the mean. Thus $P_{84} = 600 + 75 =$ **675**.

CHAPTER 11

INTRODUCTORY GEOMETRY

Assessment 11-1A: Basic Notions

1. (a) Different order implies different names; i.e., $\overrightarrow{AB}$, $\overrightarrow{AC}$, $\overrightarrow{AD}$, $\overrightarrow{BC}$, $\overrightarrow{BD}$, $\overrightarrow{CD}$, $\overrightarrow{BA}$, $\overrightarrow{CA}$, $\overrightarrow{DA}$, $\overrightarrow{CB}$, $\overrightarrow{DB}$, $\overrightarrow{DC}$, which is the permutation of four points taken two at a time. ${}_4P_2 = \frac{4!}{(4-2)!} =$ **12 ways.**

(b) Different order implies different names; i.e., a plane is determined by three non-collinear points. The number of different names $=$ ${}_5P_3 = \frac{5!}{(5-3)!} =$ **60 ways.**

3. (a) **None**. The lines are parallel.

(b) **Point *C*.** Three distinct planes intersecting at one common point.

(c) **Point *A*.** The intersection of two non-parallel lines.

(d) Answers may vary. $\overleftrightarrow{AC}$ **and** $\overleftrightarrow{BE}$ or $\overleftrightarrow{AB}$ **and** $\overleftrightarrow{CE}$ are two sets.

(e) $\overleftrightarrow{AC}$ **and** $\overleftrightarrow{DE}$ or $\overleftrightarrow{AD}$ **and** $\overleftrightarrow{CE}$ are parallel; i.e., opposite sides of a rectangle.

(f) **Planes *BCD*** or ***BEA***; i.e., planes bisecting the pyramid.

5. Answers may vary.

(a) Edges of a room; vertical and horizontal parts of a window frame; intersecting crossroads.

(b) Branches in a tree; clock hands at 7:30; angle on a yield sign.

(c) The top of a coat hanger; clock hands at 7:15.

7. (a) **(*i*)** $18°35'29'' + 22°55'41'' = 40°90'70'' = 40°91'10'' =$ **41°31′10″.**

(*ii*) $15°29' - 3°45' = 14°89' - 3°45' =$ **11°44′**

(b) **(*i*)** $0.9° = (0.9 \cdot 60)' =$ **0°54′00″.**

(*ii*) $15.13° = 15° + (0.13 \cdot 60)' = 15°7.8' = 15°7' + (0.8 \cdot 60)'' =$ **15°7′48″.**

9. (a) If $m(\angle AOB) = \frac{1}{3}m(\angle COD)$, then $m(\angle COD) = 3m(\angle AOB)$.

Let $x = m(\angle AOB)$. Then $x + 90° + 3x = 180°$ $\Rightarrow 4x = 90° \Rightarrow x = 22.5°$.
$m(\angle AOB) =$ **22.5°**; $m(\angle COD) = 3(22.5°) =$ **67.5°.**

(b) Let $x = m(\angle BOC)$. Then $x + (3x - 35°) = 90°$ $\Rightarrow 4x = 125° \Rightarrow x = 31.25°$.
$m(\angle BOC) =$ **31.25°.**
$m(\angle AOB) = 3(31.25°) - 35° =$ **58.75°.**

(c) Assume that if the position of $\overline{OE}$ is changed, the other rays will be adjusted so that all x's are congruent and all y's are congruent. $3x + 3y$ will have different values, and thus x and y will have different values. $3x + 3y$, though, will remain $180° \Rightarrow 3(x + y) = 180°$ $\Rightarrow x + y = 60° \Rightarrow m(\angle BOC) =$ **60°.**

11. (a)

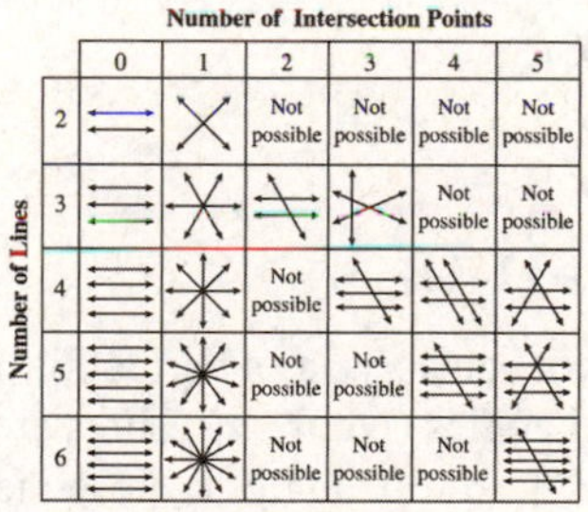

Number of Intersection Points

Number of Lines	0	1	2	3	4	5
2			Not possible	Not possible	Not possible	Not possible
3					Not possible	Not possible
4			Not possible			
5			Not possible	Not possible		
6			Not possible	Not possible	Not possible	

(b) The maximum number of intersections is the combination of n lines taken two at a time, or

$${}_nC_2 = \frac{n!}{2!(n-2)!} = \frac{n(n-1)(n-2)(n-3)\cdots}{2(n-2)(n-3)\cdots} = \mathbf{\frac{n(n-1)}{2}}.$$

13. Answers may vary; e.g.,

(a) A real-world example would be a paddle wheel:

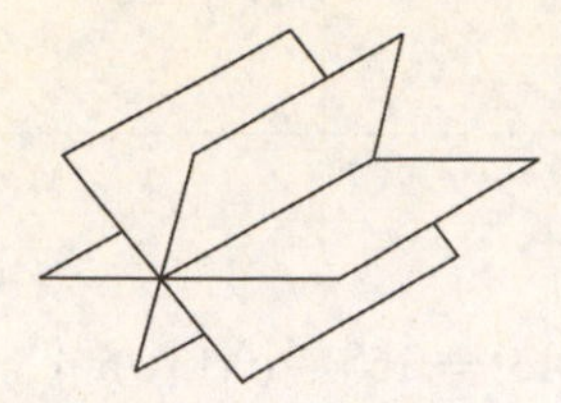

(b) A real-world example would be the intersection of floor and two adjacent walls of a house:

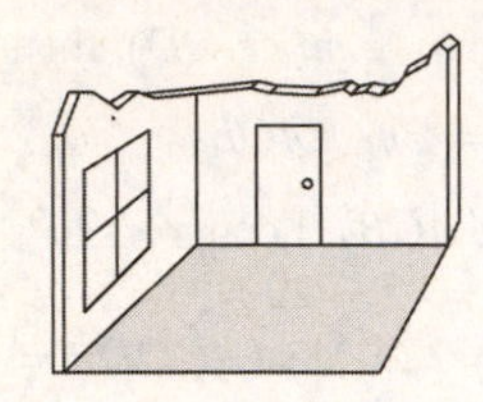

(c) A real-world example would be a field irrigated with a central pivot:

Field Irrigates with a Central Pivot

15. **(a)** $\overset{\frown}{BC}$ is the arc associated with the central angle $\angle BOC$, whose measure is 70°. Thus, $m(\overset{\frown}{BC}) = \mathbf{70°}$.

(b) $m(\overset{\frown}{CD}) = m(\angle DOC) = \mathbf{110°}$.

(c) $m(\overset{\frown}{AD}) = m(\angle AOD) = \mathbf{70°}$.

(d) **110°**.

(e) **180°**.

Assessment 11-1B

1. Each point can serve as a vertex for 6 angles, 3 normal angles (whose measures are less than 180° and 3 reflex), angles (whose measures are greater than 180°) by creating line segments from the vertex to the remaining points.

For example:

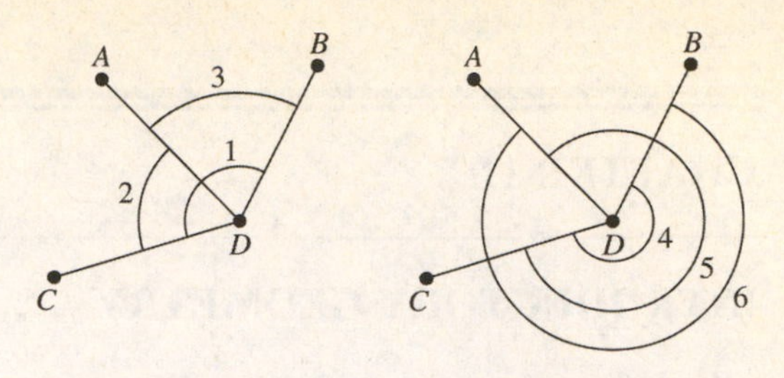

So there are 4 possible vertices times 6 angles for each vertex, and thus **24** distinct angles.
Normal: $\angle ADC, \angle BDC, \angle ADB, \angle ACB, \angle ACD, \angle BCD, \angle ABD, \angle ABC, \angle BAC, \angle BAD, \angle CAD$.
Reflex: Same list as normal.

3. **(a)** $\overline{AD}$ and $\overline{BE}$ do not intersect. Their intersection is $\boldsymbol{\emptyset}$.

(b) All points lying on $\overline{AD}$ are in this intersection. Since the planes determined are infinite, $\overleftrightarrow{AD}$ represent their intersection.

(c) $\{C\}$, a single point.

(d) Answers vary. $\overline{BE}$ and $\overleftrightarrow{CD}$ is one example.

(e) None of the lines in the drawing are parallel.

(f) Answers vary. The plane determined by *BED* is one example.

5. Answers may vary.

(a) Railroad tracks.

(b) The corners of walls in your home.

(c) Two opposite facing walls in your home.

7. **(a)** *(i)* $21°35'31'' + 49°51'32'' = 70°86'63'' = 70°87'3'' = \mathbf{71°27'3''}$.

(ii) $93°38'14'' - 13°49'27'' = 93°37'(14 + 60)'' - 13°49'27'' = 92°(37 + 60)'74'' - 13°49'27'' = \mathbf{79°48'47''}$.

(b) *(i)* $10.3° = 10° + (0.3 \cdot 60)' = \mathbf{10°18'}$.

(ii) $15.14° = 15° + (0.14 \cdot 60)' = 15°8.4' = 15°8' + (0.4 \cdot 60)'' = \mathbf{15°8'24''}$.

9. **(a)** Let $x = m(\angle BOA)$. Since $m(\angle DOC) = \frac{3}{4}m(\angle BOA)$ then $x + 120° + \frac{3}{4}x = 180° \Rightarrow \frac{7}{4}x = 60° \Rightarrow x = 34\frac{2}{7}°$.

$m(\angle DOC) = \frac{3}{4}\left(34\frac{2}{7}°\right) = \mathbf{25\frac{5}{7}°} \approx 25.71°.$

$m(\angle BOA) = \mathbf{34\frac{2}{7}°} \approx 34.29°.$

(b) Let $x = m(\angle BOC)$. Then $x + (2x - 30°) = 90° \Rightarrow 3x = 120° \Rightarrow x = 40°.$

$m(\angle BOC) = \mathbf{40°}.$

$m(\angle AOB) = 90° - 40° = \mathbf{50°}.$

(c) Let $x = m(\angle AOB)$ and $y = m(\angle BOC)$.

Then
$$\begin{aligned} x - y &= 50 \\ x + y &= 180 \\ \hline 2x &= 230 \end{aligned}$$

Or $x = 115; y = 180 - 115 = 65.$

$m(\angle AOB) = \mathbf{115°}$

$m(\angle BOC) = \mathbf{65°}.$

11. **(a)** Construction: $\angle EAF = 90°.$

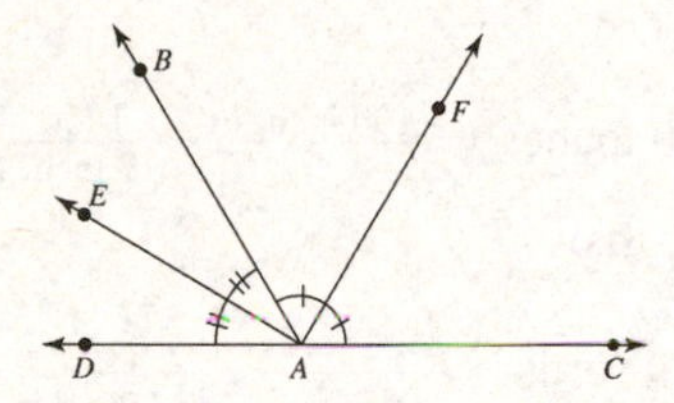

(b) Conjecture: $\angle EAF$ will always be 90°.

(c) Let the measures of the congruent angles in each pair be x and y, respectively. Then $2x + 2y = 180° \Rightarrow x + y = 90°$, thus $m(\angle EAF) = \mathbf{90°}.$

13. **(a)**

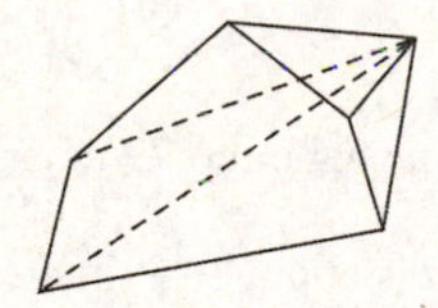

(b)

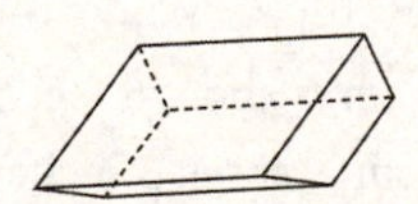

15. **(a)** $\angle ACO$ is an angle in an isosceles triangle AOC. Its measure is equivalent to $\angle CAO$, which is **35°**.

(b) $m(\widehat{AC}) = m(\angle AOC) = \mathbf{110°}.$

(c) $m(\widehat{CB}) = m(\angle COB) = \mathbf{70°}.$

(d) $m(\widehat{AB}) = m(\angle AOB) = \mathbf{180°}.$

(e) $m(\widehat{CBA}) = m(\angle COA) = 360° - 110° = \mathbf{250°}.$

Assessment 11-2A: Linear Measure

1. **(a)** $AB = 1.0$ cm $- 0$ cm $=$ **1.0 cm.**

(b) $DE = 4.5$ cm $- 3.5$ cm $=$ **1.0 cm.**

(c) $CJ = 10.0$ cm $- 2.0$ cm $=$ **8.0 cm.**

(d) $EF = 5.0$ cm $- 4.5$ cm $=$ **0.5 cm.**

(e) $IJ = 10.0$ cm $- 9.3$ cm $=$ **0.7 cm.**

(f) $AF = 5.0$ cm $- 0$ cm $=$ **5.0 cm.**

(g) $IC = 9.3$ cm $- 2.0$ cm $=$ **7.3 cm.**

(h) $GB = 6.2$ cm $- 1.0$ cm $=$ **5.2 cm.**

3. **(a)** $100 \text{ inches} = \frac{100 \text{ inches}}{1} \times \frac{1 \text{ yard}}{36 \text{ inches}} = \frac{25}{9} = \mathbf{2\frac{7}{9}}$ **yd.**

(b) $400 \text{ yards} = \frac{400 \text{ yards}}{1} \times \frac{36 \text{ inches}}{1 \text{ yard}} = \mathbf{14{,}400}$ **in.**

(c) $300 \text{ feet} = \frac{300 \text{ feet}}{1} \times \frac{1 \text{ yard}}{3 \text{ feet}} = \mathbf{100}$ **yd.**

(d) $372 \text{ inches} = \frac{372 \text{ inches}}{1} \times \frac{1 \text{ foot}}{12 \text{ inches}} = \mathbf{31}$ **ft.**

5. Answers may vary depending on your estimate; e.g.,

(a) About **65 mm**. **(b)** About **6.5 cm**.

7. **(a)** **Inches**. 19 cm is about 7.5 inches.

(b) **Inches**. 21 mm is about 0.8 or $\frac{13}{16}$ inches.

(c) **Feet**. 1.2 m is about 3 feet 11 inches.

9. **(a)** **13.50 mm**. 10 mm is about 0.4 inch.

(b) **0.770 m**. 0.77 m is about 30 inches.

(c) **10.0 m**. 10 m is about 33 feet.

(d) **15.5 cm**. 15.5 cm is about 6 inches.

11. Answers may vary; e.g.: A circle with radius about $\frac{5}{8}$ inch. $\left(r = \frac{C}{2\pi} = \frac{2}{\pi} \approx \frac{5}{8}.\right)$

13. Listing lengths of sides starting clockwise from the top of each figure (measurements are approximate):

(a) About $1.65 + 1.65 + 1.65 + 1.65 = \mathbf{6.6\ cm}$.

(b) About $3.3 + 0.7 + 1.6 + 0.9 + 1.7 + 1.7 = \mathbf{9.9\ cm}$.

15. Consider the following triangle, in which $AB \approx 20$ mm. $BC \approx 24$ mm, and $AC \approx 17$ mm. The sum of the lengths of any two sides of the triangle is greater than the length of the third side:

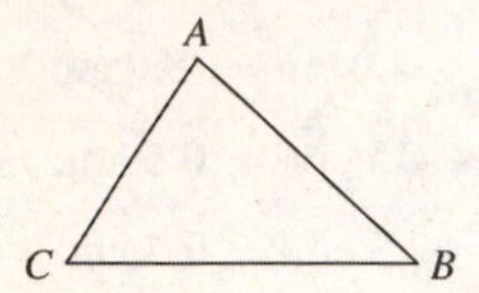

(a) $\mathbf{AB + BC} \approx 44 \text{ mm} > \mathbf{AC} \approx 17 \text{ mm}$.

(b) $\mathbf{BC + AC} \approx 41 \text{ mm} > \mathbf{AB} \approx 20 \text{ mm}$.

(c) $\mathbf{AB + AC} \approx 37 \text{ mm} > \mathbf{BC} \approx 24 \text{ mm}$.

17. The hypotenuses of the resultant right triangles are $\sqrt{4.25^2 + 11^2} \approx \mathbf{11.8\ inches}$.

(a) Form an isosceles triangle with the two congruent sides formed by the diagonals and the base formed by the two short $\left(4\frac{1}{4} \text{ inch}\right)$ sides. The perimeter is approximately **32.1 inches**.

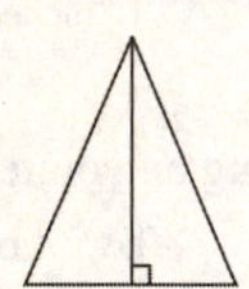

(b) Form the base with the two long (11 inch) sides. The perimeter is about **45.6 inches**.

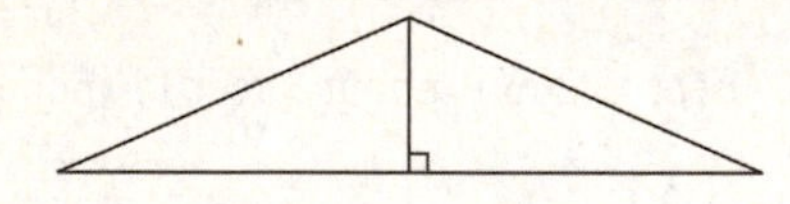

19. Circumference $(C) = \pi \cdot \text{diameter } (d)$ or $C = 2\pi \cdot \text{radius } (r)$.

(a) $C = \pi \cdot 6 = \mathbf{6\pi\ cm}$.

(b) $C = 2\pi \cdot \frac{2}{\pi} = \mathbf{4\ cm}$.

21. **(a)** Mach $2.5 \times \frac{0.344 \text{ km}}{\text{sec}} \times \frac{3600 \text{ sec}}{\text{hr}} =$ **3096 km/hr.**

(b) Mach $3 \times \frac{344 \text{ m}}{\text{sec}} = \mathbf{1032\ m/sec}$.

(c) $\text{M} = \frac{\text{speed of aircraft}}{\text{speed of sound}} = \frac{\frac{5000 \text{ km}}{\text{hr}}}{\frac{0.344 \text{ km}}{\text{sec}} \cdot \frac{3600 \text{ sec}}{\text{hr}}}$

$\approx$ **Mach 4.04.**

Assessment 11-2B

1. **(a)** $AB = 1 \text{ in.} - 0 \text{ in.} = \mathbf{1\ in.}$

(b) $DE = 3\frac{1}{4} \text{ in.} - 2\frac{1}{2} \text{ in.} = \mathbf{\frac{3}{4}\ in.}$

(c) $CJ = 6 \text{ in.} - 2 \text{ in.} = \mathbf{4\ in.}$

(d) $EF = 4 \text{ in.} - 3\frac{1}{4} \text{ in.} = \mathbf{\frac{3}{4}\ in.}$

(e) $IJ = 6 \text{ in.} - 5\frac{1}{2} \text{ in.} = \mathbf{\frac{1}{2}\ in.}$

(f) $AF = 4 \text{ in.} - 0 \text{ in.} = \mathbf{4\ in.}$

(g) $IC = 5\frac{1}{2} \text{ in.} - 2 \text{ in.} = \mathbf{3\frac{1}{2}\ in.}$

(h) $GB = 4\frac{5}{8} \text{ in.} - 1 \text{ in.} = \mathbf{3\frac{5}{8}\ in.}$

3. **(a)** $100 \text{ inches} = \frac{100 \text{ inches}}{1} \times \frac{1 \text{ foot}}{12 \text{ inches}} = \frac{50}{6} =$ $\mathbf{8\frac{1}{3}\ ft.}$

(b) $400 \text{ yards} = \frac{400 \text{ yards}}{1} \times \frac{3 \text{ feet}}{1 \text{ yard}} = \mathbf{1200\ ft.}$

(c) $300 \text{ feet} = \frac{300 \text{ feet}}{1} \times \frac{12 \text{ inches}}{1 \text{ foot}} = \mathbf{3600\ in.}$

(d) $372 \text{ inches} = \frac{372 \text{ inches}}{1} \times \frac{1 \text{ yard}}{36 \text{ inches}} =$ $\mathbf{10\frac{1}{3}\ yd.}$

5. Answers may vary depending on your estimate; e.g.,

(a) about **40 mm**. **(b)** about **4 cm**.

7. **(a)** **Inches**. 20 mm is about 0.79 or $\frac{25}{32}$ inches.

(b) **Inches**. 25 cm is about $9\frac{13}{16}$ inches.

(c) **Feet**. 1.9 m is about 6 feet 3 inches.

9. **(a)** **195.0 cm**. 195 cm is about 6 feet 5 inches.

(b) **8.100 cm**. 8.1 cm is about 3 inches.

(c) **40.0 km/hr**. 40 km/hr is about 25 mi/hr.

11. Answers may vary; e.g.:

(a) A triangle with 1, $1\frac{1}{2}$, and $1\frac{1}{2}$ inch sides.

(b) A concave kite with 3, 3, 1, and 1 cm sides.

13. Listing lengths of sides starting clockwise from the top of each figure (measurements are approximate):

(a) About $3 + 3 + 3 =$ **9 cm**.

(b) About $4 + 1 + 3 + 1 + 3 + 1 + 4 + 3 =$ **20 cm**.

15. A casual explanation is that the shortest distance between two points is a straight line. But this response assumes what is to be argued or explained. For a better explanation places four points A, B, C, and D so that C and D are not collinear with A and B and that AC plus BD is less than AB.

Assume the segments are hinged at A and B. Push AC and BD down toward AB and swing the segments to note that C and D never meet. Thus, a triangle is never formed nor is a line segment.

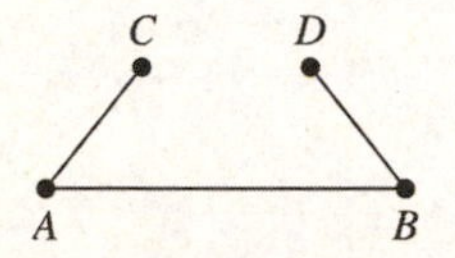

17. **(a)** Answers may vary. One way would be to form a seven-square by two-square rectangle; thus $7 + 2 + 7 + 2 = 18$; another would be to add four squares to the end of the top row.

(b) **Eight**. The minimum number of squares for a fixed perimeter occurs when the squares have the maximum number of sides exposed, such as in the figure below.

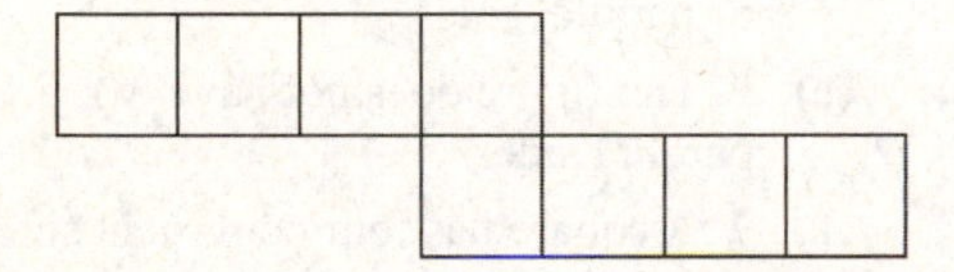

(c) **Twenty**. The maximum number of squares for a fixed perimeter occurs when the squares have the minimum number of sides exposed. I.e., from a four-squares by five-square rectangle, where the perimeter is $5 + 4 + 5 + 4 = 18$.

19. Circumference $(C) = \pi \cdot$ diameter (d) or $C = 2\pi \cdot$ radius (r).

(a) $C = 2\pi \cdot 3 = \mathbf{6\pi}$ **cm**.

(b) $C = \pi \cdot 6\pi = \mathbf{6\pi^2}$ **cm**.

21. **(a)** $\frac{300{,}000 \text{ km}}{\text{sec}} \times \frac{60 \text{ sec}}{\text{min}} \times \frac{60 \text{ min}}{\text{hr}} \times \frac{24 \text{ hr}}{\text{day}} \times \frac{365 \text{ days}}{\text{yr}} \approx \mathbf{9.5 \cdot 10^{12}}$ **km**.

(b) 4.34 light years $\times\ 9.5 \cdot 10^{12}$ km/yr $\approx$ $\mathbf{4.1 \cdot 10^{13}}$ **km**.

(c) $\frac{4.1 \cdot 10^{13} \text{ km}}{6 \cdot 10^{4} \text{ km/hr}} \approx 6.8 \cdot 10^{8}$ hours, or **about 78,000 years** at 8760 hours per year.

(d) Light travels $(8 \cdot 60 + 19)$ sec $\times$ 300,000 km/sec $= 1.497 \cdot 10^{8}$ km in 8 minutes 19 sec. $\frac{1.497 \cdot 10^{8} \text{ km}}{6 \cdot 10^{4} \text{ km/hr}} \approx 2495$ hours, or **about 104 days**.

11-2 Review Problems

9. **Yes**; see the drawing

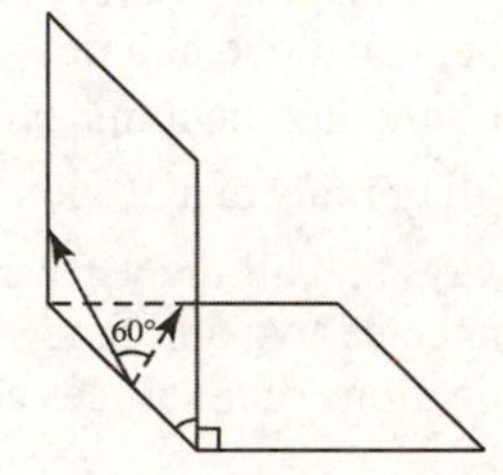

The stipulation that forces planes to be perpendicular is that rays in either plane $\perp$ to the planes' line of intersection are $\perp$. The rays we drew in the figure above are not $\perp$ to the line of intersection.

11. Any two intersecting lines determine a plane and perpendicular lines are particular intersecting lines.

Assessment 11-3A
Curves, Polygons, and Symmetry

1. **(a)** **1, 4, 6, 7, 8**. A simple closed curve does not cross itself and begins and ends at the same point.

(b) **1, 6, 7, 8**. Polygons are polygonal curves (i.e., made up entirely of line segments) which are both simple and closed.

(c) **6, 7**. All segments connecting any two points of the polygon are inside the polygon; i.e., the region is nowhere dented inwards.

(d) **1, 8**. It is possible to draw a segment between two points of the polygon such that part of the segment lies outside the polygon.

3. A **concave polygon**. The polygon is "caved in" somewhere.

5. **Square**. For a shape to be in the shaded region it must be both a rectangle and a kite. Squares are rectangles with two adjacent sides congruent, which is a property of a kite.

7. (a) The number of ways that all the vertices in a pentagon can be connected two at a time is the number of combinations of five vertices chosen two at a time, or ${}_5C_2 = 10$. This number of segments includes both diagonals and sides. Subtracting the number of sides, 5, from 10 yields **5 diagonals** in a pentagon.

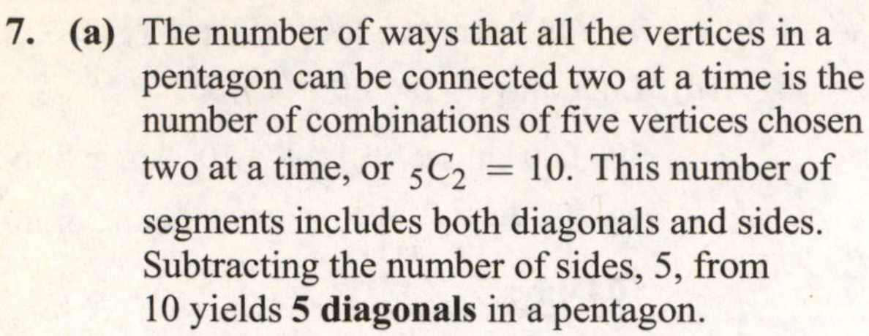

(b) The number of ways that all the vertices in a decagon can be connected two at a time is ${}_{10}C_2 = 45$. Subtracting the number of sides, 10, from 45 yields **35 diagonals** in a decagon.

(c) The number of ways that the vertices in a 20-gon can be connected two at a time is ${}_{20}C_2 = 190$. Subtracting the number of sides yields **170 diagonals** in a 20-gon.

(d) The number of ways that all the vertices in an n-gon can be connected two at a time is the number of combinations of n vertices chosen two at a time, or ${}_nC_2 = \frac{n!}{(n-2)!2!} = \frac{n(n-1)}{2}$.

This number of segments includes both the number of diagonals and the number of sides. Subtracting the number of sides n from $\frac{n(n-1)}{2}$ yields $\frac{n(n-1)}{2} - n = \frac{n(n-3)}{2}$.

9. (a) **Equilateral** and **isosceles**; three congruent sides.

(b) **Isosceles**; two congruent sides.

(c) **Scalene**; no sides congruent.

11. Answers vary.

(a)

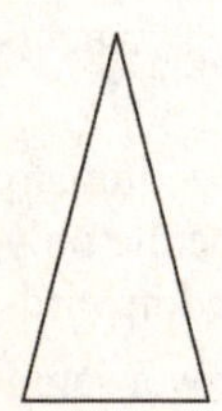

(b)

(c) The letter **N** has 180° rotational symmetry and no line of symmetry.

13.

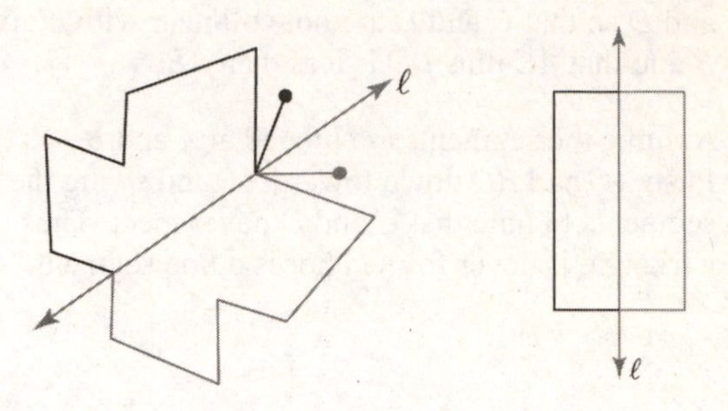

Assessment 11-3B

1. (a) **1** and **2**. Both are triangles with at least two congruent sides.

(b) **1**. 2 is isosceles but is also equilateral.

(c) **None**. By definition, if a triangle is equilateral it has three congruent sides; an isosceles triangle has at least two congruent sides.

(d) **None**. A parallelogram has two pairs of parallel sides; a trapezoid has at least one pair of parallel sides.

(e) **5**. The figure does not have two pairs of parallel sides.

(f) **7**. A square has four congruent sides.

(g) **None**. A square has four right angles, the definition of a rectangle.

(h) **None**. A square has two pairs of parallel sides; a trapezoid has at least one pair of parallel sides.

(i) **None**. A rhombus has four congruent sides; a kite has two adjacent congruent sides with the other two sides also congruent.

(j) **4** and **6**. Both have four congruent sides.

(k) **4, 6, 8,** and **9**. All have two adjacent congruent sides with the other two sides also congruent.

3. (a) There are three of the six vertices from which a diagonal cannot be drawn to any given vertex; i.e., itself and the vertices at the ends of adjacent sides. Thus **three** can be drawn.

(b) There are three of the ten vertices from which a diagonal cannot be drawn, thus **seven** can be drawn.

(c) There are three of the twenty vertices from which a diagonal cannot be drawn, thus **17** can be drawn.

(d) In any n-gon there will be three diagonals which cannot be drawn. Thus three will be $\boldsymbol{n - 3}$ which can.

5. Rectangles have four right angles while rhombuses have four congruent sides; both are parallelograms. Their intersection is therefore a **square**.

7. (a) A hexagon has six vertices and each vertex is contained in three distinct diagonals. (The vertex itself and the two consecutive vertices are excluded from forming diagonals.) This method has counted each diagonal twice, so we must divide by 2. Thus, the number of diagonals in a hexagon is $6 \cdot 3/2 = \mathbf{9}$.

(b) An 11-gon has 11 vertices, each of which can be connected to 8 non-consecutive vertices by diagonals. Again, we have counted each diagonal twice. Thus, we have $11 \cdot 8/2 =$ **44** diagonals.

(c) The key idea above is that n vertices can be connected to $n - 3$ non-consecutive vertices to form diagonals and that this method counts each vertex twice. Thus, in general, an n-gon has $n(n - 3)/2$ diagonals. Therefore, a 18-gon has $18(15)/2 =$ **135** diagonals.

9. (a) equiangular and acute.

(b) Obtuse

(c) right

11. (a) 4 lines of symmetry; 90° and 270° turn symmetry; and point symmetry, which is also 180° turn symmetry.

(b) Point symmetry, which is also 180° turn symmetry.

(c) Point symmetry; 90° and 270° turn symmetry.

(d) 4 lines of symmetry; 90° and 270° turn symmetry; and point symmetry.

13.

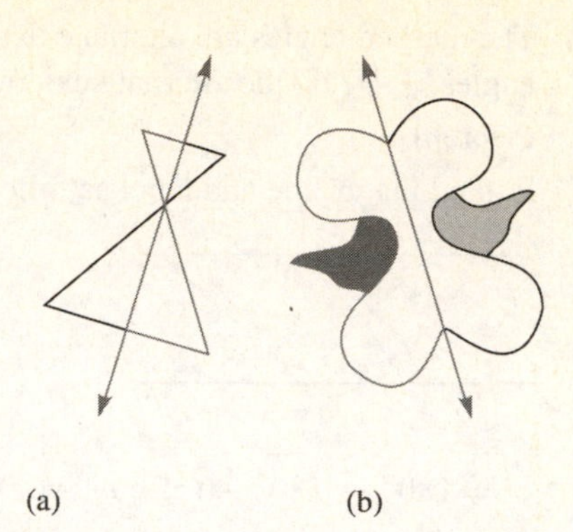

(a) (b)

Review Problems

13. **∅** if no intersection. **One**, if a ray of an angle is contained in a line. **Two**, if the line intersects both sides.
A **ray**, if the line starts at the vertex.

15. (a) $7 \text{ mm} \cdot \frac{1 \text{ meter}}{1000 \text{ mm}} = \mathbf{0.007\ m}$

(b) $17 \text{ in.} \cdot \frac{1 \text{ yard}}{36 \text{ in}} = \frac{17}{36} \text{yd} = \mathbf{0.47\overline{2}\ yd}$

(c) $4 \text{ m} \cdot \frac{100 \text{ cm}}{1 \text{ m}} = \mathbf{400\ cm.}$

(d) $1.7 \text{ yd} \cdot \frac{36 \text{ in}}{\text{yd}} = \mathbf{61.2\ in.}$

Assessment 11-4A More About Angles

1. Every pair of lines forms two pairs of vertical angles. Two times the number of combinations of three lines taken two at a time $= 2 \cdot {}_3C_2 = 2 \cdot \frac{3!}{(3-2)!2!} =$
$2 \cdot \frac{3 \cdot 2 \cdot 1}{1 \cdot (2 \cdot 1)} = 2 \cdot 3 =$ **6 pairs** of vertical angles.

3. (a) **Yes**. A pair of corresponding angles are 50° each (note that the supplementary angles formed by lines n and ℓ are 130° and 50°).

(b) **Yes**. A pair of corresponding angles are 70° each (note that the supplementary angles formed by lines n and ℓ are 110° and 70°).

(c) **Yes**. A pair of alternate interior angles are 40° each.

(d) **Yes**. A pair of corresponding angles are 90° each.

(e) The marked angles are alternate exterior angles. $k \parallel l$ by the alternate exterior angle theorem.

(f) Extend the *yx* line and label as follows:

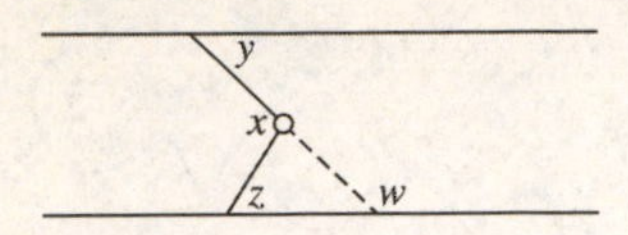

$x = 180° - [180° - (z + m\angle xwz)]$

$= z + m\angle xwz$. It is given that $x = y + z$, so $z + m\angle xwz = y + z \Rightarrow y = m\angle xwz$.

Because $\angle xwz$ and y are measures of alternate interior angles, $k \parallel l$.

5. Each exterior angle $= 180° - 162° = 18°$. Since the sum of the measurements of the exterior angles $= 360°$, then $360° \div 18° = 20$ angles, thus **20 sides**.

7. The general form of an arithmetic sequence is $a_n = (n-1)d + a_1$. Since $a_6 = 130°$, we have that $130° = 5d + a_1$. Since the sum of the interior angles of a hexagon is 720°, we have that $a_1 + (d + a_1) + (2d + a_1) + (3d + a_1) + (4d + a_1) + 130° = 720°$. Simplifying, $5a_1 + 10d = 590°$. Combining this information:

$$\begin{aligned} 5a_1 + 10d &= 590° \\ -2a_1 - 10d &= 260° \\ \hline 3a_1 \qquad &= 330° \Rightarrow a_1 = 110°. \end{aligned}$$

Thus, $130° = 5d + 110° \Rightarrow 5d = 20° \Rightarrow d = 4$. Therefore, the angles are as follows: **110°; 114°; 118°; 122°; 126°; 130°.**

9. **(a)** $x =$ **40°** (congruent vertical angles).

(b) $x + 4x = 90°$ (complementary angles). $5x = 90 \Rightarrow x =$ **18°**.

(c) Let $y = m(\angle ABC)$ and $z = m(\angle BAC)$. Then $180° - 3y = 3z$ (alternate interior angles) $\Rightarrow 3y + 3z = 180° \Rightarrow y + z = 60°$. Since $x + y + z = 180° \Rightarrow x + 60° = 180°$ $\Rightarrow \mathbf{x = 120°}$.

11. The six angles surrounding the center point sum to 360°. They can be compared to three pairs of vertical angles with the angles contained by triangles equal to those not contained.

There are two of each angle making up the circle around the intersection point (the included angle and its vertical angle). Since the included angle and its vertical angle are always congruent, the contained angles must sum to half of $360° = 180°$.

The sum of the angles in the three triangles is $3 \cdot 180° = 540°$. Thus the numbered angles sum to $540° - 180° =$ **360°**.

13. ***(i)*** $m(\angle APT) = 90°$. $m(\angle 1) = 90° - 30° =$ **60°**.

(ii) $m(\angle 2) =$ **30°** (alternate interior angles).

(iii) In ΔAPR, $m(\angle 3) = 180° - (30° + 40°)$ $=$ **110°**.

15. If $y = 2x$, then $x = \frac{1}{2}y$. If $y = \frac{1}{2}z$, then $z = 2y$. If $y = \frac{1}{3}w$, then $w = 3y$.

Thus $\frac{1}{2}y + y + 2y + 3y = 180° \Rightarrow \frac{13}{2}y = 180° \Rightarrow$

$y = \frac{360°}{13} = 27\frac{9°}{13} \approx \mathbf{27.69°}$;

$x = \frac{1}{2}\left(\frac{360}{13}\right)° = \frac{360°}{26} = 13\frac{11°}{13} \approx \mathbf{13.85°}$;

$z = 2\left(\frac{360}{13}\right)° = \frac{720°}{13} = 55\frac{5°}{13} \approx \mathbf{55.38°}$; and

$w = 3\left(\frac{360}{13}\right)° = \frac{1080°}{13} = 83\frac{1°}{13} \approx \mathbf{83.08°}$.

Assessment 11-4B

1. Answers vary. Vertical angles are created by intersecting lines and are the pair of angles whose sides are two pairs of opposite rays. Dihedral angles are formed by two rays, one from each of two intersecting half-planes, perpendicular to the line of intersection. If we extend these half-planes by adjoining their "opposite" half planes and corresponding rays, we create vertical angles as shown below:

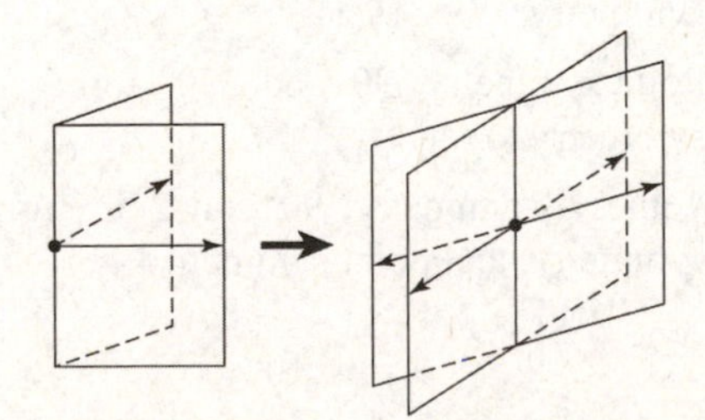

3. **(a)** Label the angle supplementary to y as x_1.

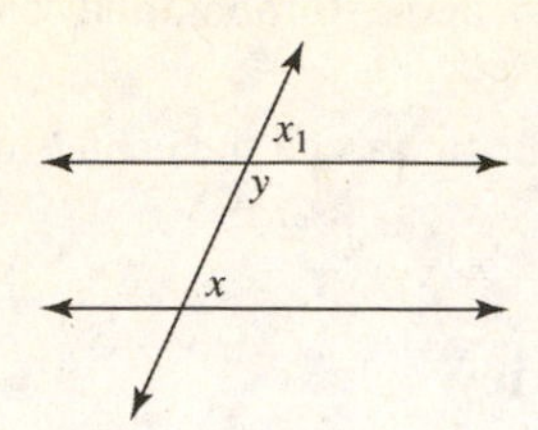

Then $x_1 + y = 180°$, and since $x + y = 180°$ it follows that $x = x_1$. Since x and x_1 are measures of corresponding angles, $k \parallel l$ and $k \parallel \ell$ if and only if $x = x_1$.

(b) Extend the yz line and label as follows:

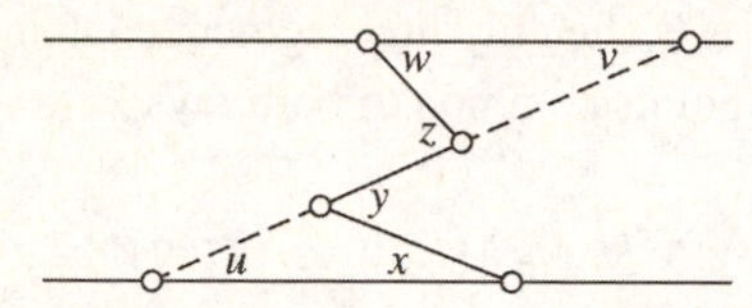

Observe that $z = w + v$ and $y = u + x$. Thus $u = y - x$ and $v = z - w$. It is given that $y = x + z - w$, so $y - x = z - w$. Therefore $u = y - x = z - w = v$. Since u and v are measures of alternate interior angles, $k \parallel l$, and $k \parallel \ell$ if and only if $u = v$.

(c) Extend the 30° ray from line a to line b; a triangle is formed. The interior angles are 20°, 30° (alternate interior angles), and 130° (total of 180° interior angles). x and 130° are supplementary, thus $\boldsymbol{x = 50°}$.

(d) Extend the 135° ray from line b to line a; a triangle is formed. The interior angles are 45° (supplement of 135°), 40° (supplement of 140°), and 95° (total of 180° interior angles). x and 95° are supplements, thus $\boldsymbol{x = 85°}$.

5. $m(\angle 1) + m(\angle 3) + m(\angle 5) = 180°$. Likewise, $m(\angle 2) + m(\angle 4) + m(\angle 6) = 180°$. $m(\text{sum of marked angles}) = 180° + 180° = \mathbf{360°}$.

7. The sum of the measures of the interior angles of a pentagon is $(5 - 2) \cdot 180° = 540°$.

The angles are $60°, 60° + d°, 60° + 2d°, 60° + 3d°$, and $60° + 4d° \Rightarrow 5 \cdot 60° + 10d° = 540° \Rightarrow d = 24°$.

The angles are **60°, 84°, 108°, 132°,** and **156°**.

9. **(a)** If $y = 3x$, then because it is an isosceles triangle, $x + 2(180° - 3x) = 180° \Rightarrow$

$^-5x = {}^-180°$. $\boldsymbol{x = 36°}$; $\boldsymbol{y} = 3(36°) = \mathbf{108°}$.

(b) One interior angle is $180° - 115° = 65°$. x is an exterior angle to the triangle, so $\boldsymbol{x} = 55° + 65° = \mathbf{120°}$.

(c) The total of the measures of the angles in a hexagon is $(n - 2) \cdot 180° = 720°$. Then starting with x, where $v = 136°$, $(136° - 4d) + (136° - 3d) + (136° - 2d) + (136° - d) + 136° + (136° + d) = 720° \Rightarrow 6(136°) - 10d + d = 720° \Rightarrow {}^-9d = {}^-96° \Rightarrow d = 10\frac{2}{3}°$.

So $x = 136° - 4\left(10\frac{2}{3}°\right) = 93\frac{1}{3}° = \mathbf{93°20'}$;

$y = 136° - 3\left(10\frac{2}{3}°\right) = \mathbf{104°}$;

$z = 136° - 2\left(10\frac{2}{3}°\right) = 114\frac{2}{3}° = \mathbf{114°40'}$;

$u = 136° - 10\frac{2}{3}° = 125\frac{1}{3}° = \mathbf{125°20'}$;

$v = \mathbf{136°}$; and

$w = 136° + 10\frac{2}{3}° = 146\frac{2}{3}° = \mathbf{146°40'}$.

11. The measure of each interior angle of a regular octagon $= \frac{(8-2) \cdot 180°}{8} = 135°$. The triangle has two equal interior angles of $180° - 135° = 45°$ (i. e., supplementary angles). Thus $m(\angle 1) = 180° - (2 \cdot 45°) = \mathbf{90°}$.

13. Let $m(\angle B) = \beta$ and $m(\angle C) = \gamma$. We have that $m(\angle D) = 180° - \beta - \gamma$ and that $\alpha = m(\angle A) = 180° - (180° - 2\beta) - (180° - 2\gamma) = 2\beta + 2\gamma - 180°$. This last equation can be written as $\alpha = 2(\beta + \gamma - 180°) + 180°$. Thus,

$\alpha = 2({}^-m(\angle D)) + 180° \Rightarrow \alpha - 180° = {}^-2m(\angle D)$

$\Rightarrow m(\angle D) = \frac{\alpha - 180°}{^-2} = \mathbf{90°} - \boldsymbol{\frac{\alpha}{2}}$.

15. **(a)** ***(i)*** The third angle in the triangle with two angles of 30° and 70° is $180° - (30° + 70°) = 80°$. x is a vertical angle, so $\boldsymbol{x = 80°}$.

(ii) The third angle in the triangle with angles of 60° and 70° is 50°. The angle supplementary to x is 100°. y is an exterior angle, thus $y = 50° + 100° = \mathbf{150°}$.

(b) If $\frac{x}{y} = \frac{3}{4}$, then $x = \frac{3}{4}y$. If $\frac{y}{z} = \frac{4}{5}$, then $z = \frac{5}{4}y$. Thus $\frac{3}{4}y + y + \frac{5}{4}y = 180°$ $\Rightarrow 3y = 180°$. $y = \mathbf{60°}$; $x = \frac{3}{4}(60°) = \mathbf{45°}$; $z = \frac{5}{4}(60°) = \mathbf{75°}$.

Review problems

17. (a) (*i*) Two sets of parallel sides: *A, B, C, D, E, F, G.*

(*ii*) One set of parallel sides: *I, J.*

(*iii*) No parallel lines: *H.*

(b) (*i*) Four right angles: *D, F, G.*

(*ii*) Two right angles: *L.*

(*iii*) No right angles: *A, B, C, E, H, J.*

(c) (*i*) Four congruent sides: *B, C, F,G.*

(*ii*) Two pairs of congruent sides: *A, D, E, H.*

(*iii*) One pair of congruent sides: *J.*

(*iv*) No congruent sides: *I.*

19. Answers vary.

(a) The figure has 180° turn symmetry but no line symmetry.

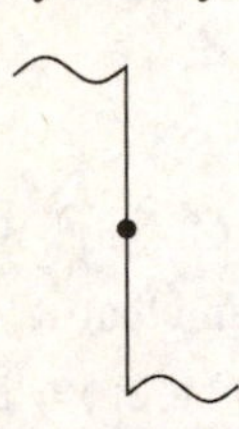

(b) The figure has point symmetry, 90° turn symmetry, and vertical and horizontal line symmetry.

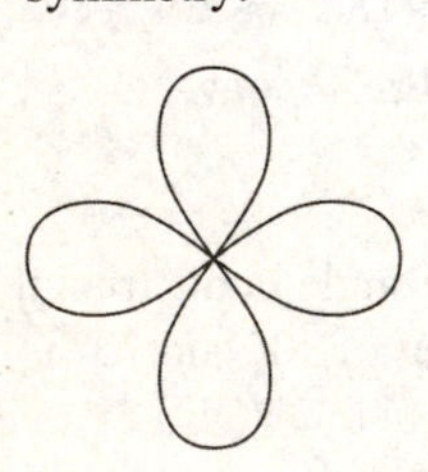

21. (a) Turn symmetries of 90°, 180° (also called point symmetry), and 270° about the center of the square.

(b) Point symmetry about the center of the square.

Chapter 11 Review

1. (a) $\overleftrightarrow{AB}, \overleftrightarrow{BC}, \overleftrightarrow{AC}$. One line can be drawn through any two given points.

(b) $\overrightarrow{BA}$ and $\overrightarrow{BC}$. A ray contains one endpoint and all points on the line on one side of the endpoint.

(c) $\overline{AB}$, the line segment between *A* and *B*.

(d) $\overline{AB}$, the only line segment containing all points common to both rays.

3. (a) $113°57' + 18°14' = 131°71' = \mathbf{132°11'}$.

(b) $84°13' - 27°45' = 83°73' - 27°45' = \mathbf{56°28'}$.

(c) $113°57' + 18.4° = 113°57' + 18°24' = 131°81' = \mathbf{132°21'}$.

(d) $0.75° = 0° + 0.75(60)' = \mathbf{0°45'0''}$.

(e) $\mathbf{35°8'35''}$.

5. Examples may vary.

(a) A simple closed curve:

(b) A closed curve that is not simple:

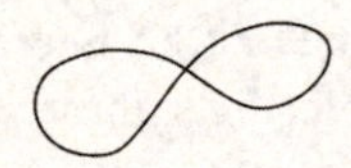

(c) A concave hexagon:

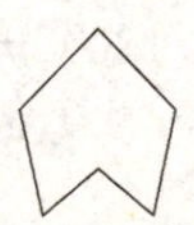

(d) A convex decagon:

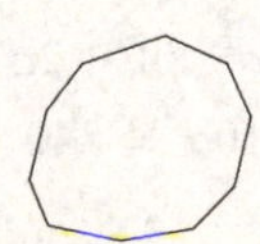

(e) Construct a regular pentagon $ABCDE$. Draw a line CD and reflect CD and DE about the line CD. The will create an equilateral pentagon that is not equiangular.

(f) Construct an rectangle that is not a square.

7. Let α be the measure of the smallest angle. Then $\alpha + 2\alpha + 7\alpha = 180° \Rightarrow 10\alpha = 180°$. $\alpha = \mathbf{18°}$, $2\alpha = \mathbf{36°}$, $7\alpha = \mathbf{126°}$.

9. $\frac{(n-2)\cdot 180}{n} = 176 \Rightarrow 180n - 360 = 176n \Rightarrow$ $4n = 360$. $n =$ **90 sides.**

11. **(a)** $m(\angle 3) = m(\angle 1) = \mathbf{60°}$ (vertical angles).

(b) $m(\angle 5) = m(\angle 3) = 60°$ (alternate interior angles) and $m(\angle 6) = 180° - m(\angle 5)$ (supplementary angles). $m(\angle 6) = 180° - 60° = \mathbf{120°}$.

(c) $m(\angle 8) = m(\angle 6) = \mathbf{120°}$ (vertical angles).

13. **(a)** $x + 30° + (180° - 70°) = 180°$
$\Rightarrow x + 30° + 110° = 180°$
$\Rightarrow \mathbf{x = 40°}$.

(b) Extend a line segment as follows.

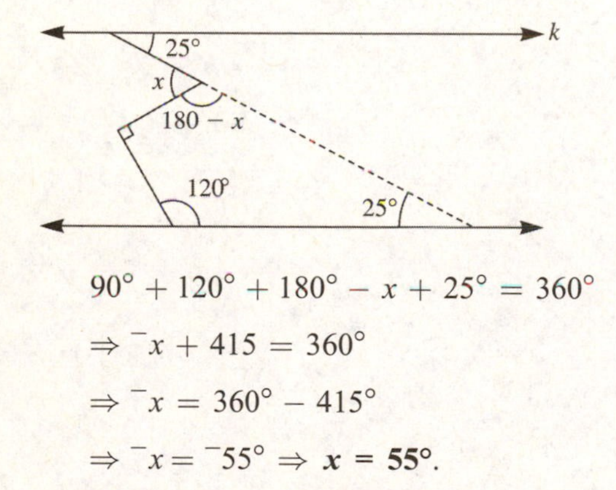

$90° + 120° + 180° - x + 25° = 360°$
$\Rightarrow {}^{-}x + 415 = 360°$
$\Rightarrow {}^{-}x = 360° - 415°$
$\Rightarrow {}^{-}x = {}^{-}55° \Rightarrow \mathbf{x = 55°}$.

15. ***(i)*** $m(\angle 1) = 180° - (70° + 45°) = \mathbf{65°}$.

(ii) $m(\angle 2) = m(\angle 1) = \mathbf{65°}$ (alt interior and corresponding angles).

(iii)
$m(\angle 3) + m(\angle 4) = 360° - (2 \cdot 65°) = 230°$.
$m(\angle 3) = m(\angle 4)$ (opposite $\angle$s of a parallelogram)
$m(\angle 3) = \frac{230°}{2} = \mathbf{115°}$.

(iv) $m(\angle 4) = m(\angle 3) = \mathbf{115°}$.

(*v*) The third angle of $\Delta BDF = 65°$ (alternate interior angle with $\angle 2$. $m(\angle 5) = 180° - (65° + 45°) = \mathbf{70°}$.

17. **(a)** Alternate interior angles are congruent by construction. $\overleftrightarrow{AB} \parallel \overleftrightarrow{BC}$ by the alternate interior angle theorem.

(b) Corresponding angles are congruent.

(c) $m(\angle B) + m(\angle C) + m(\angle BAC) = m(\angle BAD) + m(\angle DAE) + m(\angle BAC) = 180°$.

19. **Correct**. If it is assumed that a rectangle has four right angles for a total angle measurement of 360° and that a diagonal divides it into two congruent triangles, then the sum of the measures in each right triangle is 180°.

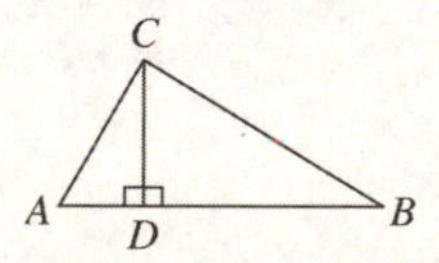

Thus the sum of the measures of the angles in ΔACD and ΔBCD is $2 \cdot 180° = 360°$. In this sum all the angles of the original triangle are included as well as the two right angles at D, so the sum of the measures of the angles in the original triangle is $360° - 2 \cdot 90° = 180°$.

21. **(a)** $\angle 3$ and $\angle 4$ are supplements of congruent angles.

(b) $m(\angle 3) = 180° - m(\angle 1)$ and $m(\angle 4) = 180° - m(\angle 2)$. If $m(\angle 1) < m(\angle 2)$ then ${}^{-}m(\angle 1) > {}^{-}m(\angle 2)$. $180° - m(\angle 1) > 180° - m(\angle 2)$ $\Rightarrow m(\angle 3) > m(\angle 4)$.

23. Example may vary.
Three planes that intersect in a point:

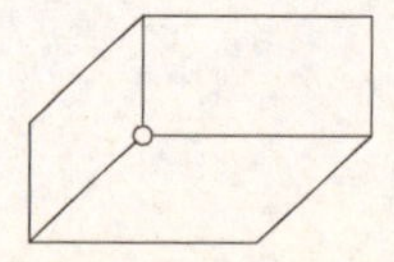

25. **(a)** $\frac{360°}{20} = 18$, so there are **18 sides**.

(b) 25 does not divide 360°; such a regular polygon does not exist.

(c) Does not exist; the sum is always 360°.

(d) Does not exist; the equation $\frac{n(n-3)}{2} = 4860$ has no natural number solution.

27. **(a)** Line and turn

(b) Line, turn, and point

(c) Line.

29. **(a)** $180° - 30° - 90° = $ **60°**.

(b) **30°**.

(c) $180° - 120° - 30° = $ **30°**.

(d) $180° - 120° = $ **60°**.

(e) $m(\overset{\frown}{CD}) = m(\angle COD) = $ **60°**.

(f) $m(\overset{\frown}{BC}) = m(\angle BOC) = $ **120°**.

31. **(a)** The triangle inequality tells us that $r + q > p$. If $p - q > r$, then $p - q + q > r + q > p$. This would say that $p > p$, which is false.

(b) If $r = p - q$, then $r + q = p$, which would form a line segment not a triangle.

33. $C = 2\pi r \Rightarrow 3m = 2\pi r \Rightarrow r = \frac{3}{2\pi}m$.

35. If a circle has radius 6 cm, then its circumference is $2\pi r = 2\pi(6 \text{ cm}) = 12\pi$ cm. Sarah is **not correct**.

Chapter 12

CONGRUENCE AND SIMILARITY WITH CONSTRUCTIONS

Assessment 12-1A: Congruence through Construction

1. If the 3 points are collinear, then it is not possible. If the 3 points are not collinear, construct a triangle whose vertices are the 3 points. Now choose any 2 sides and construct perpendicular bisectors. The intersection of these bisectors will be a point equidistant from the 3 points.

3. **(a)** The angle opposite $\overline{BC}$ is greater than the angle opposite $\overline{AC}$; i.e., $m(\angle A) > m(\angle B)$.

(b) In any triangle, the side of greater length is opposite the angle of greater measure.

5. **(a)** $\overline{AD}$ was created with a compass as shown below:

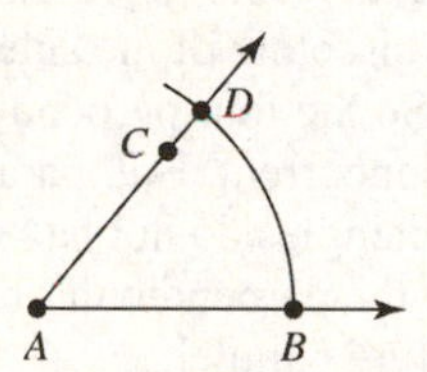

(b) One strategy is to construct a line segment for any of the three lengths, for example 2 cm. Then use a compass to draw circles of radii 3 cm and 4 cm from either endpoints. Where the circles intersect is a third vertex for a triangle with the desired properties. Alternative, cut three pieces of string of the desired lengths and arrange to form a triangle.

(c) Use a strategy similar to that in part (b).

(d) Since $4 + 5 < 10$, it is **not possible**.

(e) Use a strategy similar to that in part (b).

(f) Use protractor to create an angle whose measure is $75°$. Along the either ray of this angle draw line segment from the vertex of length 6 cm and 7 cm. Connect the end points of these line segments.

(g) Use *SSA* construction. This triangle is not unique.

(h) Use *SSA* construction. This triangle is not unique.

(i) Use a strategy similar to that in (f), i.e., *SAS* construction.

(j) (b). **Unique** by *SSS*. If the three sides of one triangle are congruent, respectively, to the three sides of a second triangle, then the triangles are congruent. I.e., all triangles with sides 2 cm, 3 cm, and 4 cm are the same. The figure below is similar to a 2 cm, 3 cm, 4 cm triangle.

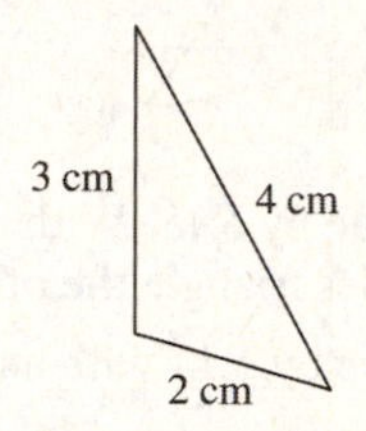

(c). **Unique** by *SSS*. A scalene right triangle.

(d) **No triangle**. The triangle inequality states that the sum of the measurements of any two sides of a triangle must be greater than the measure of the third side, and $10 > 4 + 5$.

(e) **Unique** by *SSS*.

(f) **Unique** by *SAS*. If two sides and the included angle of one triangle are congruent to two sides and the included angle of another triangle, respectively, then the two triangles are congruent. The figure below is similar to a *SAS* 6 cm-75°-7 cm triangle.

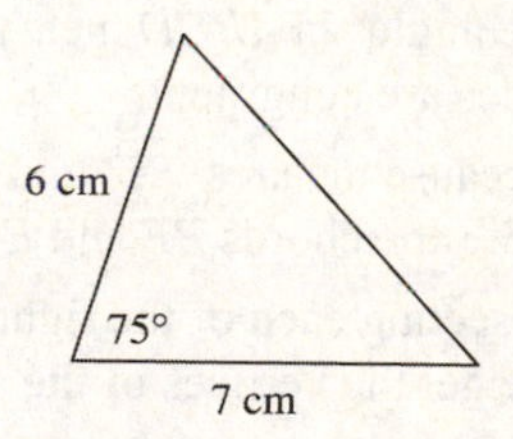

(g) **Not unique**. Three different triangles are possible; those below are representative.

(i)

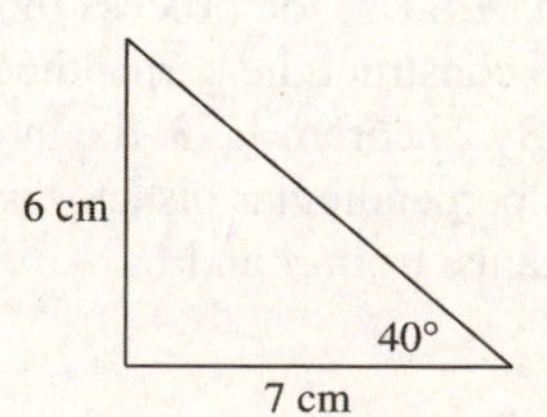

(ii)

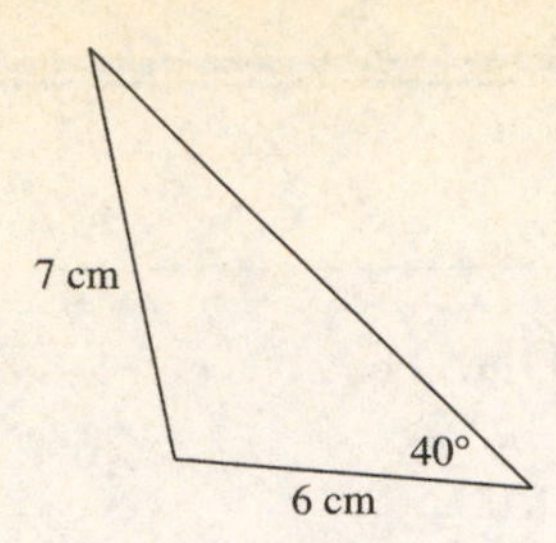

(iii)

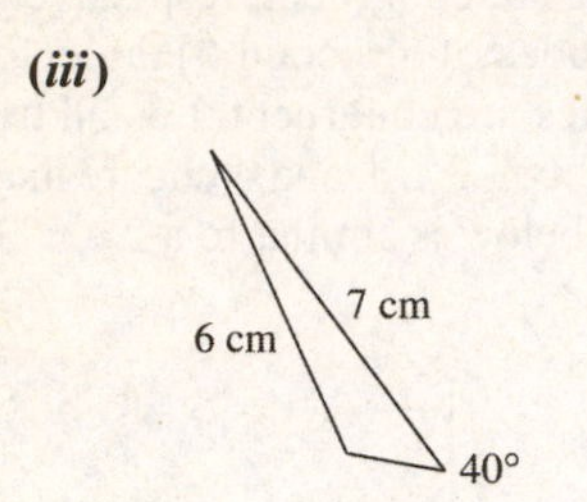

(h) **Unique** by *SAS.* With one angle given in an isosceles triangle the others are determined.

(i) **Unique** by *SAS* with an included right angle.

7. (a) Use the procedure of Figure 12-11 in the text, marking the span of $\angle B$ from the point where the span of $\angle A$ fell on the arc.

(b) Use the same procedure as in part (a), but mark the span of $\angle A$ back toward the starting point from where the span of $\angle B$ fell on the arc.

9. (a) Construction. The triangles *ABO*, *BCO*, *CDO*, and *DAO* are congruent (*SAS*) isosceles right triangles. Therefore the congruent angles in each triangle measure $45°$. Consequently all the angles in *ABCD* measure $90°$ and all the sides are congruent.

(b) Because the arcs $\overset{\frown}{BE}$ and $\overset{\frown}{EC}$ each measure $45°$, the chords *BE* and *EC* are congruent.

(c) Bisecting each of the right central angles we get the vertices of the regular octagon.

11. This point can be found only if *A* and *B* are not points on a line perpendicular to *m*. Assuming *A* and *B* have this property, construct the line segment $\overline{AB}$. Use the process illustrated in Figure 12-19 to construct the perpendicular bisector to $\overline{AB}$. By Theorem 12-3, the intersection of *m* and this perpendicular bisector is a point on *m* equidistance from *A* and *B*.

13. (a) Yes. The center of the circle is at the intersection of any two of the three perpendicular bisectors.

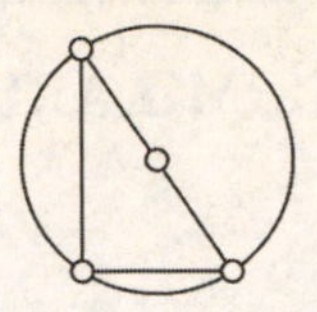

(b) Yes. Construct a regular hexagon *ABCDEF*. Bisect adjacent angles *B* and *C*. Label the intersection of these bisectors *G*. This is the center of the circle that circumscribes the hexagon. To prove that the circle circumscribes the regular hexagon, construct *AG*, *DG*, *EG*, and *FG* note that the triangles adjacent to ΔBCG are congruent to ΔBCG by SAS. Use this same approach to move around the figure noting that adjacent triangles are congruent. This shows that *AG*, for example, is congruent to *BG*.

(c) No. Construct a quadrilateral inside a circle and note that the center is equal distance from each of the vertices of the quadrilateral. By Theorem 12-3, the center is on the perpendicular bisectors of the sides of the quadrilateral. So the four perpendicular bisectors are concurrent. But, parallelograms that are not rectangles do not have this property since the perpendicular bisectors of opposite sides are parallel.

15. (a) $\Delta ACB \cong \Delta BCA$

(b) From part (a) $\angle A \cong \angle B$, by corresponding parts of congruent triangles.

Assessment 12-1B

1. No. Construct a parallelogram that is not a rectangle. Draw the $\perp$ bisectors of three adjacent sides. Points equidistant from the vertices must lie on all three of the $\perp$ bisectors. But the $\perp$ bisectors of the opposite sides do not intersect.

3. (a) Construct and angle and mark of a line segment of any length on one of the rays. Use this line segment to construct a larger angle at the opposite endpoint to the first angle. The third vertex will be the intersection of the two remaining rays.

(b) The side opposite the angle of greatest measure appears to be the one of greatest length. For all triangles, the side opposite the angle of greatest measure has the greater length than a side opposite an angle of lesser measure.

(c) The hypotenuse is longer than any of the legs. This is true because the hypotenuse is opposite the angle of measure $90°$, which is the angle of greatest measure in a right triangle.

5. (a) **Yes,** ***SAS***.

(b) **Yes,** ***SSS***.

(c) **No**. Since the angle is not between the sides (adjacent to only one side) more than one triangle can be formed.

7. (a) Use the procedure of Figure 12-11 in the text, to create two adjacent angles *A* and then an angle *B* adjacent to one of the outer most rays of one of the angles *A*.

(b) Create two angles *B* adjacent to each other. Create angle *A* inside this newly formed angle, 2*B*. The difference is $m(\angle C) = 2m(\angle B) - m(\angle A)$.

9. (a) Because $\angle BOC$ measures $60°$ and $\overline{OG}$ bisects $\overline{BC}$, $\angle BOG$ and $\angle COG$ measure $30°$. Since $\Delta BOG \cong \Delta COG$ by *SAS*, $\overline{BG} \cong \overline{CG}$. This same argument can be made for each of the adjacent vertices of the hexagon. A similar argument shows that all the constructed triangles are congruent. Since we have constructed a 12-gon with congruent sides and interior angles, we have constructed a regular 12-gon.

(b) Because the hexagon is regular, each interior angle measures $\frac{360°}{60°} = 60°$. Because ΔBOC is isosceles, the equal base angles measure $\frac{180°-60°}{2} = 60°$. Thus, ΔBOC is equilateral.

(c) Using (b) offers an approach to constructing a regular hexagon by adjoining six congruent equilateral triangles. After this step, part (a) can be used to create the 12-gon.

11. Place three points at different locations on the circle. Construct line segments connecting the points to form a triangle. Use the process described in Figure 12-19 to construct the perpendicular bisectors for two of the sides of the triangle. The intersection of these bisectors is the center of the circle. Figure 12-23 illustrates the same process.

13. (a) **Not possible** unless the rhombus is a square.

(b) **Possible**. The center *O* of the circle is the intersection of any two perpendicular bisectors of two adjacent sides. The radius is the segment from *O* to any of the vertices.

15. The three sides are congruent in any order;

$$\Delta ABC \cong \Delta ABC$$
$$\cong \Delta BAC$$
$$\cong \Delta CAB$$
$$\cong \Delta ACB$$
$$\cong \Delta BCA$$
$$\cong \Delta CBA.$$

Assessment 12-2A Additional Congruence Properties

1. Constructions may vary; those below are representative.

(a) *ASA*:

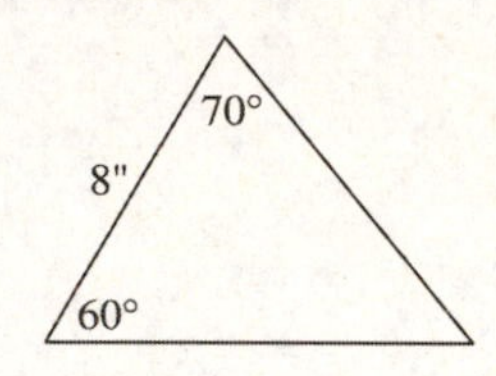

(b) *AAS*:

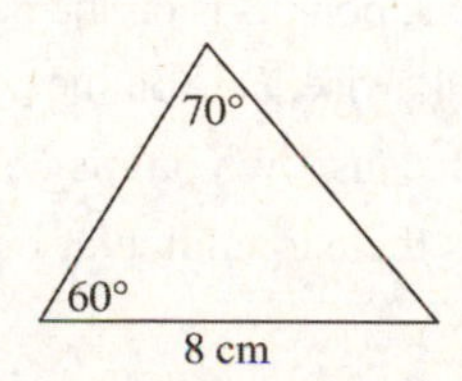

(c) **Infinitely many** are possible. All would be similar; *AAA* determines a unique shape, but not size.

3. Compare the following triangles:

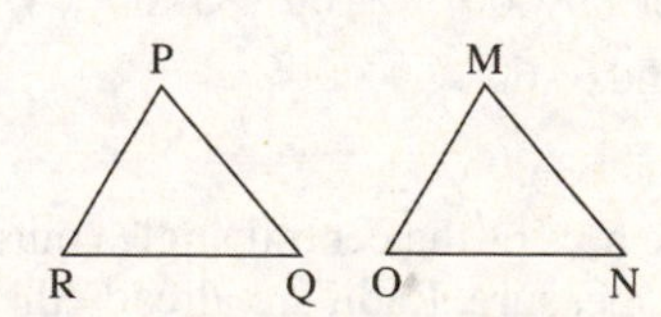

(a) **Yes**. Congruent by *ASA*.

(b) **Yes**. Congruent by *AAS*.

5. See Table 12-1 for properties.

(a) **Parallelogram**, by properties *d* and *e*.

(b) **None**. It must be known that the quadrilateral is a parallelogram before it can be known that it is a rectangle. Otherwise it could be an isosceles trapezoid.

(c) **None**. The quadrilateral could be a parallelogram, rectangle, or trapezoid even if the diagonals are not perpendicular.

7. See Table 12-1.

(a) **True**, by property.

(b) **True**, by definition.

(c) **True**, by definition.

(d) **False**. A trapezoid may have only one pair of parallel sides, while in a parallelogram each pair of opposite sides must be parallel. A trapezoid may also have two consecutive right angles with the other two not right angles.

9. *ABCD* is a kite.

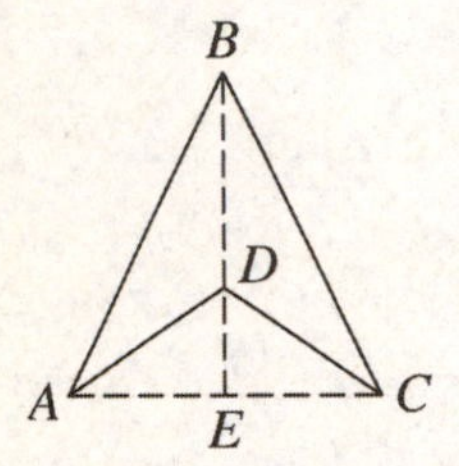

Because $AB = BC$, B is equidistant from A and C. By Theorem 12-3, point B is on the perpendicular bisector of $\overline{AC}$. Likewise, D is on the perpendicular bisector of $\overline{AC}$. Because two points determine a unique line $\overleftrightarrow{BD}$ is the perpendicular bisector of $\overline{AC} \Rightarrow \overleftrightarrow{BD} \perp \overleftrightarrow{AC}$.

11. When O, the center of the circumcircle, is connected with 6 vertices, 6 equilateral triangles result. A rotation about O by any multiple of $60°$ will map the vertices on themselves and hence there are 6 rotational symmetries by $60°$, $2 \cdot 60°$, $3 \cdot 60°$, $4 \cdot 60°$, $5 \cdot 60°$, and $6 \cdot 60°$.

13. Either the arcs or the central angles must have the same measure. Radii are already the same, since the sectors are part of the same circle.

15. Make one of the quadrilaterals a square and the other a rectangle that is not a square.

17. **Rhombus**. Use *SAS* to prove that $\Delta ECF \cong \Delta GBF \cong \Delta EDH \cong \Delta GAH$. Then $\overline{EF} \cong \overline{GF} \cong \overline{EH} \cong \overline{GH}$.

19. (a) In the figure, $\overline{BC} \parallel \overline{AD}$ and $\overline{BC} \cong \overline{AD}$.

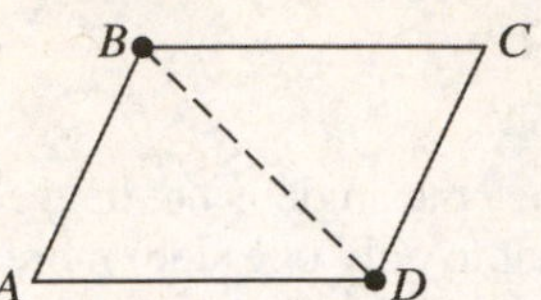

Since $\overline{BC} \parallel \overline{AD}$, $\angle ADB \cong \angle CBD$. Also, $\overline{BD} = \overline{DB}$. Thus, by SAS, $\Delta ADB \cong \Delta CBD$. Thus, corresponding parts are congruent, and $\overline{AB} \cong \overline{CD}$ and $\angle ABD \cong \angle CDB$. Since these last angles are alternate interior angles for transversal $\overline{BD}$, $\overline{AB} \parallel \overline{CD}$. Since opposite sides are parallel and congruent, the quadrilateral is a parallelogram.

(b) If the diagonals bisect each other, then $AO = OC$ and $BO = OD$ in the figure below.

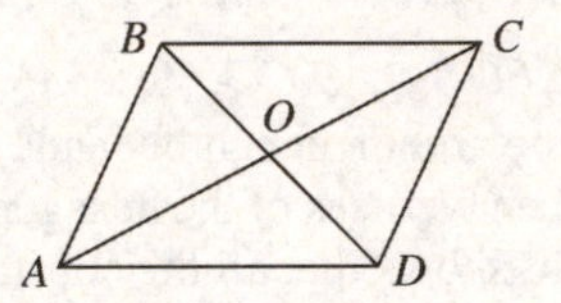

In addition, vertical angles $\angle BOC$ and $\angle DOA$ are congruent. Thus, by *SAS*, $\Delta BOC \cong \Delta DOA$. $\angle OBC \cong \angle ODA$ because they are corresponding parts of $\cong$ Δ's, and $\overline{BC} \parallel \overline{AD}$, because the angles are alternate interior angles formed by transversal $\overline{BD}$. Since quadrilateral *ABCD* has a pair of opposite sides parallel and congruent, it must be a parallelogram by part (a).

Assessment 12-2B

1. (a) *ASA*:

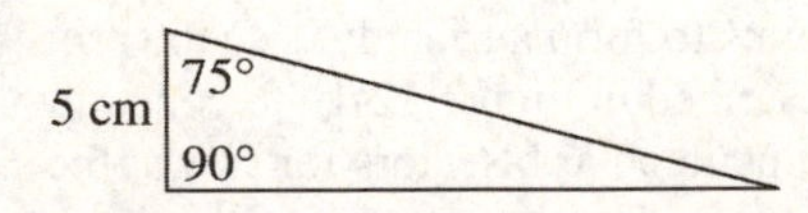

(b) **Infinitely many** are possible. All would be similar; *AAA* determines a unique shape, but not size.

3. Compare the following triangles:

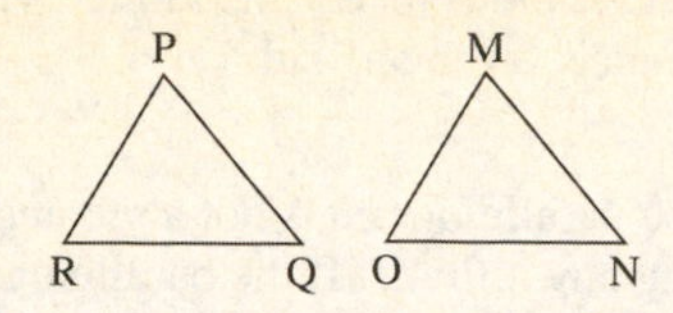

(a) **No**. *SSA* does not assure congruence.

(b) **No**. *AAA* does not assure congruence.

5. See Table 12-1 for properties.

(a) **Rectangle**, by property *d* and *e*.

(b) **Rhombus**, by property *c* and *d*.

(c) **Square**. By parallelogram property *d*, rectangle property *d and e*, and rhombus property *c* and *d*.

(d) **Parallelogram**, by definition.

7. See Table 12-1.

(a) **False**. A square is both a rectangle and a rhombus.

(b) **False**. A square satisfies all conditions of a trapezoid, thus some trapezoids must be squares.

(c) **True**. All squares, by definition, are trapezoids.

(d) **True**. A rhombus is a parallelogram with all sides congruent.

(e) **True**. A square is a rhombus with all angles congruent.

9. **Yes**. $\Delta ABD \cong \Delta CBD$ by *SSS* $\Rightarrow \angle BAD \cong \angle BCD$ by CPCTC.

11. The diagonals of a rectangle are congruent and bisect each other and therefore a rotation by 180° about the intersection of the diagonals will take each vertex to the vertex on the other end of the corresponding diagonal.

13. **(a)** Congruent sectors are ***COD* and *AOB*** as well as ***COB* and *AOD***.

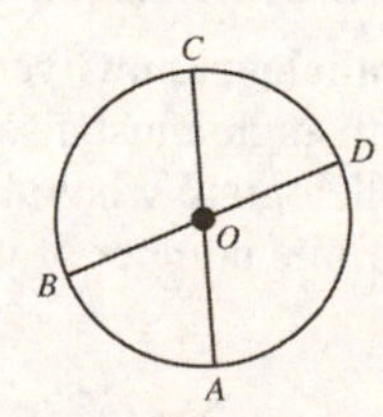

(b) Because the corresponding central angles are vertical and hence congruent.

15. Answers vary. For example, two isosceles trapezoids *AEFD* and *ABCD* as shown where $\overline{EF} \parallel \overline{AD}$.

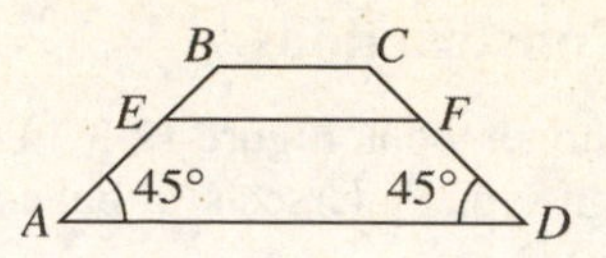

17. **(a)** **Parallelogram**.

(b) Suppose *ADCB* in part (a) is a parallelogram. By *SAS*, $\Delta EDH \cong \Delta GBF$, thus $\overline{EH} \cong \overline{GF}$.

Similarly, $\Delta EAF \cong \Delta GAH$ so $\overline{EF} \cong \overline{GH}$. If opposite sides of a quadrilateral are congruent, it is a parallelogram.

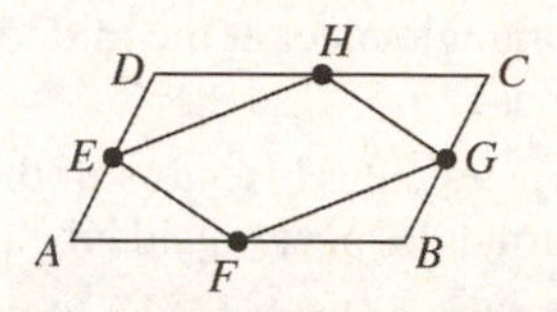

19. **(a)** $\angle ABD \cong \angle CDB$ and $\angle CBD \cong \angle ADB$ (alternate interior angles with respect to parallel lines). Then $\Delta ABD \cong \Delta CDB$ (by *ASA*).

Thus $\angle BAD \cong \angle DCB$ (CPCTC); similarly, $\angle ABC \cong \angle CDA$.

(b) $\Delta ABD \cong \Delta CDB$, thus $\overline{AB} \cong \overline{CD}$ and $\overline{AD} \cong \overline{CB}$ (CPCTC).

(c) $\angle BAC \cong \angle DCA$ and $\angle ABD \cong \angle CDB$ (alternate interior angles), and $\overline{AB} \cong \overline{DC}$ [from part (b)]. Then $\Delta BAF \cong \Delta DCF$, so $\overline{AF} \cong \overline{CF}$ and $\overline{BF} \cong \overline{DF}$ (CPCTC).

(d) In ΔABD, $m(\angle BAD) + m(\angle ABD) + m(\angle ADB) = 180°$.

From part (a), $\angle CBD \cong \angle ADB$. Substitution yields $m(\angle BAD) + m(\angle ABD) + m(\angle CBD) = 180°$.

Since $m(\angle ABC) = m(\angle ABD) + m(\angle CBD)$, then $m(\angle BAD) + m(\angle ABC) = 180°$. $\angle ABC$ and $\angle BAD$, therefore, are supplementary.

Review Problems

15. Triangles ***BCD*, *CDE*, *DEA***, and ***EAB***. By definition, all five sides and angles are congruent. Therefore the triangles are congruent by *SAS*.

17. Follow the procedure of Figure 12-10, using the given segment for all three sides.

Assessment 12-3A
Additional Constructions

1. (a) Refer to textbook Figure 12-37. Draw a line through P that intersects ℓ, but copy α as an alternate interior angle (i.e., vertical to α as shown in the Figure).

(b) Refer to Figure 12-36. Draw the line through P that intersects ℓ, and use the construction shown to create a rhombus.

3. (a) The perpendicular bisectors of the sides of an acute triangle meet inside the triangle.

(b) The perpendicular bisectors of the sides of a right triangle meet at the midpoint of the hypotenuse.

(c) The perpendicular bisectors of the sides of an obtuse triangle meet outside the triangle.

(d) Use the point of intersection of the perpendicular bisectors as the center of the circle, then use the distance from the intersection to any of the vertices as the radius of the circle.

5. (a) If the rectangle is not a square it is **impossible** to construct an inscribed circle. The angle bisectors of a rectangle do not intersect in a single point.

(b) Possible. The center of the circle is the intersection of the diagonals (which are also the angle bisectors of the vertices). The radius of the circle is the distance from the center to any of the sides.

(c) Possible. The intersection of the three longest diagonals is the center of the circle.

7. (a) Make an arc of radius BC with center A and one with radius AB and center C so that the two intersect. This intersection is the location of the fourth vertex.

Another technique would be to construct a parallel to $\overline{BC}$ through A, then mark off the distance BC from A.

(b) Since one use of a compass or straight edge is not sufficient to find the fourth vertex, we must use the tools twice to locate the fourth vertex. Then, we must use the straight edge twice to construct the remaining sides of the parallelogram after the fourth vertex is located. The cheapest way is **40¢**.

9. The incenter is located where the perpendicular bisectors of the sides meet. The radius is the distance from the incenter to any side.

11. (a) If the parallelogram is not a rectangle, cut along any altitude. If the parallelogram is a rectangle, cut along any line through the point where the diagonals meet. The line must not be a diagonal nor parallel to any side.

(b) Make a copy of the given trapezoid and put it upside down next to $\overline{CD}$; i.e.,

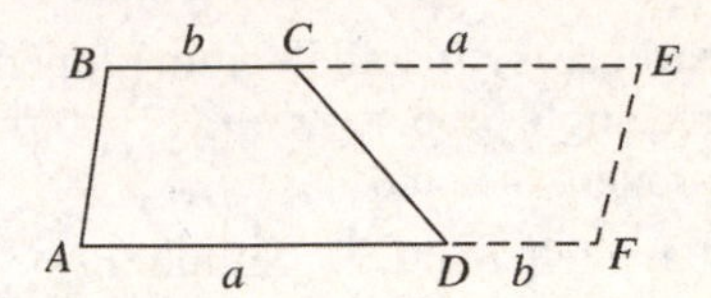

extend $\overline{BC}$ so that $CE = a$ and extend $\overline{AD}$ so that $DF = b$. Since $\overline{BE} \parallel \overline{AF}$ and $\overline{BE} = \overline{AF}$ (the length of each is $a + b$), $ABEF$ is a parallelogram.

13. (a) Possible. Draw line $\overline{AB}$ the length of a side of the square, then construct a right angle at A using the method of Figure 12-45. Measure AB with a compass and use it to mark off side AC along the perpendicular.

From C and B, mark off the length of $\overline{AB}$ by drawing compass arcs at the approximate location of the final vertex. Call the intersection of the two arcs point D.

Construct $\overline{AD}$ and $\overline{CD}$ to from square $ABCD$.

(b) No unique rectangle. The endpoints of two segments bisecting each other and congruent to the given diagonal determine a rectangle, but since the segments may intersect at any angle there are infinitely many such rectangles.

(c) Not possible. The sum of the measurements of the angles would be greater than 180°.

(d) No unique parallelogram. Given three right angles, the fourth angle must also be a right angle. The parallelogram would be a rectangle or square, an infinite number of which could be constructed..

15. Make arcs of the same radius from A and B above $\overline{AB}$ and label their intersection C. Repeat the process with a new radus, labeling this intersection D. $\overleftrightarrow{CD}$ is the perpendicular bisector of $\overline{AB}$.

17. Answers vary. One method is to construct a circle with center O and radius r. Then draw a diameter and label one of the intersections of the diameter and the circle as D. Construct an arc with radius r and center O. This will determine an equilateral triangle whose vertex O determines a central angle of 60°.

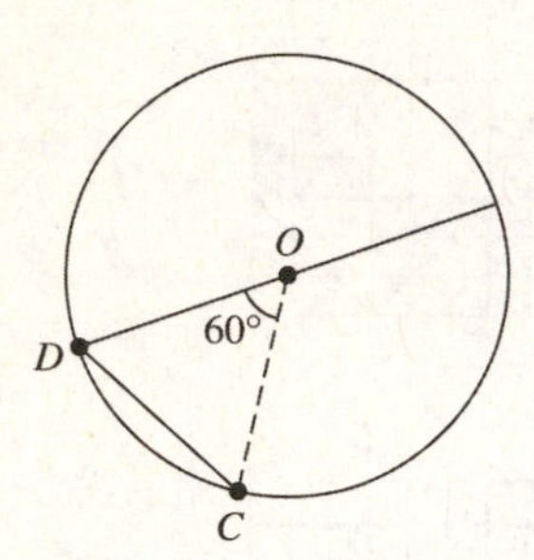

This method can be continued to construct six congruent triangles as shown below. Two adjacent triangles determine a central angle of 120° with vertex O. Because $\Delta AOB, \Delta BOC,$ and ΔAOC are congruent by SAS, $\overline{BC}, \overline{BA},$ and $\overline{AC}$ are congruent and hence we have constructed an equilateral inscribed triangle.

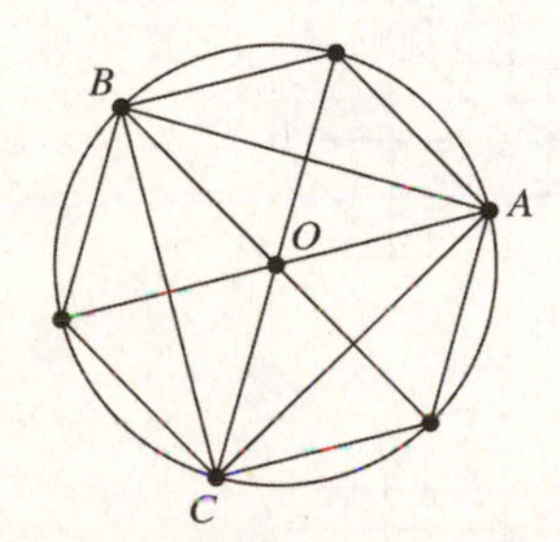

Assessment 12-3B

1. **(a)** Construct line m through P perpendicular to ℓ; then a line through P perpendicular to m.

(b) Draw any line through P that intersects ℓ. Let Q be the point where the line intersects line ℓ. Find the midpoint M of $\overline{PQ}$. Draw any line through M intersecting ℓ at Q'. Find P' so M is the midpoint of $\overline{P'Q'}$. Then the line $\overleftrightarrow{PP'}$ is the desired figure.

3. The intersection of the two perpendicular bisectors will be the center of the circumscribing circle. The center will be outside the triangle.

5. Yes, because the angle bisectors are concurrent. One of the diagonals of the kite is on the angle bisectors of opposite angles. For the other angle opposite angles, the angle bisector intersects on that diagonal. The distances from this intersection point to the sides are all equal. Thus, the distance from the intersection of the angle bisectors to any one of the sides defines the radius of an inscribing circle.

7. ***(i)*** With a compass construct a circle with center A and radius r (10¢).

(ii) Choose any point B on the circle and construct a circle with radius r and center B. Choose one of the two points where the two circles intersect and name it point C (10¢).

(iii) Use a straight edge to construct line segments $\overline{AB}, \overline{BC},$ and $\overline{CA}$ (30¢).

Total: **50¢**.

9. The center is the intersection of any two angle bisectors. The radius of the incircle is the distance from the incenter to any side, measured along the perpendicular bisector of that side.

11. The following figure shows two such congruent trapezoids. For different values of a we get different trapezoids.

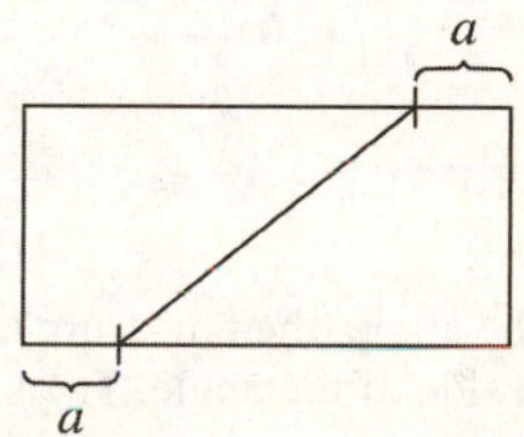

13. **(a)** **Possible**. Construct two perpendicular segments bisecting each other and congruent to the given diagonal. Then connect the endpoints.

(b) **No unique parallelogram**. Without the angle between the sides, there are infinitely many parallelograms which could be constructed from the two given sides.

(c) **Possible**. Construct two perpendicular segments bisecting each other and congruent to the given diagonals. Then connect the endpoints.

(d) **No unique parallelogram**. Given a side and all the angles, the other pair of congruent sides can be of any length.

15. Make the edge of the ruler coincide with ℓ. Let one of the legs of the right triangle slide along the edge of the ruler until the other leg goes through P.

The line along the edge containing P is the required perpendicular.

17. Construct an equilateral triangle PQS inscribed in the circle (see Assessment 12-3A problem 17) with center O. Next construct perpendiculars to $\overline{OP}$, $\overline{OQ}$ and $\overline{OS}$ at P, Q, and S respectively. The points of intersection of the perpendiculars are the vertices of a required triangle. To prove that the triangle is equilateral, let A be the intersection of the perpendiculars at P and Q. Because $m(\angle QOP) = 120°$, $m(\angle OPA) = m(\angle OQA) = 90°$, and $APOQ$ forms a quadrilateral, $m(\angle A) = 60°$. Similarly,the other angles of ΔABC measure $60°$ and hence the triangle is equilateral.

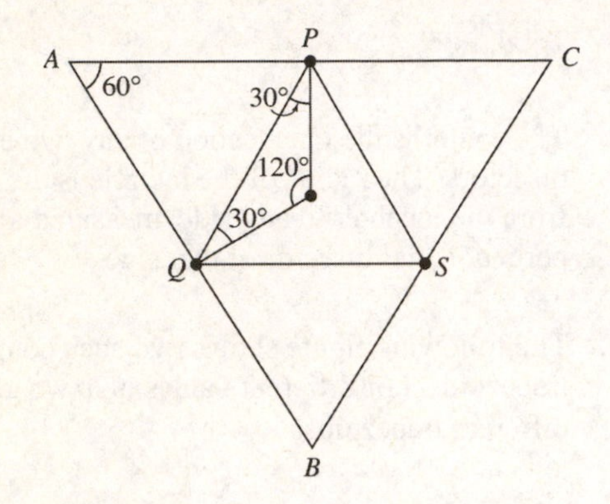

Review Problems

17. **(a)** Copy the angle, then measure off each side along a side of the angle. Then connect the end points.

(b) Copy $\overline{AB}$. Make arcs from A and B, one with radius AC and the other with radius BC. Their intersection is C.

(c) Copy the side. Copy the angles at opposite ends, extending their sides until they meet to from the triangle.

Assessment 12-4A: Similar Triangles and Other Similar Figures

1. In a rhombus, adjacent angles are supplementary. Since $\angle BAD \cong \angle B_1A_1D_1$ we know that $\angle ABC \cong \angle A_1B_1C_1$. Thus, the corresponding angles are congruent. Since, in a rhombus, adjacent sides are congruent, $\frac{AD}{AB} = 1 = \frac{A_1D_1}{A_1B_1}$.

This implies that $\frac{AD}{A_1D_1} = \frac{AB}{A_1B_1}$, in other words, corresponding sides are proportional. Therefore, the rhombuses are similar.

3. Make all dimensions three times as long. For example, in part (c) each side would be three diagonal units long. One possible solution set is below.

(a)

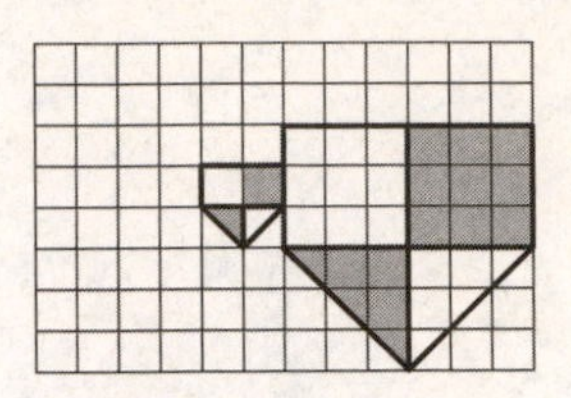

(b)

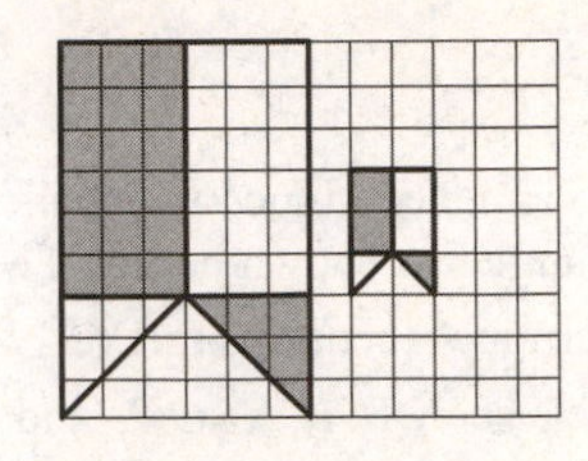

(c)

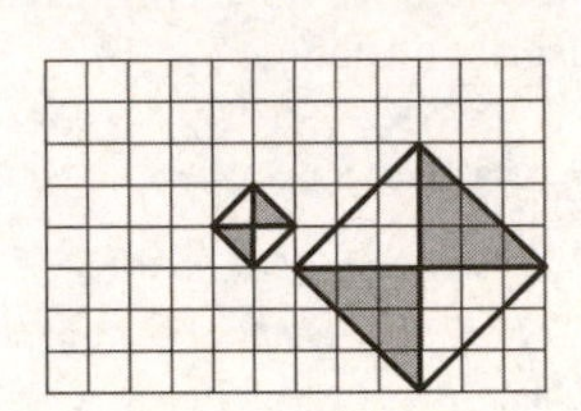

(d)

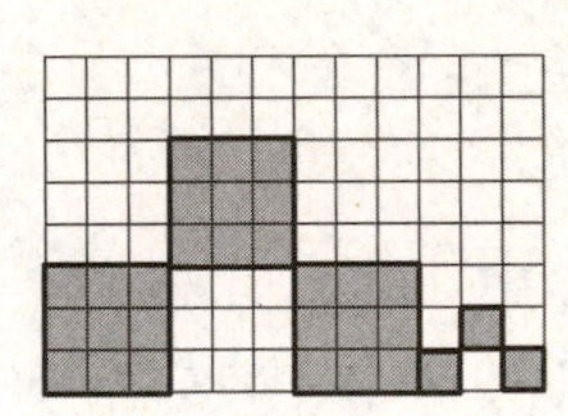

5. **(a)** $\frac{\text{Short side}}{\text{Long side}} = \frac{5}{10} = \frac{x}{x+7} \Rightarrow 5(x+7) = 10x \Rightarrow 5x + 35 = 10x$. $x = 7$.

(b) $\frac{3}{x} = \frac{7}{8} \Rightarrow 7x = 3 \cdot 8$. $x = \frac{24}{7}$.

7. If $\frac{a}{b} = \frac{c}{x}$, then $x = \frac{bc}{a}$. Construct as below, using the technique of Figure 12-56.

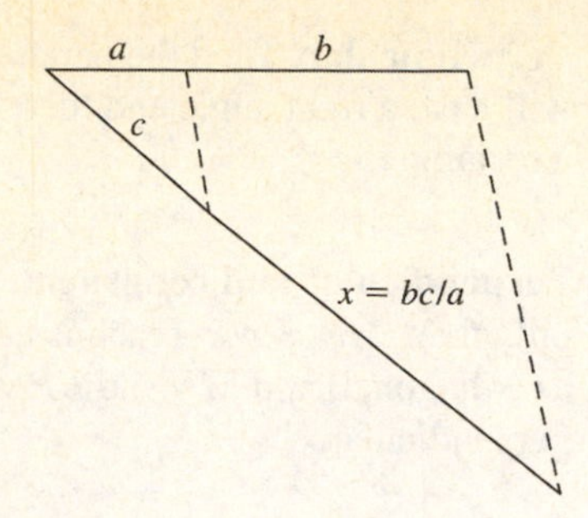

Note that the two dashed lines are parallel.

9. Sketches may vary. Make the sides proportional but interior angles not congruent; e. g.:

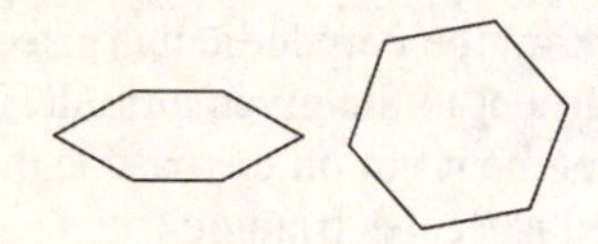

11. Let the height above ground be h and convert all measurements to inches (3 feet $=$ 36 inches: 7 feet$=$ 84 inches) and use similar triangles. $\frac{36}{13} = \frac{h}{84} \Rightarrow 13h = 36 \cdot 84.\ h \approx 232.6$ inches or **about 19.38 feet.**

13. **(a)** A perimeter is the sum of the lengths of the sides, so the ratio of the perimeters is the same as the ratio of the sides.

(b) If $a, b, c,$ and d are the sides of one quadrilateral and $a_1, b_1, c_1,$ and d_1 are the corresponding sides of a similar quadrilateral, then $\frac{a}{a_1} = \frac{b}{b_1} = \frac{c}{c_1} = \frac{d}{d_1} = r$ the scale factor.

Then $\frac{a+b+c+d}{a_1+b_1+c_1+d_1} = \frac{a_1r+b_1r+c_1r+d_1r}{a_1+b_1+c_1+d_1}$

$= \frac{(a_1+b_1+c_1+d_1)r}{a_1+b_1+c_1+d_1} = r$

So the ratio of the perimeters is r. An analgous proof works for any two similar n-gons.

15. **(a)** $ABCD$ is a rhombus $\Rightarrow \overline{AB} = \overline{BC} = \overline{CD} = \overline{DA}$, thus all angles of $MNPQ$ are right and it is a **rectangle**.

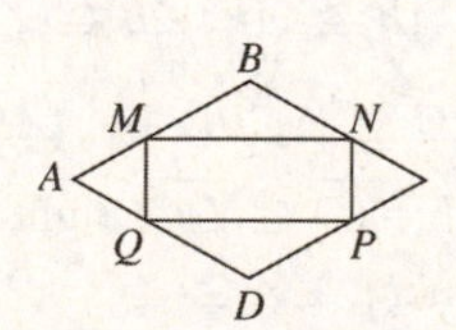

(b) If $ABCD$ is a kite, $MNPQ$ is a **rectangle**.

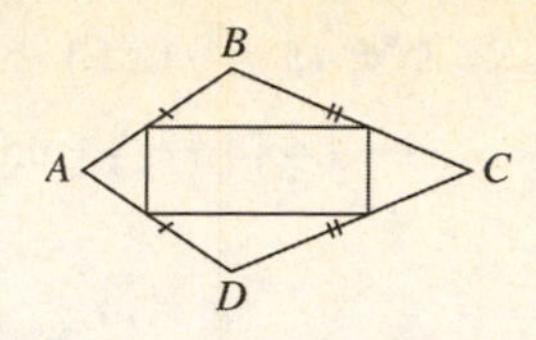

(c) If $ABCD$ is an isosceles trapezoid, $MNPQ$ is a **rhombus**.

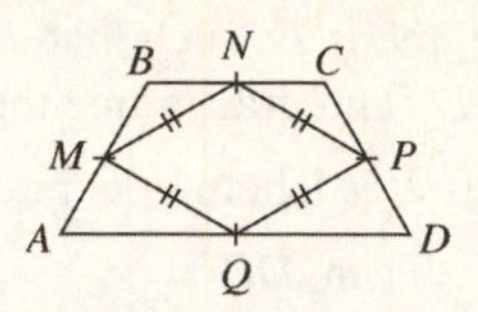

(d) If $ABCD$ is quadrilateral which is not a rhombus nor a kite but whose diagonals are perpendicular to each other, $MNPQ$ is a **rectangle**.

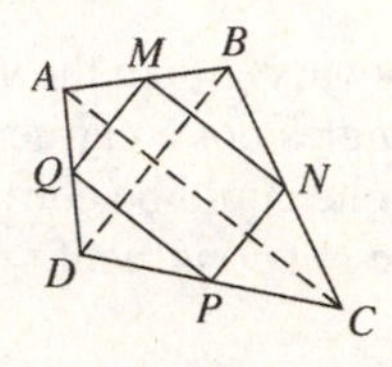

Assessment 12-4B

1. **Not necessarily** similar. Consider a parallelogram $ABCD$ in which $BC = 2AB$, and the rhombus $ABEF$ where E and F are the midpoints of $\overline{BC}$ and $\overline{AD}$ respectively.

3. Answers vary. For example, choose any natural number a and draw $a, a, a\sqrt{2}$ by starting at a corner of one grid and constructing vertical and horizontal line segments. Note that all right isosceles triangles are all similar.

5. **(a)** Similar by SAS because $\frac{5}{5+5} = \frac{7}{7+7} = \frac{1}{2}$ $\Rightarrow x = 9\left(\frac{1}{2}\right) = \mathbf{4.5}$.

(b) Similar by AA. Scale factor is $\frac{7}{3} \Rightarrow$ $x = \left(\frac{8}{6}\right)5 = \mathbf{\frac{20}{3}}$.

(c) Similar by SAS because $\frac{4+3.2}{3.2} = \frac{4+5}{4} = \frac{9}{4}$. $x = \left(\frac{9}{4}\right)3 = \mathbf{\frac{27}{4}}$.

(d) Similar by AA because each has a right angle and $m(\angle CBA) = 180° - 90° - m(\angle DBE)$ and in ΔBDE, $m(\angle E) = 90° - m(\angle DBE)$

$\Rightarrow m(\angle CBA) = m(\angle E)$. Scale factor is $\frac{2}{3} \Rightarrow x = \left(\frac{2}{3}\right)4 = \frac{8}{3}$ and $y = \left(\frac{2}{3}\right)5 = \frac{10}{3}$.

7. $\frac{a}{b} = \frac{x}{c} \Rightarrow \frac{b}{a} = \frac{c}{x}$. Let b and a be end-to-end on a side of an arbitrary angle. On the other side of the angle, mark D such that $DA = c$. Connect B with D, and draw a line through C that is parallel to $\overline{BD}$, labeling its intersection with $\overrightarrow{AD}$ as C. Then $DE = x$.

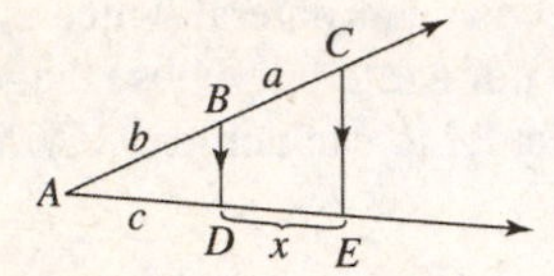

9. Sketches may vary. Make the sides proportional but interior angles not congruent. One could be concave and one could be convex to make sure the shapes are not congruent.

11. The figure shows similar triangles with proportional sides. Convert the girl's height to 1.5 m and let h be the height of the tree. Then $\frac{1.5}{3} = \frac{h}{15+3} \Rightarrow$ $3h = 1.5(15+3)$. $h = \mathbf{9\ m}$.

13. **(a)** The ratio of corresponding heights is the same as the ratio of corresponding sides.

(b) Let $ABC \sim A_1B_1C_1$. Construct altitudes BD and B_1D_1.

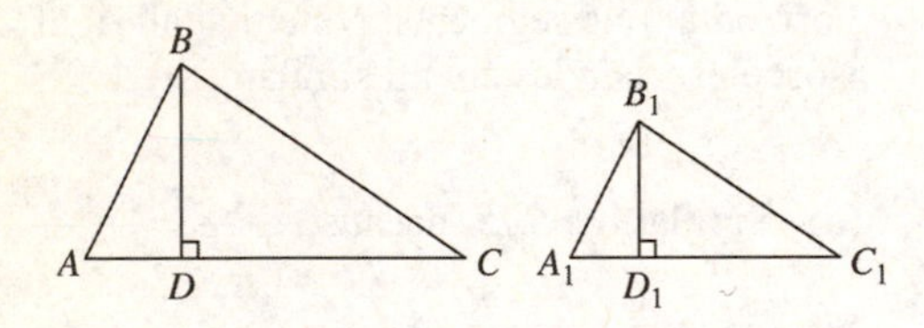

By the AA Similarity Theorem, $\Delta ABD \sim \Delta A_1B_1D_1$. Thus, $\frac{AB}{A_1B_1} = \frac{BD}{B_1D_1}$.

15. **(a)** $\overline{NP}$ and $\overline{MN}$ are midsegments, so $\overline{NP} \parallel \overline{BD}$ and $\overline{MN} \parallel \overline{AC} \Rightarrow NSTV$ is a parallelogram. $\angle VNS \cong \angle VTS$ since these are opposite angles in parallelogram.

(b) **(*i*)** **Perpendicular**. Then the parallelogram will have a right angle and therefore be a rectangle.

(*ii*) **Perpendicular and congruent**. Then and only then $MN = NP$ (each is half as long as a diagonal) and MN and NP will be perpendicular.

Review Problems

15. Start with the given base and construct its perpendicular bisector. The vertex of the required triangle must be on this perpendicular bisector. Starting at the point where the perpendicular bisector intersects the base, mark on the perpendicular bisector a segment congruent to the given altitude. The endpoint of the segment not on the base is the vertex of the required isosceles triangle.

17. Answers may vary. E.g.: construct a perpendicular line at one of the endpoints of the hypotenuse, giving a 90° angle. Bisect this angle to yield a 45° angle. Copy the 45° angle at the other endpoint of the hypotenuse, extending until it meets the bisector which formed the first 45° angle. These will meet at a 90° angle across from the given hypotenuse.

19. Two quadrilaterals may be congruent by *SASAS* or by *ASASA*.

Chapter 12 Review

1. **(a)** $\Delta ABD \cong \Delta CBD$ by ***SAS***. $\overline{BD}$ is common to both; right angles are congruent; $\overline{AD} \cong \overline{DC}$.

(b) $\Delta GAC \cong \Delta EDB$ by ***SAS***. $\overline{AC} \cong \overline{DB}$; right angles are congruent; $\overline{AG} \cong \overline{DE}$.

(c) $\Delta ABC \cong \Delta EDC$ by ***AAS***. Right angles are congruent; vertical angles; $\angle ACB \cong \angle DCE$; $\overline{AC} \cong \overline{CE}$.

(d) $\Delta BAD \cong \Delta EAC$ by ***ASA***. $\angle A$ is common to both; $\overline{AB} \cong \overline{AE}$; $\angle B \cong \angle E$.

(e) $\Delta ABD \cong \Delta CBD$ by ***AAS***. $\overline{BD}$ is common to both; $\overline{AD} \cong \overline{DC}$; right angles are congruent; $\angle A \cong \angle C$.

(f) $\Delta ABD \cong \Delta CBD$ by ***SAS***. $\overline{AB} \cong \overline{BC}$; $\angle ABD \cong \angle DBC$; $\overline{BD}$ is common to both.

(g) $\Delta ABD \cong \Delta CBE$ by **SSS**. $\overline{BD} \cong \overline{BE}$; $\overline{AB} \cong \overline{BC}$; $\overline{AD} \cong \overline{EC}$.

(h) $\Delta ABC \cong \Delta ADC$ by **SSS**; $\Delta ABE \cong \Delta ADE$ by **SSS** or **SAS**; $\Delta EBC \cong \Delta EDC$ by **SSS** or **SAS**. $\overline{BC} \cong \overline{CD}$; $\overline{AB} \cong \overline{AD}$; properties of kites contribute to most of congruency.

3. (a) (*i*) Use the method illustrated by Figure 12-40 in the text.

(*ii*) Fold the angle down the middle so the sides match and the crease passes through A.

(b) (*i*) See text Figure 12-45.

(*ii*) Fold the line on top of itself so that the crease passes through B.

(c) (*i*) See text Figure 12-43.

(*ii*) Same as in part (b)(*ii*).

(d) (*i*) See text Figures 12-36 or 12-37.

(*ii*) Make line $k \perp \ell$ through P as in part (b) and (c), then make $m \perp k$ through P as in part (b) $\Rightarrow m \parallel \ell$.

5. Use the method illustrated by Figure 12-57 and 12-58 in the text.

7. $\overline{AB}$ must be a chord of the circle, and the perpendicular bisector of a chord passes through the center. Construct this line to locate the center on ℓ, measure the radius to A or B, and draw the circle with your compass.

9. If h is the height of the building: $\frac{2 \text{ m tall}}{1 \text{ m shadow}} = \frac{h \text{ m high}}{6 \text{ m shadow}} \Rightarrow 1h = 2 \cdot 6$. $\boldsymbol{h = 12}$ **m**.

11. $\frac{h}{8} = \frac{1.5}{2} \Rightarrow 2h = 1.5 \cdot 8$. $\boldsymbol{h = 6}$ **m**.

13. In the figure below:

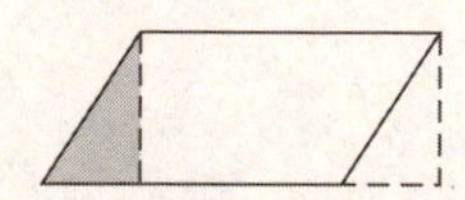

Slide the shaded triangle to the right.

15. The statement is false since in (*i*) we have a counterexample.

(*i*) Quadrilateral $ABCD$ below is not a square, even through its diagonals are perpendicular and congruent (i.e., $\overline{AB} \cong \overline{BD}$).

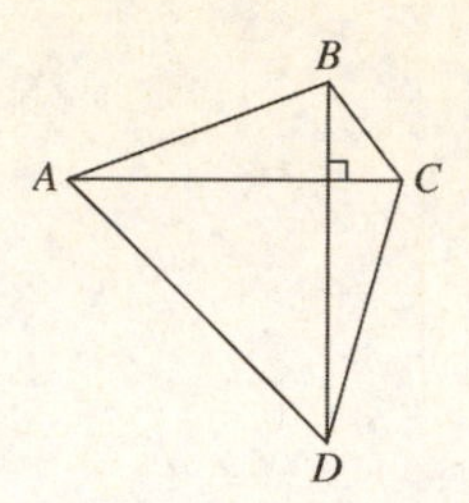

(*ii*) If the diagonals bisect each other, the quadrilateral is a square.

17. Congruent triangle have congruent corresponding angles. Thus they are similar by AA.

Chapter 13

CONGRUENCE AND SIMILARITY WITH TRANSFORMATIONs

Assessment 13-1A: Translations and Rotations

1. (a) Each corner of the trapezoid moves two dots to the right.

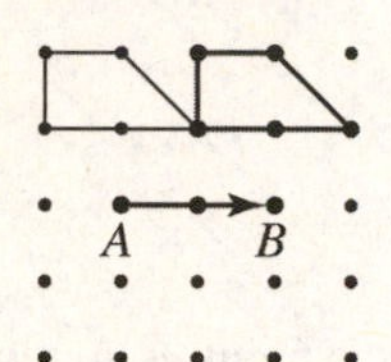

(b) Each corner of the trapezoid moves one dot down and one dot to the right.

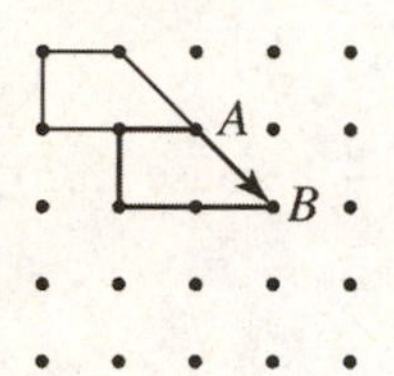

3. (a) $(0, 0) \to (0 + 3, 0 - 4) = \mathbf{(3, {}^-4)}$.

(b) $({}^-3, 4) \to ({}^-3 + 3, 4 - 4) = \mathbf{(0, 0)}$.

(c) $({}^-6, {}^-9) \to ({}^-6 + 3, {}^-9 - 4) = \mathbf{({}^-3, {}^-13)}$.

5. (a) $A({}^-4, 2) \to ({}^-4 + 3, 2 - 4) = A'({}^-1, {}^-2)$;

$B({}^-2, 2) \to ({}^-2 + 3, 2 - 4) = B'(1, {}^-2)$;

$C(0, 0) \to (0 + 3, 0 - 4) = C'(3, {}^-4)$;

$D({}^-2, 0) \to ({}^-2 + 3, 0 - 4) = D'(1, {}^-4)$.

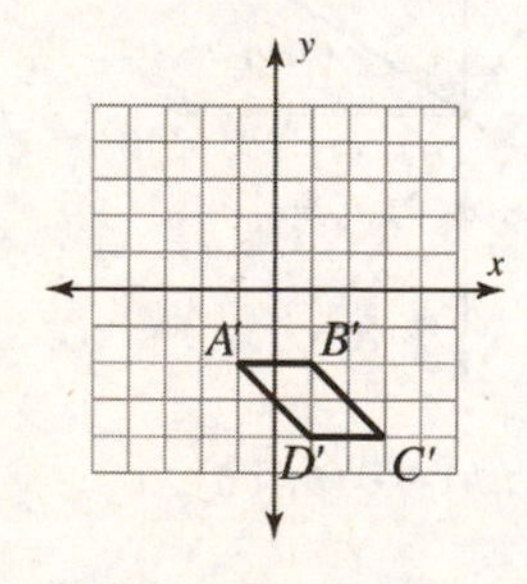

(b) $A({}^-1, 1) \to ({}^-1 + 3, 1 - 4) = A'(2, {}^-3)$;

$B(1, 4) \to (1 + 3, 4 - 4) = B'(4, 0)$;

$C(3, {}^-1) \to (3 + 3, {}^-1 - 4) = C'(6, {}^-5)$.

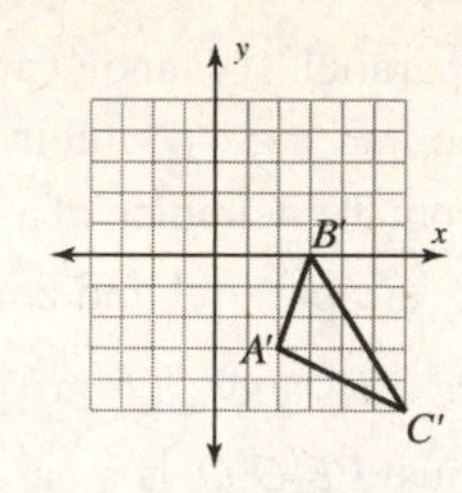

7. Use the technique demonstrated in Example 13-3 of the text, where each corner is rotated 90° counterclockwise around O to obtain:

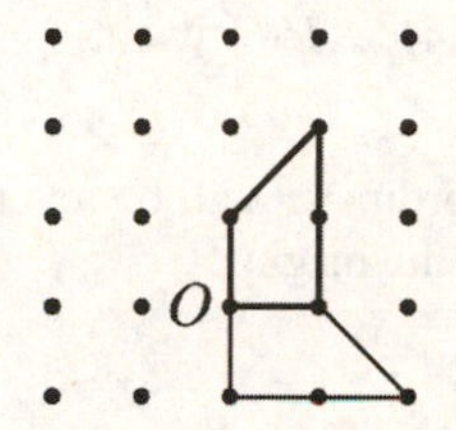

9. If y is the image of k under $(x + 3, y - 2)$, k is the translation of the image under $(x - 3, y + 2)$. Two sample points on the image are $A'(0,3)$ and $B'(1, 1)$.

$A = (0 - 3, 3 + 2) = ({}^-3, 5)$ and $B = (1 - 3, 1 + 2) = ({}^-2, 3)$.

Then $m = \frac{3-5}{{}^-2-{}^-3} = {}^-2$. Using x and y values from A in the general form, $5 = {}^-2({}^-3) + b \Rightarrow b = {}^-1$.

Line k is $\mathbf{y = {}^-2x - 1}$.

11. Reverse the rotation [i.e., counterclockwise] to locate $\overline{AB}$, which is the pre-image.

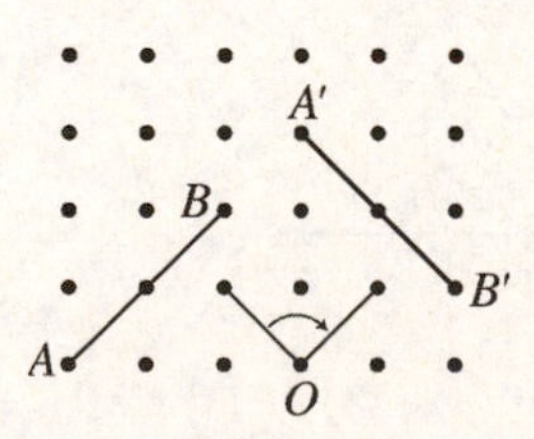

13. (a) Since $\overline{AB}$ is parallel to line ℓ, the image is the line ℓ itself.

(b) Sketches may vary.

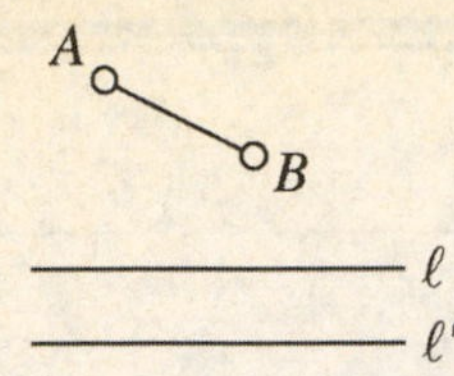

(c) ℓ and ℓ' are parallel. If P and Q are any two points on ℓ and P' and Q' their respective images, then from the definition of a translation $\overline{PP'}$ and $\overline{QQ'}$ are parallel and congruent to $\overline{AB}$. Therefore $\overline{PP'}$ and $\overline{QQ'}$ are parallel and congruent. Thus $PP'Q'Q$ is a parallelogram and by definition $\ell \parallel \ell'$.

(d) Ten successive rotations of $36° = 360°$. Thus the image of $\angle ABC$ is an identity transformation and is $\angle ABC$ itself.

15. Signs of the x and y coordinates will be reversed under a half-turn about the origin.

(a) $(4, 0) \to$ $\mathbf{(^-4, 0)}$.

(b) $(2, 4) \to$ $\mathbf{(^-2, ^-4)}$.

(c) $(^-2, ^-4) \to$ $\mathbf{(2, 4)}$.

(d) $(a, b) \to$ $\mathbf{(^-a, ^-b)}$.

17. (a) $\ell' = \ell$.

(b) $\ell' \perp \ell$.

19. (a) The following figures show 90° counterclockwise rotation about the origin.

(*i*) $m_\ell = \frac{3}{2} \Rightarrow m_\perp = \frac{2}{^-3}$ because $(2, 3) \to$ $\mathbf{(^-3, 2)}$.

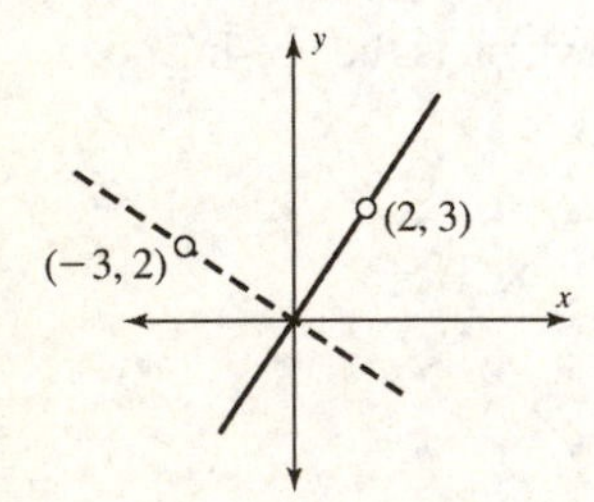

(*ii*) $m_\ell = \frac{2}{^-1} \Rightarrow m_\perp = \frac{^-1}{^-2}$ because $(^-1,2) \to$ $\mathbf{(^-2, ^-1)}$.

(*iii*) Under 90° counterclockwise rotation about the origin, $(m,n) \to$ $\mathbf{(\text{-}n, m)}$ or $m_\perp = \frac{m}{^-n}$.

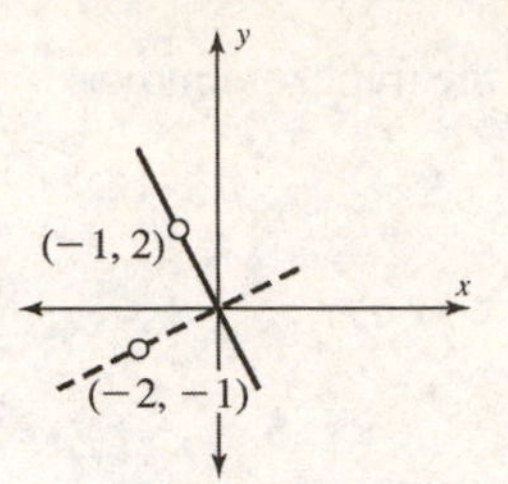

(b) In problem 19(a)(*iii*) the point (a, b) moves to $(^-b, a)$. Repeating this 90° counterclockwise rotation would produce a full half-turn and the point would end at $(^-a, ^-b)$.

(c) $m_\ell = \frac{b}{a} \Rightarrow m_\perp = \frac{^-a}{b}$ because $(a,b) \to$ $\mathbf{(b, ^-a)}$.

(*i*) (a, b) rotated 90° counterclockwise:

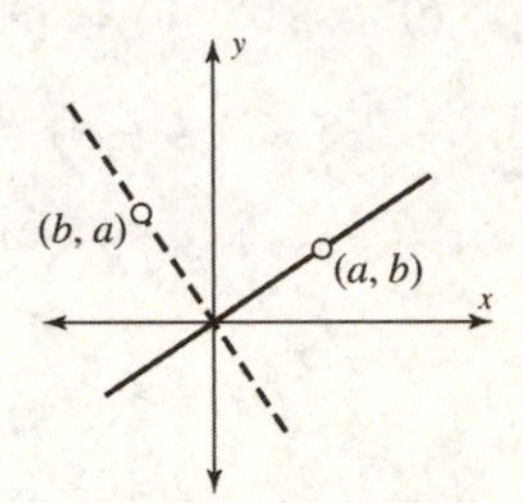

(*ii*) $(^-b, a)$ under a half-turn about the origin:

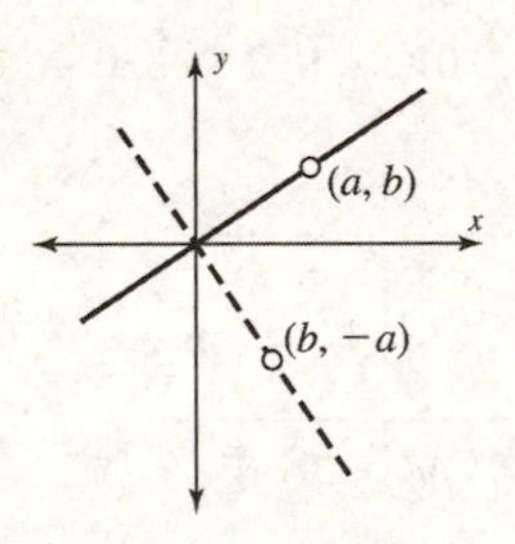

Or $(a,b) \to$ $\mathbf{(b, ^-a)}$.

21. (a) The image of $A(h, k)$ under the translation from O to C is $B(h + a, k + 0) = B(h + a, k)$.

(b) The distance formula yields $(OA)^2 = h^2 + k^2$. $OA = OC$, thus $h^2 + k^2 = a^2$.

(c) $m_{\overline{OB}} = \frac{k-0}{(h+a)-0} = \frac{k}{h+a}$. $m_{\overline{AC}} = \frac{k-0}{h-a} = \frac{k}{h-a}$. $m_{\overline{OB}} \cdot m_{\overline{AC}} = \frac{k}{h+a} \cdot \frac{k}{h-a} = \frac{k^2}{h^2-a^2}$ $= \frac{k^2}{{}^-k^2} = {}^-1$. Thus the diagonals are perpendicular.

(d) There is a turn symmetry of 180 degrees about the point of intersection of the two diagonals.

23. The image is a line **parallel** to the original line. To see this let ℓ represent the original line. Choose two points on the original line A and B. Let C be the center. If C is on ℓ, then A and B remain on ℓ. If C is not on ℓ, then A' and B' determine a new line ℓ' under the half turn.

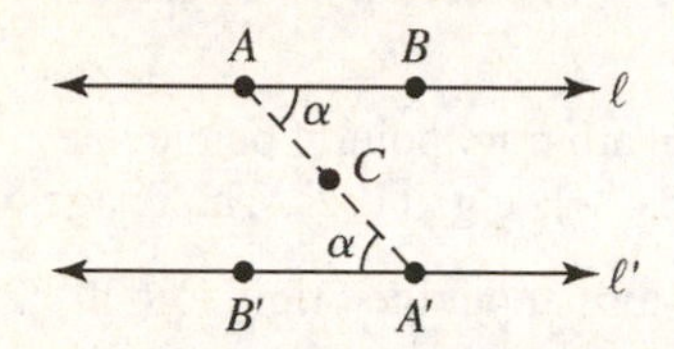

Since $\Delta ABC \cong \Delta A'B'C'$, $\angle CA'B' = \alpha = \angle CAB$. Thus, $\ell \parallel \ell'$.

25. Parallelograms can be formed by translating lines AB, AC, and BC to form parallels:

(i) $A(1,3) \to A'(1+3, 3+2) = C \Rightarrow C(4,5) \to C'(4+1, 5+3) =$ $\mathbf{(5,8)} = D_1$

(ii) $C(4,5) \to C'(4-2, 5+1) = B$ $A(1,3) \to A'(1-2, 3+1) =$ $\mathbf{({}^-1,4)} = D_2$

(iii) $B(2,6) \to B'(2-1, 6-3) = A \Rightarrow C(4,5) \to C'(4-1, 5-3) =$ $\mathbf{(3,2)} = D_3$.

See below:

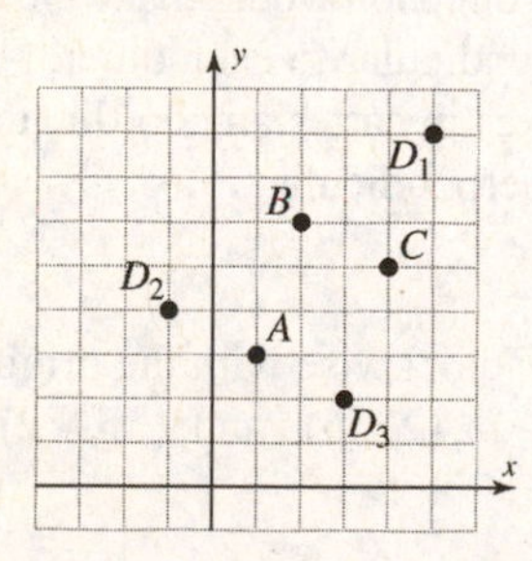

AD_2BC, ABD_1C, and $ABCD_3$ all then form parallelograms.

Assessment 13-1B

1. Reverse the translation so that the image completes a slide from X' to X (i.e., to its pre-image); then check by carrying out the given motion in the "forward" direction (i.e., to see if $\overline{AB}$ maps to $\overline{A'B'}$).

(a)

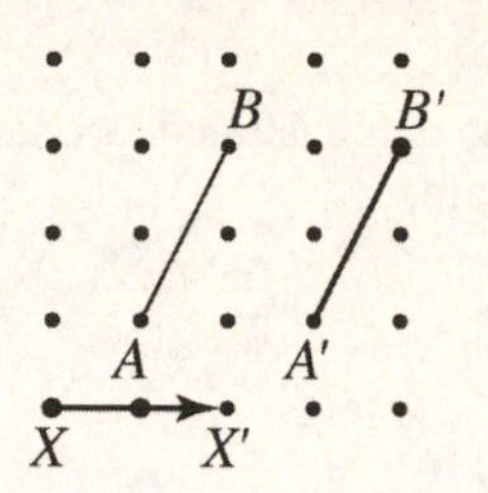

(b)

X'

X

3. **(a)** $(7,14) \to (7+3, 14-4) = \mathbf{(10,10)}$.

(b) $({}^-3, {}^-5) \to ({}^-3+3, {}^-5-4) = \mathbf{(0, {}^-9)}$.

(c) $(h,k) \to \mathbf{(h+3, k-4)}$.

5. **(a)** $A({}^-2,3) \to ({}^-2+3, 3-4) = A'(1, {}^-1)$; $B({}^-2,1) \to ({}^-2+3, 1-4) = B'(1, {}^-3)$; $C({}^-2, {}^-1) \to ({}^-2+3, {}^-1-4) = C'(1, {}^-5)$.

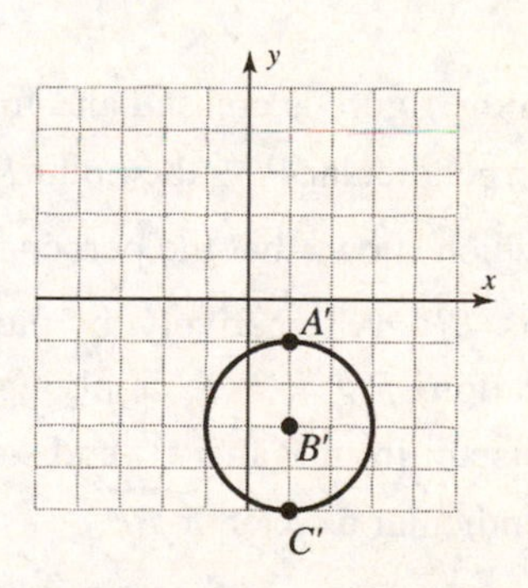

(b) $A({}^-4, {}^-1) \to ({}^-4+3, {}^-1-4) = A'({}^-1, {}^-5)$; $B({}^-3,3) \to ({}^-3+3, 3-4) = B'(0, {}^-1)$; $C(2,2) \to (2+3, 2-4) = C'(5, {}^-2)$; $D(1, {}^-1) \to ({}^-1+3, {}^-1-4) = D'(2, {}^-5)$.

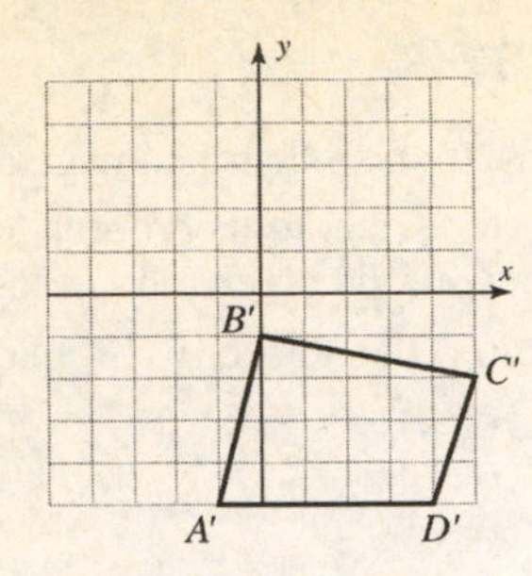

7. Use the technique demonstrated in Example 13-3 of the text to obtain:

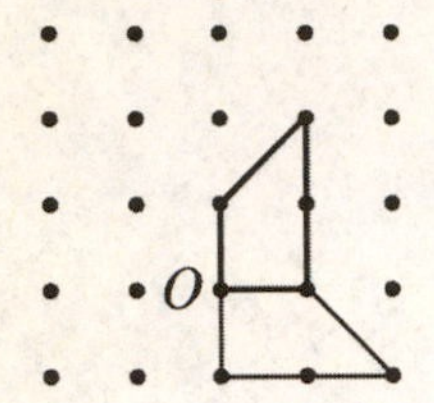

9. **P', P, and O are collinear** because the measure of $\angle POP'$ under a half-turn must be 180°.

11. Reverse the rotation in either direction to locate $\overline{AB}$; i.e., the pre-image.

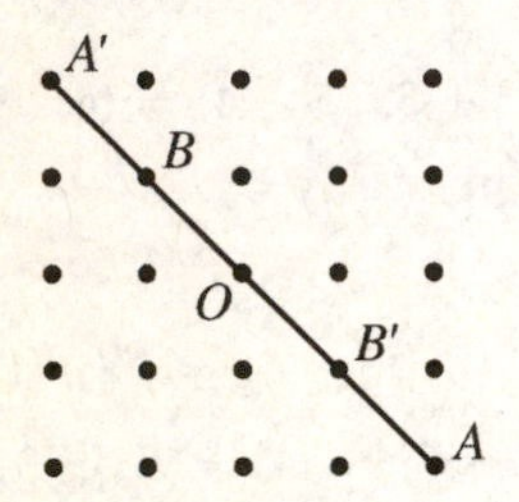

13. **(a)** When the figure is creased and folded along the perpendicular $\overline{PP'}$, the point P falls on $\overline{P'}$, which shows that the perpendicular also bisects $\overline{PP'}$. Alternatively, by the definition of rotation, $PO = PO'$, implying that O is equidistant from P and P' and so is on the perpendicular bisector of $\overline{PP'}$.

(b) From part (a), O is on the perpendicular bisector of $\overline{AA'}$ and $\overline{BB'}$, as well as $\overline{CC'}$. Then O may be found by determining the point at which any two of the perpendicular bisectors intersect. $\angle AOA'$ is the angle of rotation.

(c) The perpendicular bisectors of the corresp-onding sides are not concurrent; i.e., they do not contain the same point; although the perpendicular bisectors of two sides may intersect.

15. Signs of the x and y coordinates will be reversed under a half-turn about the origin.

(a) $(0, 3) \rightarrow$ **$(0, {}^{-}3)$**.

(b) $({}^{-}2, 5) \rightarrow$ **$(2, {}^{-}5)$**.

(c) $({}^{-}a, {}^{-}b) \rightarrow$ **(a, b)**.

17. **(a)** $\ell' \parallel \ell$.

(b) ℓ' and ℓ intersect at a $60°$ angle.

19. **(a)** Pick an arbitrary point A on the line $y = 3x - 1$; e.g., $(1, 2)$. The image of A under a half-turn about the origin is $({}^{-}1, {}^{-}2)$. The line passing through $({}^{-}1, {}^{-}2)$ with $m = 3$ is **$y = 3x + 1$**.

(b) The image of the arbitrary point A under a 90° counterclockwise turn about the origin is $({}^{-}2, 1)$. The line passing through $({}^{-}2, 1)$ with $m_{\perp} = \frac{{}^{-}1}{3}$ is **$y = \frac{{}^{-}1}{3}x + \frac{1}{3}$**.

21. O must be the midpoint of segments $\overline{AA'}$ and $\overline{BB'}$. So A' must lie on the x-axis and B' must lie on the y-axis. Since A' lies on the x-axis, A' has coordinates $(x, 0)$. But $AB = \sqrt{a^2 + b^2} = BA' = \sqrt{x^2 + b^2} \Rightarrow a^2 + b^2 = x^2 + b^2 \Rightarrow x^2 = a^2$. Since x is positive $x = a$. So, A' has coordinates $(a, 0)$. Similarly, B' has coordinates $(0, -b)$..

23. A 90° turn in either direction would make the original and its image perpendicular to each other. Note that the x- and y-axes in cartesian coordinates are 90° apart and are perpendicular.

25. Rotate $({}^{-}2, 5)$ 90° clockwise with the origin as the center to obtain $(5, 2)$, $(2, {}^{-}5)$, and $({}^{-}5, {}^{-}2)$. See below:

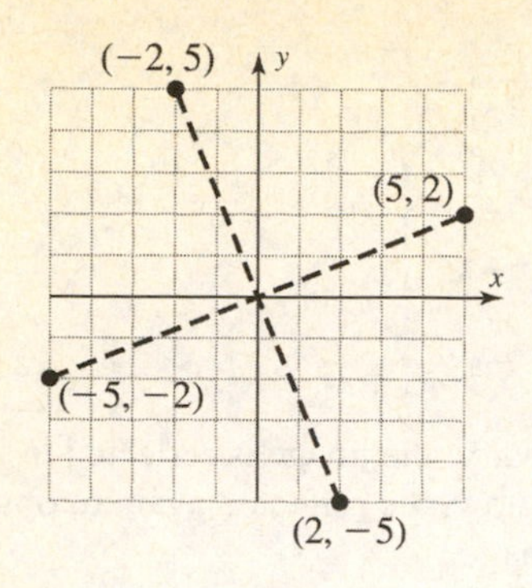

Assessment 13-2A: Reflections and Glide Reflections

1. Locate the image of vertices directly across (perpendicular to) ℓ on the geoboard.

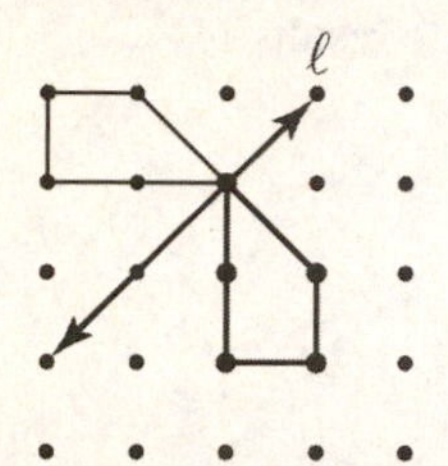

3. The original figure is reflected back upon itself. I.e., if the first reflection yields $\Delta A'B'C'$, the second reflection brings $\Delta A'B'C'$ back to ΔABC.

5. (a) **No**. The final images are congruent but in different locations, thus they are not the same.
   (b) A translation determined by a slide arrow from P to R. Let P be any point on ℓ and Q on m such that $\overrightarrow{PQ} \perp \ell$. Point R is on $\overrightarrow{PQ}$ such that $PQ = QR$.

7. The line of reflection is the **perpendicular bisector** of $\overline{AA'}$, $\overline{BB'}$, or $\overline{CC'}$.

9. Given the line $y = 2x + 1$, two sample points are $A(0, 1)$ and $B(1, 3)$.
   (a) $P(x, y) \to P(x, {}^-y)$ when reflected in the x-axis $\Rightarrow A(0, 1) \to A'(0, {}^-1)$ and $B(1, 3) \to B'(1, {}^-3)$.

   Then $m = \frac{{}^-3 - {}^-1}{1-0} = {}^-2$.

   Using A', ${}^-1 = {}^-2(0) + b \Rightarrow b = {}^-1$.

   So the equation of the line is $\boldsymbol{y = {}^-2x - 1}$.

   (b) $P(x, y) \to P({}^-x, y)$ when reflected in the y-axis $\Rightarrow A(0, 1) \to A'(0, 1)$ and $B(1, 3) \to B'({}^-1, 3)$.

   Then $m = \frac{3-1}{{}^-1-0} = {}^-2$.

   Using A', $1 = {}^-2(0) + b \Rightarrow b = 1$.

   So the equation of the line is $\boldsymbol{y = {}^-2x + 1}$.

   (c) $P(x, y) \to P(y, x)$ when reflected in the line $y = x \Rightarrow A(0, 1) \to A'(1, 0)$ and $B(1, 3) \to B'(3, 1)$.

   Then $m = \frac{1-0}{3-1} = \frac{1}{2}$.

   Using A', $0 = \frac{1}{2}(1) + b \Rightarrow b = \frac{{}^-1}{2}$.

   So the equation of the line is $\boldsymbol{y = \frac{1}{2}x - \frac{1}{2}}$.

11. None of the images has a reverse orientation, so there are no reflections or glide reflections involved. Thus

   1 to 2 can be viewed as a counterclockwise rotation;

   1 to 3 can be viewed as a clockwise rotation;

   1 to 4 is a translation down;

   1 to 5 is a clockwise rotation followed by a translation down or a rotation about an exterior point;

   1 to 6 is a translation or a clockwise rotation; and

   1 to 7 is a translation or a clockwise rotation.

13. (a) $P(x, y) \to \boldsymbol{P'(x, {}^-y)}$ under reflection in the x-axis; $P(x, y) \to \boldsymbol{P'(y, x)}$ under reflection about $y = x$;
   (b) $\boldsymbol{({}^-x, {}^-y)}$. From part (a) the reason can be seen as below:

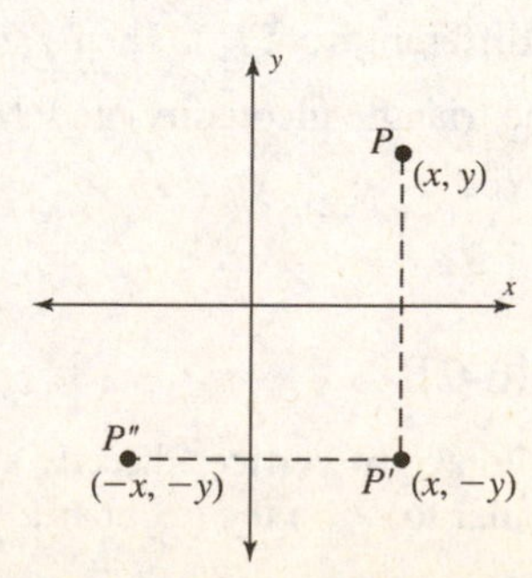

15. $P(x, y) \to P(^{-}x, y)$ when reflected in the y-axis. Thus:

(a) $y = {}^{-}x + 3 \to y = {}^{-}({}^{-}x) + 3 \Rightarrow$ $\mathbf{y = x + 3}$.

(b) $y = 0 \to \mathbf{y = 0}$.

17. Given the circles below:

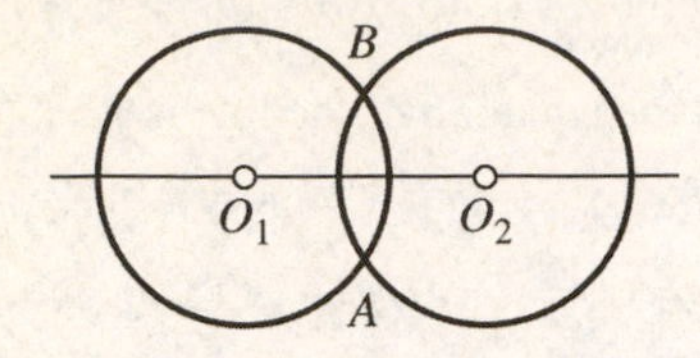

(a) The line $\overleftrightarrow{AB}$, since the circles are congruent.

(b) **Yes**, a translation taking O_1 to O_2.

19. The **intersection of $\overline{HT'}$ and r** in the figure below determines the point on the road at which the pole should be placed.

If H represents the house, P the pole, and T the other house, then the shortest path connecting H and T' (the reflection of T in r) is **the segment $\overline{HT'}$**.

Use the figure below to prove that the path from H to P_1 to T is the shortest possible.

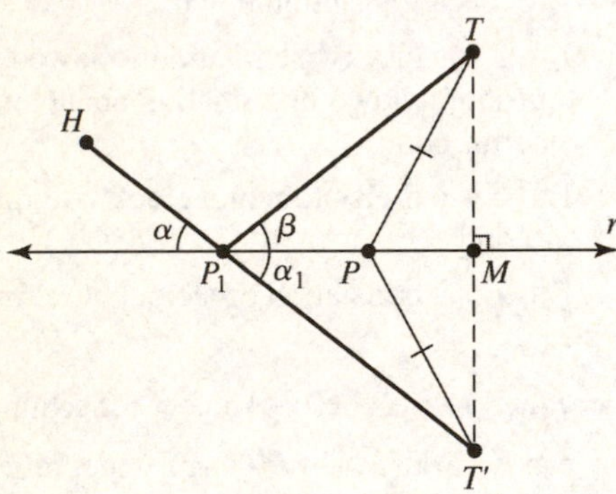

$HP_1 + P_1T = HP_1 + P_1T' = HT'$ (P_1T and P_1T' are the equal legs of the isosceles triangle TP_1T') and $HP + PT = HP + PT'$ (where P is any point on r different from P_1). Then $HT' < HP + PT'$ (by the triangle inequality) and $HP_1 + P_1T < HP + PT$.

Assessment 13-2B

1. Locate the image of vertices directly across (perpendicular to) ℓ on the geoboard.

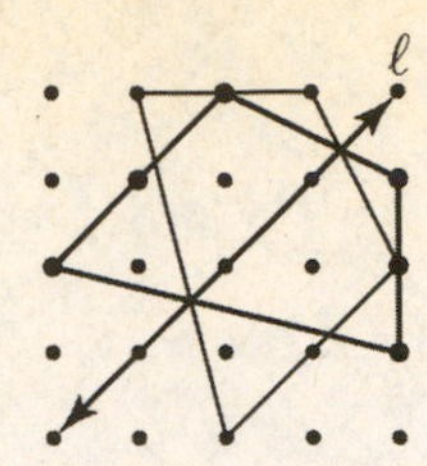

3. Methods will vary; the image is reflected in line ℓ as shown below, the reflected again to obtain the original image.

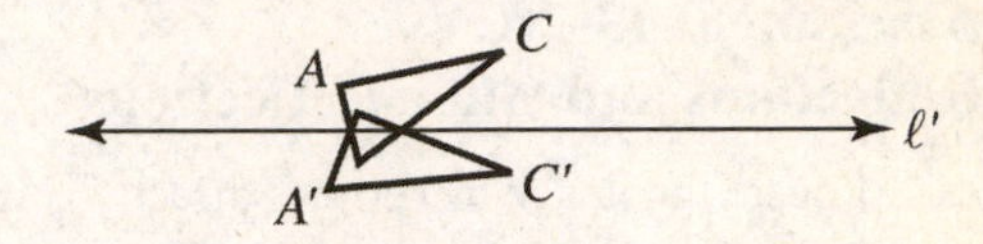

5. **(a)** The following image results:

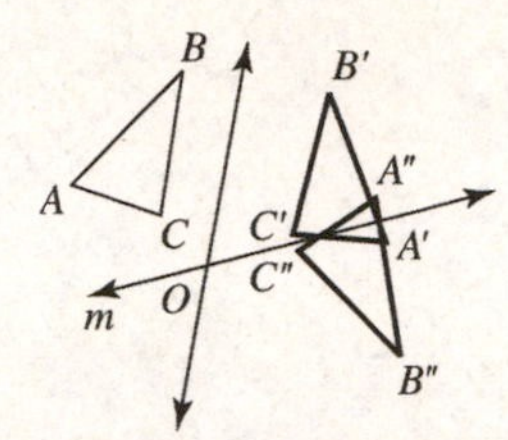

(b) A rotation about O through 2α, where α is the measure of the angle between ℓ and m in the direction from ℓ to m as shown.

(c) A half-turn about O:

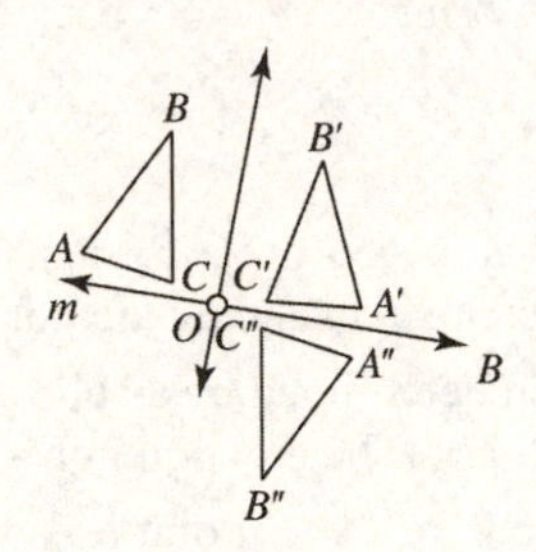

7. **(a)**

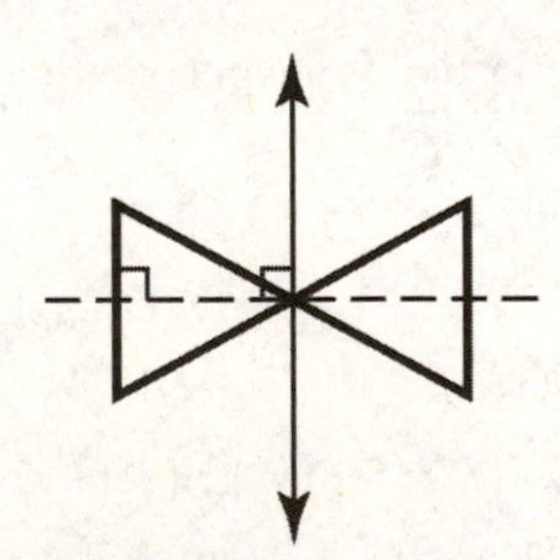

(b) Use the line perpendicular to an altitude of the original triangle. The bow-tie may be skewed, depending on the type of the triangle, but the two bases will be parallel.

(c) At least three reflecting lines are possible. Two other than the one used in part (a) could be found in the same way at the other vertices.

9. **(a)** The x-axis and all lines perpendicular to the x-axis.

(b) The y-axis and all lines perpendicular to the y-axis.

(c) The line $y = x$ and all lines perpendicular to the line $y = x$.

11. **(a)** Answers may vary. Part (a) could be moved to part (b) by:

(*i*) A reflection in the line through the common point of the parts and containing a diagonal of the large square.

(*ii*) A 90° counterclockwise rotation about the common point of the parts.

(b) Answers may vary. A 90° clockwise rotation about the center of the large square would take part (a) to part (b).

13. $P(x, y) \to \mathbf{P'(^{-}x, y)}$ under reflection in the y-axis; $P(x, y) \to \mathbf{P'(^{-}y, ^{-}x)}$ under reflection about the line $y = {^{-}x}$.

15. $P(x, y) \to P(^{-}x, y)$ when reflected in the y-axis. Thus:

(a) $y = 3x \to y = 3(^{-}x) \Rightarrow \mathbf{y = {^{-}3x}}$.

(b) $y = {^{-}x} \to y = {^{-}(^{-}x)} \Rightarrow \mathbf{y = x}$.

(c) $x = 0 \to {^{-}x} = 0 \to \mathbf{x = 0}$.

17. Given:

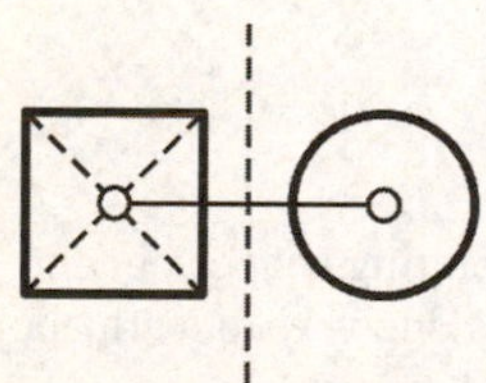

The diameter of the circle must be congruent to the distance between opposite sides of the square. Then the line of reflection is the perpendicular bisector of the segment connecting the center of the circle with the point of intersection of the diagonals of the square.

19. **(a)** All the **points on** ℓ are points with images that are the points themselves.

(b) The **point O**.

(c) **None**. There are no fixed points under translation.

(d) **None**. There are no fixed points under glide reflection.

Review Problems

15. $(^{-}a, ^{-}b)$. See below, where, for example, (a, b) is in the first quadrant. The relationship is the same regardless of the quadrant in which (a, b) is located.

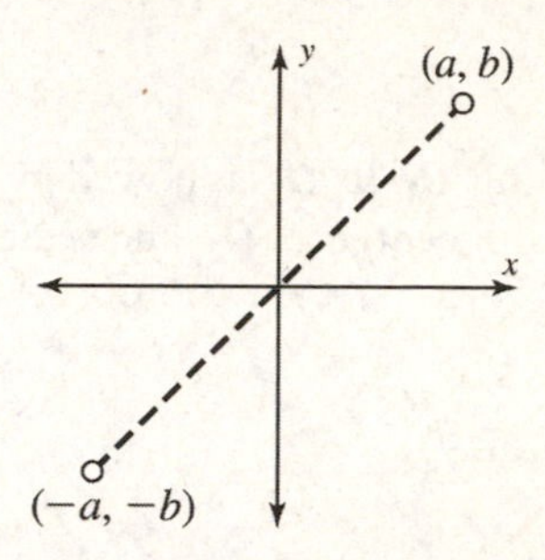

17. Construct $\overline{BE}$ perpendicular to $\overline{AD}$ as shown below:

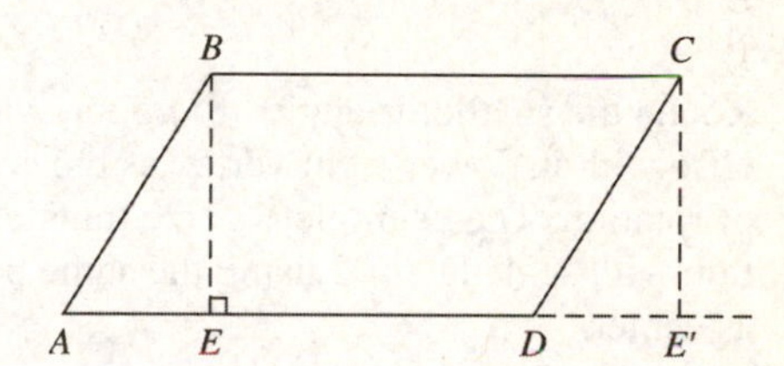

Translate ΔABE by the slide arrow from B to C. The image of ΔABE is $\Delta DCE'$, thus the rectangle $BCE'E$ is the required rectangle.

Assessment 13-3A: Dilations

1. **(a)** Slide the small triangle down three units (translation). Then complete the dilation with scale factor 2 (i.e., the larger triangle has sides twice as long) using the top right vertex as center.

(b) Slide the smaller triangle right 5 and up 1. Then complete the size transformation with scale factor 2 using the top right vertex as center.

3. Answers may vary.

(a) Translate B to B' followed by a size transformation with center B' and scale factor $\frac{A'B'}{AB}$.

(b) Rotate a half-turn with the midpoint of $\overline{AA'}$ as the center followed by a size transformation with center A'.

5. $\frac{3 \text{ cm}}{10 \text{ cm}} = \frac{BA \text{ cm}}{40 \text{ cm}} \Rightarrow 10 \cdot BA = 120 \Rightarrow BA =$ the height of the candle $=$ **12 cm**.

7. The dilation with center O and scale factor $\frac{1}{r}$ (equivalent to using division to reverse multiplication).

9. A translation taking O_1 **to** O_2 followed by a size transformation with **center at** O_2 and **scale factor** $\frac{3}{2}$.

11. **(*i*)** Scale factor of $\frac{7}{15}$.

(*ii*) $x = \frac{7}{15}(14) = \mathbf{\frac{98}{15}}$. $y = 6 \div \frac{7}{15} = \mathbf{\frac{90}{7}}$.

Assessment 13-3B

1. **(a)** Rotate the smaller triangle 90° counterclockwise with its lower-right vertex as the center of rotation. Then complete a size transformation with scale factor 2 using the same point as center.

(b) Assume the lower left corner of the grid has coordinates (0, 0). Translate the smaller triangle using $(x, y) \to (x - 5, y - 1)$. Then use a size transformation with (1, 2) as center and scale factor of 2.

3. Answers may vary.

(a) Rotate 90° counterclockwise using center B. Then translate to take B to B'. Finally complete a dilation with center B' and scale factor $\frac{A'B'}{AB} = \frac{1}{2}$.

(b) Rotate a half turn about C followed by dilation with center C and scale factor $\frac{A'B'}{AB} = \frac{3}{2}$.

5. **(*i*)** Scale factor of $\frac{6}{3} = $ **2**.

(*ii*) $x + 5 = 2x \Rightarrow \mathbf{x = 5}$..

$y + 4 = 2y \Rightarrow \mathbf{y = 4}$.

7. The dilation with **center at the origin** and **scale factor** $\frac{3}{4}$.

9.

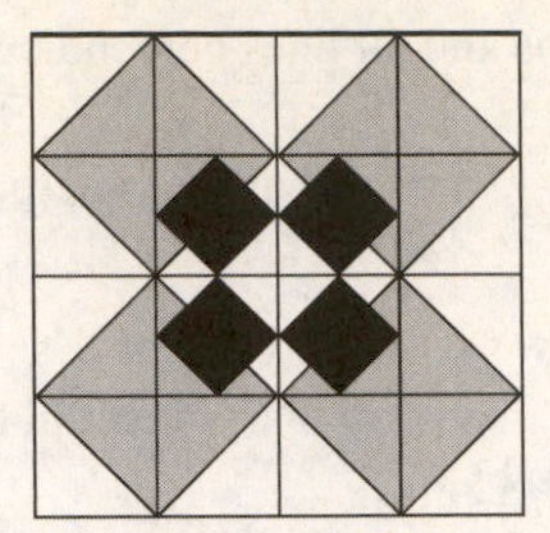

11. The center would remain at O; the undoing scale factor would be the reciprocal of $\frac{3}{4}$, or $\frac{4}{3}$.

Review Problems

11. **(a)** The translation given by slide arrow from N to M.

(b) A counterclockwise rotation of 75° about O.

(c) A clockwise rotation of 45° about A.

(d) A reflection in m and translation from B to A.

(e) A second reflection in line n.

13. **(a)** The angle will reflect to the angle itself.

(b) The square will reflect to the square itself.

Assessment 13-4A: Tessellations of the Plane

1. Forming rectangles will tessellate the plane.

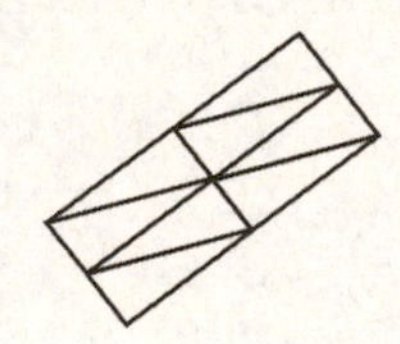

3. Experimentation by cutting shapes out and moving them about is one way to learn about these tessellations.

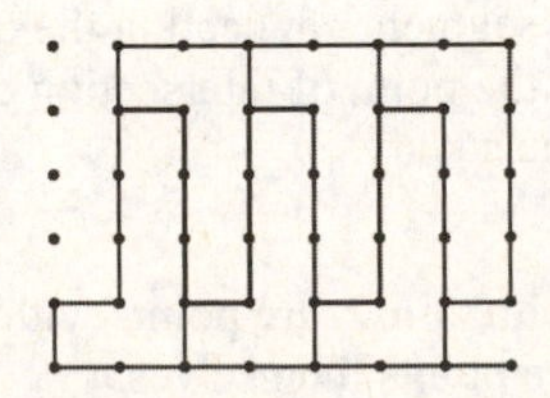

5. The shape will tessellate a plane as shown below, provided that it has a symmetry line as shown below:

7. (a) The dual is another tessellation of squares (congruent to those given).

(b) A tessellation of equilateral triangles.

(c) The tessellation of equilateral hexagons is illustrated in the statement of the problem.

9. A semiregular tessellation uses more than one kind of regular polygon.

(a) **Not Semiregular**; arrangement is different at adjacent vertexes.

(b) **Semiregular**; squares and equilateral triangles.

(c) **Semiregular**; squares, hexagons, and 12-gons.

Assessment 13-4B

1. Rotate the trapezoid 180° and place to form a parallelogram. Place these together to cover the plane.

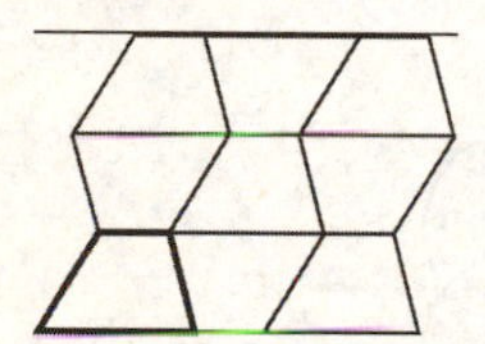

3. Experimentation by cutting shapes out and moving them about is one way to learn about these tessellations.

(a)

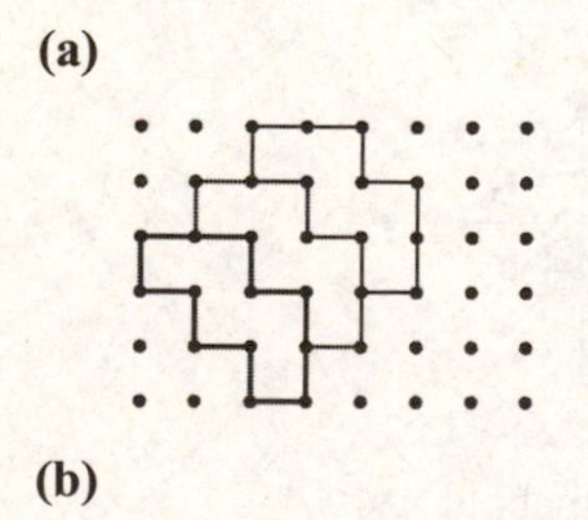

(b)

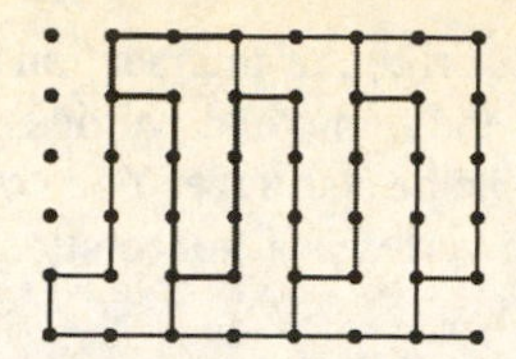

5. The shape shown is a "bow-tie" which can tessellate the plane using only copies of that shape and translations. See below:

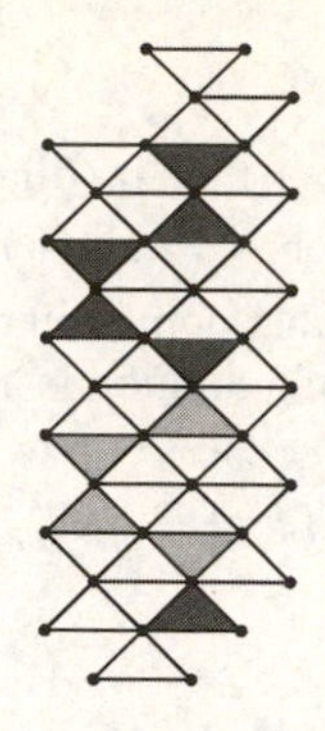

7. (a) Figure 17: A translation from A to B.

(b) Figure 18: A translation would move lizard C to lizard D; a rotation would move lizard C to lizard E.

9. (a) **Semiregular**; equilateral triangles and 12-gons.

(b) **Semiregular**; equilateral triangles and hexagons.

Review Questions

13. (a) **90°**, **180°** (point), and **270°** symmetries.

(b) **(*i*)** Two **line** symmetries; one through each of the congruent parallel sides.

(*ii*) **Point** symmetry about the center of the square.

15.

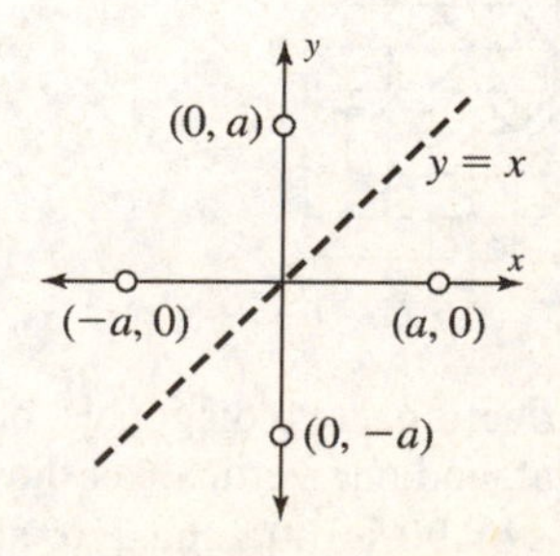

(a) $d(a)$ from the origin, or intersection of the diagonals, is the same to all vertices. Thus the figure must be a **square**. You could also say that each vertex is a 90° rotation of a previous vertex.

(b) $P(x, y) \rightarrow P(y, x)$ under reflection in the line $y = x$. Thus $(a, 0) \rightarrow$ $\mathbf{(0, a)}$, $(0, a) \rightarrow$ $\mathbf{(a, 0)}$, $(^{-}a, 0) \rightarrow$ $\mathbf{(0, {}^{-}a)}$, and $(0, {}^{-}a) \rightarrow$ $\mathbf{({}^{-}a, 0)}$.

17. Answers will vary. If $k \neq 1$, then no dilation exists since the center, a point on $y = x$ other than the center, and the image of that point will not be collinear. One example of a dilation is a dilation of $k=1$ with a center O on the line $y = x$. Any such choice of O will suffice if $k=1$.

Chapter 13 Review

1. **(a)** Reflect across ℓ:

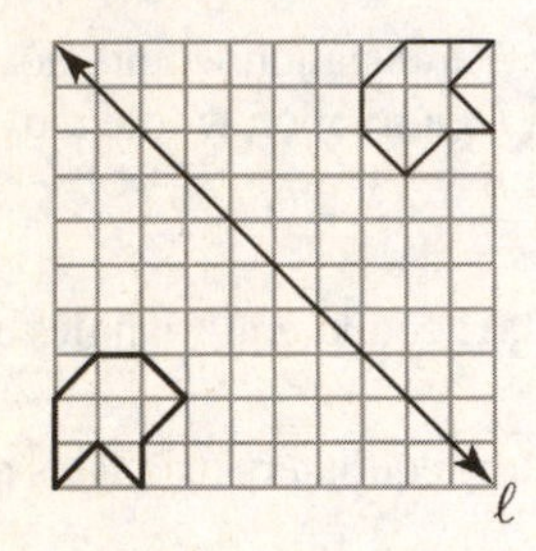

(b) Rotate 90° clockwise with O as a center:

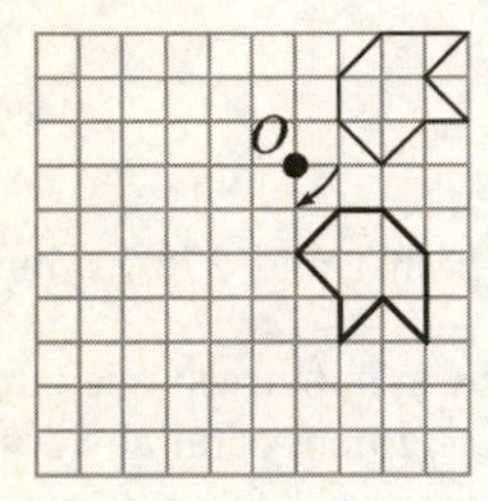

(c) Translate three units down and three units left:

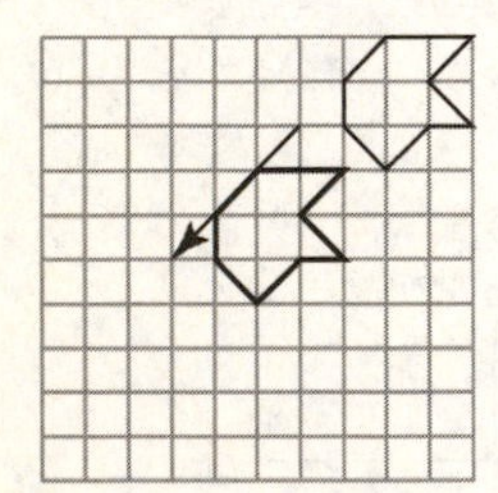

3. **(a)** **Four reflections**, two diagonals, one horizontal, and one vertical, as shown below. Rotations of 90°, 180°, and 270° also work.

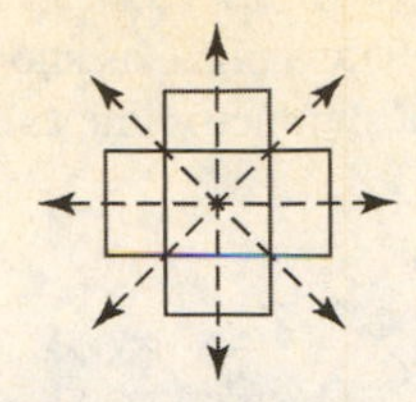

(b) **One**, the diameter bisecting the central angle.

(c) **One**, the bisector of the point angle.

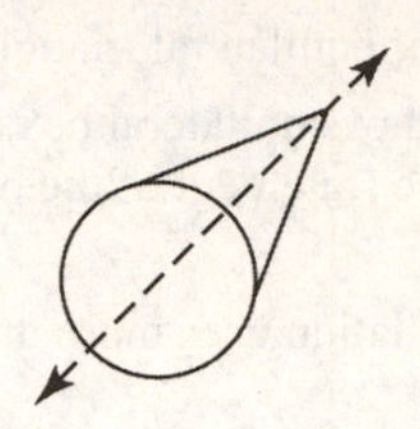

(d) **One:** a half turn through center 0.

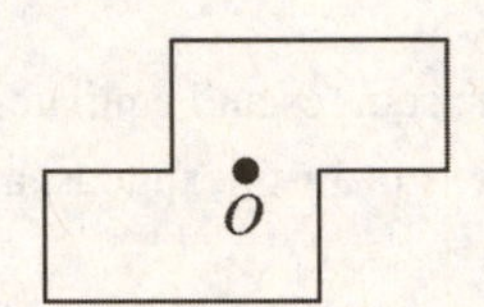

(e) **Three**: A horizontal reflection through the points of intersection, a vertical reflection through the centers, and a half turn through center 0.

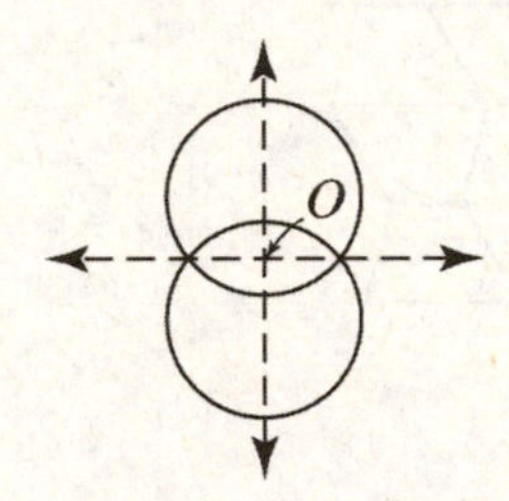

(f) **Three**: Vertically and horizontally reflections through the center and a half turn through the center.

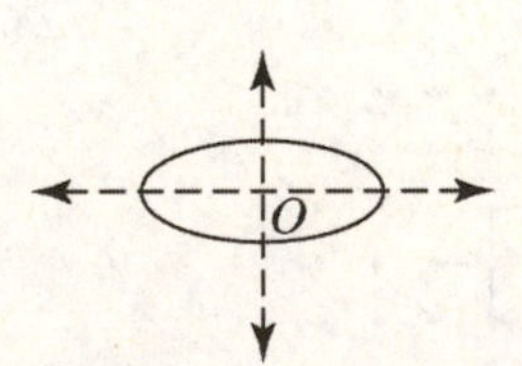

5. **(a)** A **half-turn** about x.
(b) A **half-turn** about x.

7. A **reflection** in $\overline{SO}$.

9. Rotate ΔPIG 180° (i.e., a half-turn) about the mid-point of $\overline{PT}$, then perform a size transformation with scale factor 2 and center $P'(=T)$.

11. $(x, y) \to (x, y)$. This is an "identity" translation; i.e., $(x-3, y+2)$ reverses $(x+3, y-2)$ so that every point is its own image.

13. **(a)** ***(i)*** A translation from A **to** C.

(ii) **Same** as (i). Translations with the same center are commutative.

(b) A rotation about O by $90° - 30° = $ **60° counterclockwise**.

(c) ***(i)*** A size transformations with center O and scale factor $3 \cdot 2 = \mathbf{6}$.

(ii) **Same** as (i). Size transformations are commutative as long as the center does not change.

15 **(a)** If $(x, y) \to (x+2, y-3)$ then each point on the line is translated. Thus $(0, 3) \to (2, 0)$ and $(3, 0) \to (5, {}^{-}3)$.

$m = \frac{{}^{-}3-0}{5-2} = {}^{-}1$ and $0 = {}^{-}1(2) + b \Rightarrow$ $b = 2$. The equation of the image is $\mathbf{y = {}^{-}x + 2}$.

(b) $P(x, y) \to P(x, {}^{-}y)$ under reflection in the x-axis. Thus $P(0, 3) \to P'(0, {}^{-}3)$ and $P(3, 0) \to P'(3, 0)$.

The equation of the reflected line is $\mathbf{y = x - 3}$.

(c) $P(x, y) \to P'({}^{-}x, y)$ under reflection in the y-axis. Thus $P(0, 3) \to P'(0, 3)$ and $P(3, 0) \to P'({}^{-}3, 0)$.

The equation of the reflected line is $\mathbf{y = x + 3}$.

(d) $P(x, y) \to P'(y, x)$ under reflection in the line $y = x$. Thus $P(0, 3) \to P'(3, 0)$ and $P(3, 0) \to P'(0, 3)$.

The equation of the reflected line is $\mathbf{y = {}^{-}x + 3}$.

(e) $P(x, y) \to P'({}^{-}x, {}^{-}y)$ under a half-turn about the origin. Thus $P(0, 3) \to P'(0, {}^{-}3)$ and $P(3, 0) \to P'({}^{-}3, 0)$.

The equation of the reflected line is $\mathbf{y = {}^{-}x - 3}$.

(f) $P(x, y) \to P(2x, 2y)$ under a scale factor of 2. Thus $P(0, 3) \to P'(0, 6)$ and $P(3, 0) \to P'(6, 0)$.

The equation of the scaled line is $\mathbf{y = {}^{-}x + 6}$.

17. The measure of each exterior angle of a regular octagon is $\frac{360°}{8} = 45°$. Thus the measure of each interior angle is $180° - 45° = 135°$, and $135° \nmid 360°$. Therefore a regular octagon does not tessellate the plane.

19. **Yes**. If the shape fits exactly into the object being measured, the area is the number of tessellating shapes. If the shape does not fit exactly the measuring may be difficult

21. **(a)** **Yes**, it will tessallate. Below is a start.

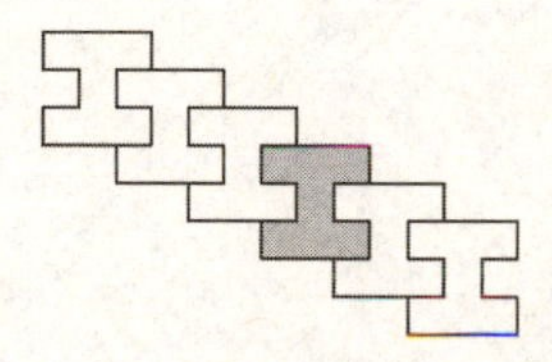

(b) **Yes**, it will tessallate. Below is a start.

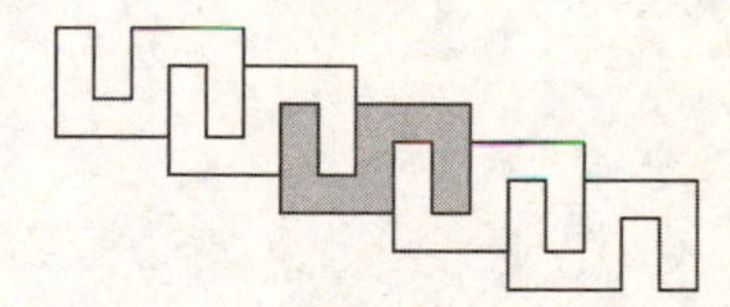

(c) **No**, it will not tessellate.

Chapter 14

Area, Pythagorean Theorem, and Volume

Assessment 14-1A: Areas of Polygons and Circles

1. Answers may vary. E.g., for a 30-inch by 60-inch desktop, and if the $8\frac{1}{2}$ by 11-inch notebook papers were to be oriented in portrait fashion, the desktop would be about 7 papers by $2\frac{3}{4}$ papers $= 19\frac{1}{4}$ square notebook papers.

3. Answers may vary. Some possible approximate measures are:

(a) A 30 in. by 78 in. door (approximately 0.8 m by 2.0 m), **about 1.6 m^2.**

(b) A 30 in. by 60 in. desktop (approximately 0.76 m by 1.5 m), **about 1.1 m^2.**

5. (a) $\frac{4000\text{ ft}^2}{1} \cdot \frac{1\text{ yd}^2}{9\text{ ft}^2} = \mathbf{444\frac{4}{9}\text{ yd}^2}$.

(b) $\frac{10^6\text{ yd}^2}{1} \cdot \frac{1\text{ mi}^2}{3.0976 \cdot 10^6\text{ yd}^2} \approx \mathbf{0.32\text{ mi}^2}$.

(c) $\frac{10\text{ mi}^2}{1} \cdot \frac{640\text{ acre}}{1\text{ mi}^2} = \mathbf{6400\text{ acres}}$.

(d) $\frac{3\text{ acre}}{1} \cdot \frac{4840\text{ yd}^2}{1\text{ acre}} \cdot \frac{9\text{ ft}^2}{1\text{ yd}^2} = \mathbf{130{,}680\text{ ft}^2}$.

7. (a) $49\text{ m} \cdot 100\text{ m} = \mathbf{4900\text{ m}^2}$.

(b) $\frac{2\text{ fields}}{1} \cdot \frac{4900\text{ m}^2}{1\text{ field}} \cdot \frac{1\text{ a}}{100\text{ m}^2} = \mathbf{98\text{ a}}$.

(c) $\frac{98\text{ a}}{2\text{ fields}} \cdot \frac{1\text{ ha}}{100\text{ a}} = \mathbf{0.98\text{ ha}}$.

9. The greatest area of the triangle would occur if drawn as below:

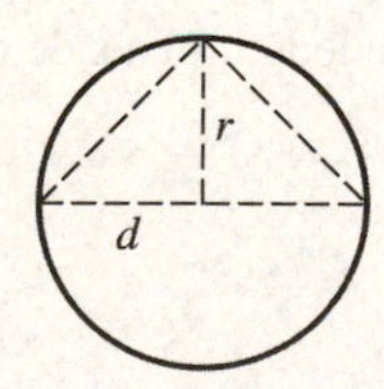

In this case, $A = \frac{1}{2}(2r)(r) = \mathbf{r^2}$.

11. (a) $A = 3 \cdot 3 = \mathbf{9\text{ cm}^2}$.

(b) $A = 8 \cdot 12 = \mathbf{96\text{ cm}^2}$.

(c) $A = 5 \cdot 4 = \mathbf{20\text{ cm}^2}$.

(d) $A = \frac{1}{2} \cdot 7 \cdot (14 + 10) = \mathbf{84\text{ cm}^2}$.

13. (a) **True**. It is not possible to determine a height when only side lengths are known.

(b) The area could be 60cm^2 if the parallelogram is a rectangle; however the assertion is **false**.

(c) **False**. The area cannot be greater than 60cm^2 since the maximum is 60cm^2 if the parallelogram is a rectangle.

(d) **False**. The area could equal 60cm^2.

15. (a) $6.5\text{ m} \times 4.5\text{ m} = 29.25\text{ m}^2$. $29.25\text{ m}^2 \times \$13.85\text{ per m}^2 = \mathbf{\$405.11}$.

(b) $15\text{ ft} \times 11\text{ ft} = 165\text{ ft}^2$. $\left(\frac{165\text{ ft}^2}{1}\right) \cdot \left(\frac{1\text{ yd}^2}{9\text{ ft}^2}\right) = \frac{165}{9}\text{ yd}^2$. $18\frac{1}{3}\text{ yd}^2 \times \$30\text{ per yd}^2 = \mathbf{\$550}$.

17. Bathroom area $= 300\text{ cm} \times 400\text{ cm} = 120{,}000\text{ cm}^2$. Each tile is $10\text{ cm} \times 10\text{ cm} = 100\text{ cm}^2$, thus $\frac{120{,}000\text{ cm}^2}{100\text{ cm}^2} =$ **1200 tiles** (assuming no waste).

19. (a) $C = 2\pi r \Rightarrow r = \frac{C}{2\pi} = \frac{8\pi}{2\pi} = 4$.
$A = \pi r^2 = \pi(4)^2 = \mathbf{16\pi\text{ cm}^2}$.

(b) $A_{circle} = A_{square} \Rightarrow \pi r^2 = s^2 \Rightarrow r^2 = \frac{s^2}{\pi}$.
$\mathbf{r = \frac{s}{\sqrt{\pi}}}$.

21. The flower bed with its encircling sidewalk forms a circle with radius $(3 + 1) = 4$ m. $A_{encircled\ bed} = \pi \cdot 4^2 = 16\pi\text{ m}^2$. $A_{flower\ bed} = \pi \cdot 3^2 = 9\pi\text{ m}^2$. $A_{side\ walk} = A_{encircled\ bed} - A_{flower\ bed} = 16\pi - 9\pi = \mathbf{7\pi\text{ m}^2}$.

23. **(a)** Area is **quadrupled**. Sides of length s mean area $= s^2$. Sides of length $2s$ mean area $= (2s)^2 = 4s^2$, which is quadruple the original area.

(b) **1 : 25**. Sides of first square $= 1s$, so area $= (1s)^2 = s^2$. Sides of second square $= 5s$, so area $= (5s)^2 = 25s^2$—or a ratio of 1 : 25.

25. **(a)** The area of the large rectangle $= a(b + c)$. The total of the areas of the separate rectangles $= ab + ac = a(b + c)$.

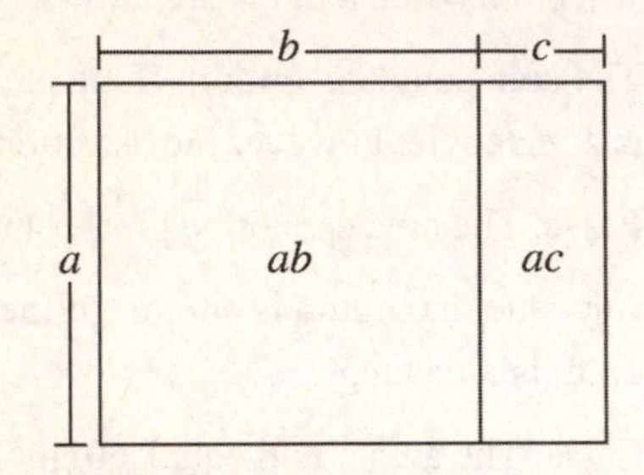

27. Label the triangle as shown:

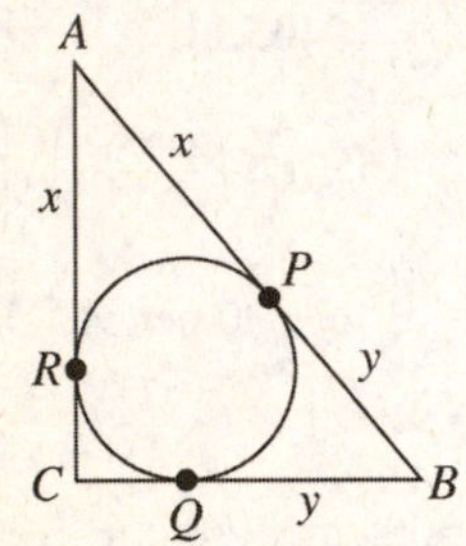

If two tangents are drawn to a circle from the same point on the exterior of the circle, the distances from the common point to the points of tangency are equal. Thus $RC = 4 - x$ and $QC = 3 - y$. Since $QC = RC$, then $4 - x = 3 - y$, or $x - y = 1$. $AB = 5 \Rightarrow x + y = 5$.
Solving this system of equations:

$$\begin{aligned} x - y &= 1 \\ \underline{x + y} &\underline{= 5} \\ 2x &= 6 \\ x &= 3 \end{aligned}$$

Since $x = 3$, then RC, and the radius of the inscribed circle, $= 4 - 3 = 1$. The radius of the circular glass is **1 ft**.

29. Draw altitudes $\overline{BE}$ and $\overline{DF}$ of triangles BCP and DCP, respectively. $\Delta ABE \cong \Delta CDF$ by AAS, thus $\overline{BE} \cong \overline{DF}$.

Because $\overline{CP}$ is a base of ΔBCP and ΔDCP, and because their heights are the same, the areas must be equal.

Assessment 14-1B

1. Answers may vary. E.g., for a 60-inch by 30-inch desktop and a hand measuring 4 inches by 8 inches (thumb closed) and oriented in portrait fashion, the desktop would be about 15 hands by $3\frac{3}{4}$ hands, or about $56\frac{1}{4}$ hands.

3. Answers may vary. Some possible approximate measures are:

(a) An 18 in. by 18 in. chair seat (approximately 46 cm by 46 cm), **about 2100 cm²**.

(b) A 4 ft by 6 ft white-or chalkboard (approximately 1.2 m by 1.8 m), **about 2.2 m²**.

5. **(a)** $\frac{99\text{ ft}^2}{1} \cdot \frac{1\text{ yd}^2}{9\text{ ft}^2} = \mathbf{11\,yd^2}$.

(b) $\frac{10^6\text{ yd}^2}{1} \cdot \frac{9\text{ ft}^2}{1\text{ yd}^2} \approx \mathbf{9 \cdot 10^6\,ft^2}$.

(c) $\frac{6.5\text{ mi}^2}{1} \cdot \frac{640\text{ acre}}{1\text{ mi}^2} = \mathbf{4160\ acres}$.

(d) $\frac{3\text{ acre}}{1} \cdot \frac{4840\text{ yd}^2}{1\text{ acre}} = \mathbf{14{,}520\,yd^2}$.

7. **(a)** $I = 3$; $B = 8$. $A = 3 + \frac{1}{2} \cdot 8 - 1 = \mathbf{6\ units^2}$.

(b) $I = 0$; $B = 11$. $A = 0 + \frac{1}{2} \cdot 11 - 1 = \mathbf{4\frac{1}{2}\ units^2}$.

9. Answers may vary. One possibility is as below:

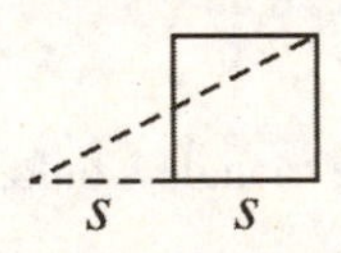

Where s is the length of a side of the square. Then $A_{triangle} = \frac{1}{2}(2s)(s) = s^2$, which is the area of the square.

11. **(a)** $A = 9 \cdot 9 = \mathbf{81\,cm^2}$.

(b) $A = \frac{1}{2} \cdot 6 \cdot (27 + 8) = \mathbf{105\,cm^2}$.

13. **(a)** Area $= L \cdot W$. If the perimeter is increased for a constant area, length must increase while width decreases.

Thus the rectangle will elongate and become less like a square.

(b) For the area to change, the dimensions must change. For the perimeter to remain constant, as one side increases the adjacent side must decrease. If we start with a rectangle that is not a square, then the area will increase as we change the dimensions toward a square.

15. The plot is $22 \text{ m} \times 28 \text{ m} = 616 \text{ m}^2$. $\frac{616 \text{ m}^2}{85 \text{ m}^2 \text{ per bag}}$ ≈ 7.25 bags, so **8 bags** must be purchased.

17. $A_{rectangle} = 64 \cdot 25 = 1600 \text{ m}^2$. Length of each side of a square with area of $1600 \text{ m}^2 =$ $\sqrt{1600} = \mathbf{40\,m}$.

19. **(*i*)** Square peg inside circular hole: Diagonal of square $= 2r$. Then $(2r)^2 = s^2 + s^2$ (by the Pythagorean theorem), where r is the radius of the circle and s is the length of a side of the square, so $s^2 = 2r^2$. $A_{square} = s^2 = 2r^2$.

Percentage of wasted space $= \frac{A_{circle} - A_{square}}{A_{circle}} =$

$\frac{\pi r^2 - 2r^2}{\pi r^2} = \frac{\pi - 2}{\pi} \approx 36.34\%$.

(*ii*) Circular peg inside square hole: Length of side of square $= 2r$. $A_{square} = (2r)^2 = 4r^2$.

Percentage of wasted space $= \frac{A_{square} - A_{circle}}{A_{square}} =$

$\frac{4r^2 - \pi r^2}{4r^2} = \frac{4 - \pi}{4} \approx 21.46\%$.

The **circular peg inside the square hole** has less wasted space.

21. $A_{complete\ target} = \pi(5^2) = 25\pi \text{ in.}^2$;

$A_{out\,side\,shaded\ region} = 25\pi - \pi(4^2) = 9\pi \text{ in.}^2$; and

$A_{in\,side\,sheded\ region} = \pi(3^2) = 9\pi \text{ in.}^2$.

The areas of the inside and outside shaded regions are **both 9π in.^{2}**

23. **(a)** Area is **quadrupled**. Diameter doubled means radius doubled. Area $= \pi(2r)^2 = 4\pi r^2$, which is quadruple the original area.

(b) Area is 1.1^2, or **1.21 times** as great. New radius is 110% of old radius $= 1.1r$, so area $= \pi(1.1r)^2 = 1.21\pi r^2$ (or area is increased by 21%).

(c) Area will **increase by** factor of **9**. $C = 2\pi r$ $\Rightarrow r = \frac{C}{2\pi} \Rightarrow$ area $= \pi\left(\frac{C}{2\pi}\right)^2$. If circumference is increased by a factor of 3, then area $= \pi\left(\frac{3C}{2\pi}\right)^2$, or area will increase by 3^2, a factor of 9.

25. The area of the large rectangle is $(a + b)(c + d)$ square units. The areas of the four small rectangles that make up the large rectangle are ac, ad, bc, and bd square units. The sum of the areas of the four small rectangles equals the area of the large rectangle they construct. Thus,
$\mathbf{(a + b)(c + d) = ac + ad + bc + bd}$.

27. Rotate square B clockwise around the center of square A until one side is perpendicular to the corresponding side of square A:

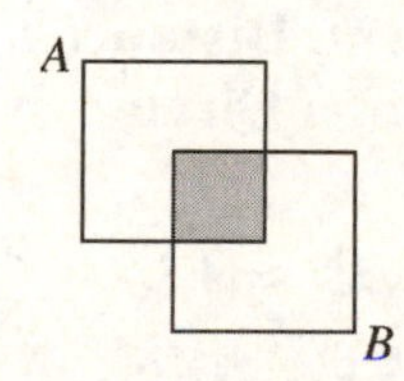

The shaded area is then $\frac{1}{4}$ **the area of square *A*.**

The shaded area will remain $\frac{1}{4}$ the area of square A regardless of the degree of rotation.

29. In the figure of problem 28, let h be the height between each pair of parallel lines (and between $\overline{DE}$ and B). Since the segments are congruent, $FG = 2DE$ and $AC = 3DE$.

(a) $\frac{\text{Area } DBE}{\text{Area } DEGF} = \frac{\frac{1}{2}(DE)h}{\frac{1}{2}(DE+FG)h} = \frac{DE}{DE+2DE} = \frac{1}{3}$.

(b) $\frac{\text{Area } DBE}{\text{Area } FGCA} = \frac{\frac{1}{2}(DE)h}{\frac{1}{2}(FG+CA)h} = \frac{DE}{2DE+3DE} = \frac{1}{5}$.

(c) $\frac{\text{Area } DEGF}{\text{Area } FGCA} = \frac{\frac{1}{2}(DE+FG)h}{\frac{1}{2}(FG+CA)h} = \frac{DE+2DE}{2DE+3DE} = \frac{3}{5}$.

(d) $\frac{\text{Area } DEGF}{\text{Area } ABC} = \frac{\frac{1}{2}(DE+FG)h}{\frac{1}{2}(AC)3h} = \frac{DE+2DE}{3(3DE)}$

$= \frac{3}{9} = \mathbf{\frac{1}{3}}.$

(e) $\frac{\text{Area } FGCA}{\text{Area } ABC} = \frac{\frac{1}{2}(FG+AC)h}{\frac{1}{2}(AC)3h} = \frac{2DE+3DE}{3(3DE)} = \mathbf{\frac{5}{9}}.$

(f) $\frac{\text{Area } ABC}{\text{Area } DECA} = \frac{\frac{1}{2}(AC)3h}{\frac{1}{2}(DE+AC)2h} = \frac{3(3DE)}{2(DE+3DE)} = \mathbf{\frac{9}{8}}.$

Review Problems

19. Each rounded circular corner has perimeter $\frac{1}{4}(2\pi \cdot 3) = \frac{3}{2}\pi$ in. Thus the total perimeter $= 4\left(\frac{3}{2}\pi\right) + 4(24) = (6\pi + 96)$ inches $\approx$ 114.8 inches or **about 9.6 feet.**

21. The perimeter of a regular hexagon is $6s$, where s is the length of each side. In a regular hexagon the radius r of the circumscribed circle equals s and its circumference is thus $2\pi s$. The ratio of the circumference to the perimeter is then $\frac{2\pi s}{6s} = \frac{\pi}{3} \approx$ 1.04720, making the circumference about **4.7%** longer. Alternatively, the difference in length is $2\pi s - 6s \approx$ **0.28** s.

Assessment 14-2A: The Pythagorean Theorem, Distance Formula, and Equation of a Circle

1. (a) $d = \sqrt{4^2 + 2^2} = \mathbf{\sqrt{20} \approx 4.5}.$

(b) $d = \sqrt{4^2 + 1^2} = \mathbf{\sqrt{17} \approx 4.1}.$

3. (a) Make a right triangle with **sides 2 and 3**.

$d = \sqrt{2^2 + 3^2} = \sqrt{13}.$

(b) Make a right triangle with **sides 1 and 2**.

$d = \sqrt{1^2 + 2^2} = \sqrt{5}.$

5. The hypotenuse has length $\sqrt{4^2 + 4^2} = 4\sqrt{2}$.

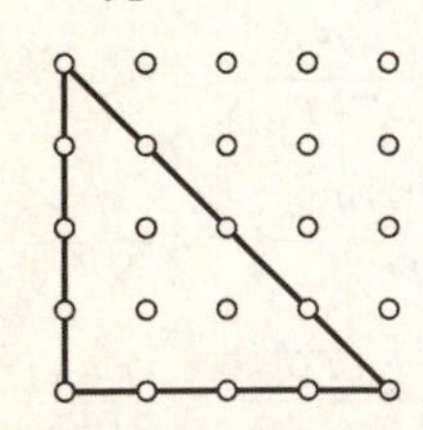

$P = 8 + 4\sqrt{2} \approx$ **13.66 units.**

7. For the answer to be yes, the number must satisfy the Pythagorean theorem.

(a) No. $24^2 \neq 10^2 + 16^2$.

(b) Yes. $34^2 = 16^2 + 30^2$.

(c) Yes. $2^2 = (\sqrt{2})^2 + (\sqrt{2})^2$.

9. The boat is 10 miles south and 5 miles east of A. The distance from A is $d = \sqrt{10^2 + 5^2} = \sqrt{125}$, or **about 11.2 miles.**

11. The tall pole stands 10 m above the short pole. Draw a horizontal line from the top of the short pole to form a right triangle, where d is the distance between the poles. $d^2 + 10^2 = 14^2 \Rightarrow d^2 = 14^2 - 10^2 \Rightarrow d = \sqrt{196 - 100} = \sqrt{96} = \sqrt{16 \cdot 6} = 4\sqrt{6}$, or **about 9.8 m.**

13. (a) $A = \frac{1}{2} \cdot$ (length of diagonal 1) $\cdot$ (length of diagonal 2) $= \frac{1}{2}(8)\left(2 \cdot \sqrt{10^2 - 4^2}\right) = \mathbf{8\sqrt{84}}.$

Alternatively, observe that the figure is made of 4 triangles, each of which have area $\frac{1}{2}(4) \cdot \sqrt{10^2 - 4^2} = 2\sqrt{84}$.

(b) The figure is made of four triangles, each having area $\frac{1}{2}(2)\sqrt{10^2 - 2^2} = \sqrt{96}$. Thus, the area is $\mathbf{4\sqrt{96}}$.

15. Let ℓ be the length of the lake. $\ell^2 + 150^2 = 180 \Rightarrow \ell^2 = 180^2 - 150^2 \Rightarrow \ell = \sqrt{9900} = \sqrt{100 \cdot 99} = 10\sqrt{99}$, or **about 99.5 ft.**

17. The area of the large square equals the sum of the areas of the small square and the four right triangles:

$(a + b)^2 = c^2 + 4\left(\frac{ab}{2}\right) \Rightarrow a^2 + 2ab + b^2 = c^2 + 2ab \Rightarrow a^2 + b^2 = c^2.$

The inside quadrilateral is in fact a square, since each side is the hypotenuse of a triangle with the same length sides, and since at each of its vertices there are three angles whose measures sum to 180º— two of which are complementary. Thus the angles of the quadrilateral are right angles.

19. We can position the figure on a coordinate system with the figure positioned so that the origin is at the midpoint of the two right angles, as shown below.

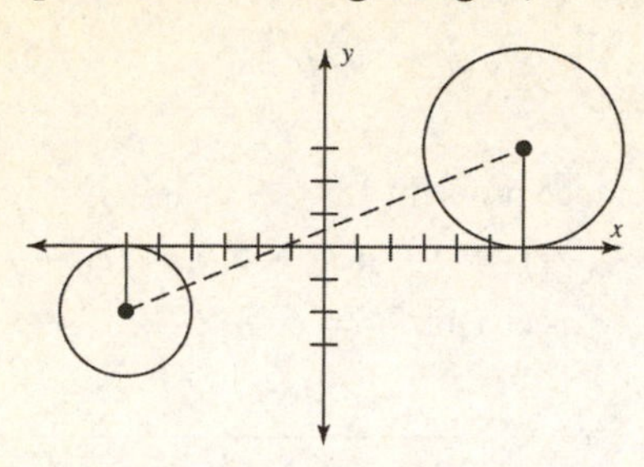

Thus, the distance is $\sqrt{(6-(-6))^2+(3-(-2))^2}$ $=\sqrt{12^2+5^2}=\sqrt{144+25}=\mathbf{13}$.

21. From the special properties of a 30°-60°-90° right triangle, the short leg (opposite the 30° angle) is half the hypotenuse, or $\frac{1}{2}\cdot\frac{c}{2}=\frac{c}{4}$. The longer leg (opposite the 60° angle) is $\sqrt{3}$ times the short leg. Thus the side opposite the 60° angle $=\sqrt{3}\cdot\frac{c}{4}=$ $\frac{c\sqrt{3}}{4}$.

23. Given the triangle, draw and label it as follows:

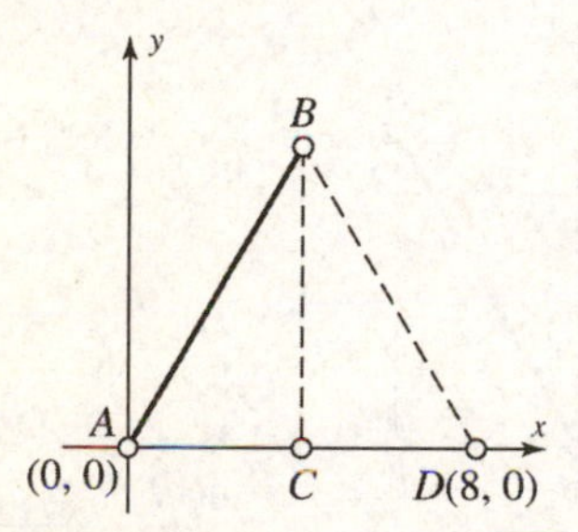

(a) Draw the line segment $\overline{BC}$ so that it is perpendicular to the x-axis. Reflect the triangle ABC about $\overline{BC}$ to form isosceles triangle ABC with third vertex at **(16, 0)**.

(b) $AB^2=AC^2+BC^2\Rightarrow AB=\sqrt{8^2+8^2}$ $=8\sqrt{2}$. Thus the sides are $\mathbf{8\sqrt{2}, 8\sqrt{2},}$ and **16**.

(c) $(8\sqrt{2})^2+(8\sqrt{2})^2=128+128=256=\mathbf{16^2}$.

25. **(a)** If $(x-0)^2+(y-0)^2=4^2$, then the center of the circle is **(0,0)** and the radius is **4**.

(b) If $(x-3)^2+(y-2)^2=10^2$, then the center of the circle is **(3, 2)** and the radius is **10**.

(c) If $(x-{}^-2)^2+(y-3)^2=(\sqrt{5})^2$, then the center of the circle is **(⁻2, 3)** and the radius is $\mathbf{\sqrt{5}}$.

(d) If $(x-0)^2+(y-{}^-3)^2=3^2$, the center of the circle is **(0, ⁻3)** and the radius is **3**.

Assessment 14-2B

1. **(a)** $d=\sqrt{2^2+2^2}=\sqrt{8}\approx\mathbf{2.8}$.

(b) $d=\sqrt{3^2+2^2}=\sqrt{13}\approx\mathbf{3.6}$.

3. **(a)** Make a right triangle with **sides 1 and 3**. $d=\sqrt{1^2+3^2}=\sqrt{10}$.

(b) Make a right triangle with **sides 1 and 4**. $d=\sqrt{1^2+4^2}=\sqrt{17}$.

5. The sides of a square made in this fashion (i.e. the sides not unit numbers) will always be hypotenuses of right triangles [see part (a)]. As such, the relationship $s^2=a^2+b^2$ must hold, with a and b whole numbers (i.e. they are spaces between dots). Only number representable in this way can be areas of squares, or s^2.

(a) $s^2=1^2+2^2=5$.

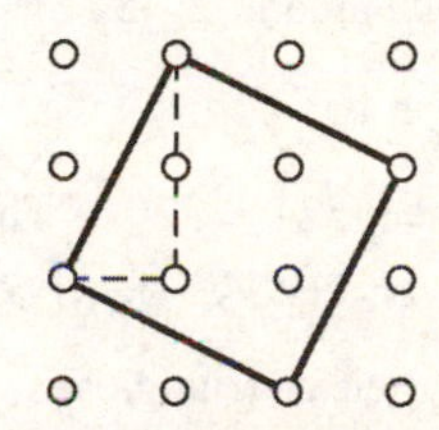

(b) **Not possible.** No sums of squares of whole numbers equals 7.

(c) $s^2=2^2+2^2=8$.

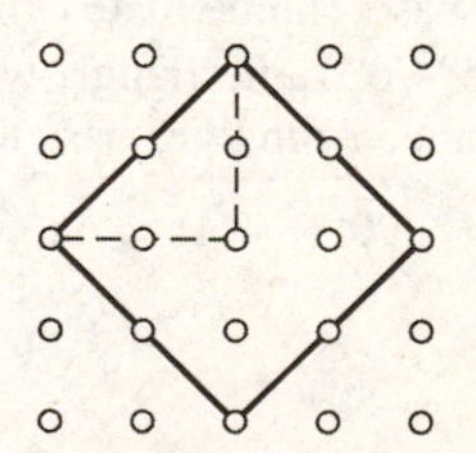

(d) **Not possible.** No sums of squares of whole numbers equals 14.

(e) **Not possible.** No sums of squares of whole numbers equals 15.

7. For the answer to be yes, the numbers must satisfy the Pythagorean theorem.

(a) **Yes.** $\left(\frac{5}{2}\right)^2 = \left(\frac{3}{2}\right)^2 + \left(\frac{4}{2}\right)^2$.

(b) **Yes.** $(\sqrt{5})^2 = (\sqrt{2})^2 + (\sqrt{3})^2$.

(c) **Yes.** $30^2 = 18^2 + 24^2$.

9. Let d be the distance along the highway intersected by a 6.1 mile circle with center C. $\left(\frac{1}{2}d\right)^2 + 3^2 = 6.1^2 \Rightarrow \frac{1}{4}d^2 = 6.1^2 - 3^2 \Rightarrow d^2 \approx 4(37.21 - 9) \Rightarrow d = \sqrt{112.84}$, or **about 10.6 mi.**

11. Label the lower right corner D. Then $BD = CD = d$; $d^2 + d^2 = 12^2 \Rightarrow 2d^2 = 144 \Rightarrow d = \sqrt{72} = \sqrt{36 \cdot 2} = 6\sqrt{2}$. Each side $= 2d = 12\sqrt{2}, \approx 16.97$ in., or **17 in.** to the nearest $\frac{1}{10}$ in.

13. **(a)** Let d be the distance from the intersection of the diagonals to the vertical apex. $d = \sqrt{5^2 - 2^2} = \sqrt{21}$.

The area of the top triangle is: $A = \frac{1}{2} \cdot (2 + 2) \cdot \sqrt{21} = 2\sqrt{21}$. There are two triangles, thus total area is $2 \cdot 2\sqrt{21} = \mathbf{4\sqrt{21}\ cm^2}$.

(b) $d = \sqrt{5^2 - 1^2} = \sqrt{24} = 2\sqrt{6}$. The area of each triangle is: $A = \frac{1}{2} \cdot 2 \cdot 2\sqrt{6} = 2\sqrt{6}$, thus the total of the two triangles is: $A = 2 \cdot 2\sqrt{6} = \mathbf{4\sqrt{6}\ cm^2}$.

15. The distance from home plate to third base is the same as the distance from home plate to first base (90 feet).The third base - home plate - first base triangle is a 45°-45°-90° right triangle with 90-feet legs. Then the distance from third base to first base $= \sqrt{90^2 + 90^2} = \sqrt{90^2(1 + 1)} = 90\sqrt{2}$, or **about 127.28 feet.**

17. The outside quadrilateral is a square of area c^2. The inside quadrilateral is a square of area $(b - a)^2 = b^2 - 2ab + a^2$.

Each of the four triangles have are $\frac{1}{2}ab$.

Thus $c^2 = (b^2 - 2ab + a^2) + 4\left(\frac{1}{2}ab\right)$;
$c^2 = b^2 - 2ab + a^2 + 2ab$; so
$c^2 = a^2 + b^2$.

19. Reference problem 18. $t + a_1 + a_2 = (c_1 + a_1) + (c_2 + a_2) \Rightarrow t + (a_1 + a_2) = (c_1 + c_2) + (a_1 + a_2)$. Thus $t = c_1 + c_2$.

21. **(*i*)** $AB = \sqrt{(1 - {}^-2)^2 + ({}^-1 - {}^-5)^2} = \sqrt{9 + 16} = 5$.

(*ii*) $BC = \sqrt{(5 - 1)^2 + (2 - {}^-1)^2} = \sqrt{16 + 9} = 5$.

(*iii*) $AC = \sqrt{({}^-2 - 5)^2 + ({}^-5 - 2)^2} = \sqrt{49 + 49} = 7\sqrt{2}$.

$AB = BC$, so the triangle is isosceles.

23. Given the triangle, draw and label it as follows:

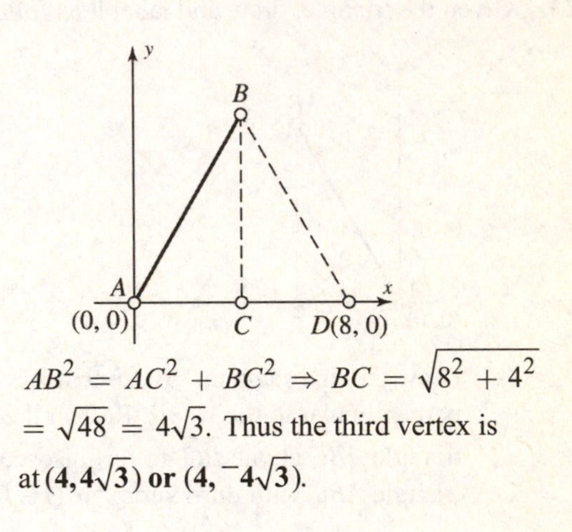

$AB^2 = AC^2 + BC^2 \Rightarrow BC = \sqrt{8^2 + 4^2} = \sqrt{48} = 4\sqrt{3}$. Thus the third vertex is at **$(4, 4\sqrt{3})$ or $(4, {}^-4\sqrt{3})$**.

25. **(a)** If $(x - 3)^2 + (y - {}^-2)^2 = 3^2$, then the center of the circle is **$(3, {}^-2)$** and the radius is **3**.

(b) $3x^2 + 3y^2 = 9 \Rightarrow x^2 + y^2 = 3$. Then if $(x - 0)^2 + (y - 0)^2 = (\sqrt{3})^2$, then the center of the circle is **(0, 0)** and the radius is **$\sqrt{3}$**.

27. The intersection of the radius of the circle with the point of tangency is the right angle APO, so

ΔAPO is a right triangle.

Thus $AP = \sqrt{d^2 - r^2}$.

Review Problems

15. Change all values to meters. Then $0.032 \text{ km}^2 = 32000 \text{ m}^2 > 3.2 \text{ m}^2 > 322 \text{ cm}^2 = 0.0322 \text{ m}^2 > 3020 \text{ mm}^2 = .00302 \text{ m}^2$.

17. In each case: circumference $= \pi d$ or $2\pi r$; area $= \pi r^2$. If given A, $r = \sqrt{\frac{A}{\pi}}$ and if given C, $r = \frac{C}{2\pi}$.

	Radius	Diameter	Circumference	Area
(a)	5 cm	**10 cm**	**10π cm**	**$25\pi \text{ cm}^2$**
(b)	**12 cm**	24 cm	**24π cm**	**$144\pi \text{ cm}^2$**
(c)	**$\sqrt{17}$ m**	**$2\sqrt{17}$ m**	**$2\pi\sqrt{17}$ m**	$17\pi \text{ m}^2$
(d)	**10 cm**	**20 cm**	20π cm	**$100\pi \text{ cm}^2$**

Assessment 14-3A: Geometry in Three Dimensions

1. (a) **Quadrilateral pyramid;** possibly square pyramid. This is a polyhedron determined by a simple closed polygonal region, a point not in the plane of the region, and triangular regions determined by the point and each pair of consecutive vertices of the polygonal region.

(b) **Quadrilateral prism;** possibly trapezoidal or right trapezoidal prism. This is a polyhedron in which two congruent polygonal faces lie in parallel planes and the other faces are bounded by parallelograms.

(c) **Pentagonal pyramid**. The closed polygonal region is a pentagon.

3. (a) Vertices: ***A, D, R, W***.

(b) Edges: $\overline{AR}, \overline{RD}, \overline{AD}, \overline{AW}, \overline{WR}, \overline{WD}$.

(c) Faces: $\Delta ARD, \Delta DAW, \Delta AWR, \Delta DRW$.

(d) Intersection of ΔDRW and $\overline{RA}$: point ***R***.

(e) Intersection of ΔDRW and ΔDAW: $\overline{DW}$.

5. (a) **True**. This is the definition of a right prism.

(b) **False**. No pyramid is a prism; i.e., a pyramid has one base and a prism two bases.

(c) **True**. The definition of a pyramid starts with the fact that it is a polyhedron.

(d) **False**. They lie in parallel planes.

7. (a) **Right regular hexagonal pyramid**. Assume the base is a regular hexagon and the triangles are congruent. Points, if folded up, would meet at a vertex.

(b) **Right square pyramid**. Assume the base is a square and triangles are congruent. Points would meet at a vertex.

(c) **Cube**. All six faces are squares; assume they are congruent.

(d) **Right square prism**. Assume the lateral bases are congruent and the ends congruent. Lateral faces of the prism would all be bounded by rectangles.

(e) **Right regular hexagonal prism**. Assume both bases are congruent regular hexagons and the lateral faces are congruent.

9. (a) ***i*** (end view), ***ii*** (top view), ***iii*** (side view).

(b) ***i, ii, iii, iv*** (top views).

11. (a) There will be three pairs of parallel faces, determined by any three pairs of opposite sides of the hexagonal base. E.g., $AA'B'B$ and $EE'D'D$.

(b) **120°**. E.g., $\angle FAB$ is a dihedral angle between two adjacent faces (each of the interior angles of a regular hexagon measure 120°).

Assessment 14-3B

1. (a) triangular prism

(b) quadrilateral pyramid

(c) cylinder

3. (a) A, B, C, D, E, F

(b) $\overline{AB}, \overline{BC}, \overline{CE}, \overline{BE}, \overline{EF}, \overline{CD}, \overline{AD}, \overline{DF}, \overline{AF}$

(c) $\Delta BCE, \Delta ADF$, quadrilateral $CDFE$, quadrilateral $ABEF$, quadrilateral $ABCD$.

(d) E.

(e) $\varnothing$.

5. (a) **False**. They base can be any simple closed curve.

(b) **False**. It has two bases.

(c) **False**. They are parallelograms; if they were rectangles it would be a right prism.

(d) **True,** by definition.

7. (a) **Triangular right prism**. Assuming the squares and triangles are congruent, the two congruent triangular bases lie in parallel planes.

(b) **Right pyramid** with square base. I.e., four triangular faces, assuming the triangles are congruent and the rectangle is a square.

9. (a) **Object 2**. Note the relationship between numbered faces and the orientation of the numbers.

   (b) **Object 4**. The two designs cannot be on adjoining faces.

11. (a) **None**. A pentagon has no parallel edges, thus there are no parallel lateral faces.

   (b) The measure of any of the five congruent interior angles of either of the regular pentagon bases is $\frac{(5-2)\cdot 180^\circ}{5} = 108^\circ$ [per theorem 11-4, the sum of measure of the interior angles of any convex n-gon is $(n-2)\cdot 180^\circ$ and there are five interior angles in a pentagon]. Each of these angles is a dihedral angle between adjacent faces $\Rightarrow$ **108°**.

13.

Prism	Vertices per Base	Diagonals per Vertex	Total Number of Diagonals
Quadrilateral	4	1	4
Pentagonal	5	2	10
Hexagonal	6	3	18
Heptagonal	7	4	28
Octagonal	8	5	40
⋮	⋮	⋮	⋮
n-gon	n	$n-3$	$n(n-3)$

Review Problems

13. (a) $\frac{1}{2}bh = \frac{1}{2}\cdot 9\cdot 12 = \mathbf{54\ m^2}$

   (b) $bh = 15\cdot 20 = 300\text{ in.}^2$

Assessment 14-4A: Surface Areas

1. (b) and (d) can form cubes. The other figures have one of the faces out of order.

3. Area of walls $= 2(6\text{ m})(2.5\text{ m}) + 2(4\text{ m})(2.5\text{ m}) = 50\text{ m}^2$. Paint needed is $\left(\frac{50\text{ m}^2}{1}\right)\cdot\left(\frac{1\text{ L}}{20\text{ m}^2}\right) =$ **2.5 liters**, so buy 3 liters.

5. $SA = 4\pi(6370\text{ km})^2 = \mathbf{162{,}307{,}600\pi\ km^2}$.

7. SA of a right pyramid is $B + \frac{1}{2}p\ell$, where B is the area of the base, p is the perimeter of the base, and ℓ is the slant height from the base to the apex.

   $B = \frac{1}{2}ap$, where a is the apothem (the height of the triangles forming the base of a regular polygon). Since a hexagon is composed of equilateral triangles about the centre, the distance from each vertex to the center is the same as the length of each edge (see below).

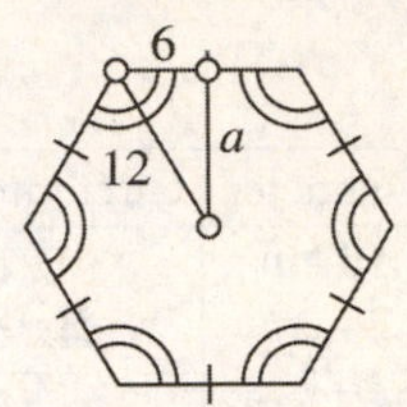

Regular hexagon

$a = \sqrt{12^2 - 6^2} = \sqrt{108} = 6\sqrt{3}$ and $B = \frac{1}{2}(6\sqrt{3})(6\cdot 12) = 216\sqrt{3}$.

The apothem $= 6\sqrt{3}$ and altitude $= 9$.

$\ell = \sqrt{9^2 + (6\sqrt{3})^2} = \sqrt{189} = 3\sqrt{21}$.

Thus $SA = 216\sqrt{3} + \frac{1}{2}(6\cdot 12)(3\sqrt{21}) = \mathbf{(216\sqrt{3} + 108\sqrt{21})\,m^2}$.

9. (*i*) The top could be $(1\cdot 88)\text{ cm}^2, (2\cdot 44)\text{ cm}^2, (4\cdot 22)\text{ cm}^2$, or $(8\cdot 11)\text{ cm}^2$.

   (*ii*) One side could be $(1\cdot 32)\text{ cm}^2, (2\cdot 16)\text{ cm}^2$ or $(4\cdot 8)\text{ cm}^2$. The other side could be $(1\cdot 44)\text{ cm}^2, (2\cdot 22)\text{ cm}^2$, or $(4\cdot 11)\text{ cm}^2$.

   The only dimensions shared by two groups each are $(8\cdot 11)\text{ cm}^2, (4\cdot 8)\text{ cm}^2$, and $(4\cdot 11)\text{ cm}^2$. Thus the box is **4 cm by 8 cm by 11 cm**.

11. $SA = B + \frac{1}{2}p\ell$, where $B = 100, p = 4\sqrt{100} = 40$ (i.e., the length of each of the four sides of the square base is the square root of the area), and $\ell = \sqrt{20^2 + 5^2} = \sqrt{425} = 5\sqrt{17}$. $SA = \left(100 + \frac{1}{2}\cdot 40\cdot 5\sqrt{17}\right) = \mathbf{(100 + 100\sqrt{17})\,cm^2}$, or about 512.3 cm^2.

13. **(a)** $2\pi r = 6\pi \Rightarrow \frac{6\pi}{2\pi} = $ **3 units.**

(b) $\ell =$ sector radius $=$ **5 units.**

(c) $h = \sqrt{\ell^2 - r^2} = \sqrt{25 - 9} = $ **4 units.**

(d) $C_{full\ circle} = 2\pi(5) = 10\pi$ units. Then $m(\angle_{sector}) = \frac{6\pi}{10\pi}(360°) = $ **216°.**

15. Surface area is proportional to the square of the radius, thus their ratio would be $\left(\frac{1}{2}\right)^2 = \frac{1}{4}$.

17. $A_{cube} = 6s^2$, where s is the length of a side. Thus $s = \sqrt{\frac{A}{6}}$. The ratio of the edges of the cubes would be $\sqrt{\frac{\left(\frac{64}{6}\right)}{\left(\frac{36}{6}\right)}} = \sqrt{\frac{16}{9}} = \frac{4}{3}$.

19. If h is the height of the completed cone, $\frac{h}{100} = \frac{h-40}{60} \Rightarrow h = 100$ cm. Slant height $= \sqrt{100^2 + 100^2} = 100\sqrt{2}$ cm.

Slant height of the missing piece $= \sqrt{60^2 + 60^2} = 60\sqrt{2}$ cm.

SA of the total cone $= \pi r^2 + \pi r\ell = \pi(100^2) + \pi(100)(100\sqrt{2}) = 10{,}000(1 + \sqrt{2})\pi$ cm^2.
Cutting off the top eliminates the lateral surface area $\pi r\ell = \pi(60)(60\sqrt{2}) = 3600\sqrt{2}\pi$ cm^2 and adds the circular area $\pi r^2 = \pi(60^2) = 3600\pi$ cm^2.

Thus total $SA = 10{,}000(1 + \sqrt{2})\pi - 3600\sqrt{2}\pi + 3600\pi = $ **$(6400\sqrt{2} + 13{,}600)\pi$ cm^2.**

Assessment 14-4B

1. (a) and (b) can form rectangular prisms. (c) when folded would be missing some faces.

3. Area of walls $= 2(8\text{ m})(2.5\text{ m}) + 2(5\text{ m})(2.5\text{ m}) = 65\text{ m}^2$. Paint needed is $\left(\frac{65\text{ m}^2}{1}\right)\cdot\left(\frac{1\text{ L}}{20\text{ m}^2}\right) =$ **3.25 liters**, so buy 4 liters.

5. **(a)** SA of a sphere is proportional to the square of the radius. If radius is doubled, surface area is **multiplied by 4**.

(b) SA of a sphere is proportional to the square of the radius. If radius is tripled, surface area is **multiplied by 9**.

7. Slant heights are given by $s = \sqrt{\left(2\frac{1}{2}\right)^2 + 3^2} = \sqrt{\frac{61}{4}} = \frac{\sqrt{61}}{2}$ ft.

Area of each side $= 6\left(\frac{\sqrt{61}}{2}\right)$ ft^2. Area of bottom $= 5 \cdot 6 = 30$ ft^2. Area of each end $= \frac{1}{2}\cdot 5\cdot 3 = 7\frac{1}{2}$ ft^2.

Amount of material $=$ total area $= 2\left[6\left(\frac{\sqrt{61}}{2}\right)\right] + 30 + 2\left(7\frac{1}{2}\right) = 12\left(\frac{\sqrt{61}}{2}\right) + 45$, or **about 91.86 ft^2**.

9. **(a)** SA of the given structure is 40 units^2 (i.e., each face has surface area of 1 unit^2 and there are 40 faces showing). Adding one cube could increase surface area by 4 units^2 (i.e., adding five faces but eliminating one). Thus the maximum would be **44 units^2**.

(b) Placing a cube in the center hole eliminates four faces while adding only two, for **38 units^2**.

(c) **Yes**. Answers may vary. E.g., arrange five cubes in the shape of a C. Filling the hole with a sixth cube would add no surface area.

11. The surface area is the sum of the area of the four triangle faces and the area of the base. It is $4\left(\frac{1}{2}\cdot\sqrt{169}\cdot 13\right) + 169 = 2\cdot 169 + 169 = 3\cdot 169 = $ **507 cm^2**.

13. **(a)** $SA = n\left(\frac{1}{2}b\ell\right) + B$, where n is the number of faces $= 4$, b is the length of each side of the base $= 10$, ℓ is slant height $= \sqrt{5^2 + 10^2} = 5\sqrt{5}$, and B is the area of the base $= 10\cdot 10 = 100$. $SA = 4\left(\frac{1}{2}\cdot 10\cdot 5\sqrt{5}\right) + 100 = 100(\sqrt{5} + 1)$in.^{2}, or **about 323.6 in.^{2}**.

(b) $SA = \pi r^2 + \pi r\ell$, where $r = 5$ and $\ell = 5\sqrt{5}$. $SA = \pi(5^2) + \pi(5)(5\sqrt{5}) = 25(1 + \sqrt{5})\pi$ in.^{2}, or **about 254.2 in.^{2}**.

15. One way to address the problem is by noting that the ratio of lengths in the smaller cone to lengths in the larger cone is $\frac{2}{3}$. If we denote lengths in the smaller cone by a subscript of 1 and lengths in the larger cone by a subscript of 2, we have $\ell_1 = \frac{2}{3}\ell_2$. Thus, $S.A._1 = \pi r_1^2 + \pi r_1 \ell_1 =$

$\pi\left(\frac{2}{3}r_2\right)^2 + \pi\left(\frac{2}{3}r_2\right)\left(\frac{2}{3}\ell_2\right) = \frac{4}{9}\left[\pi r_2^2 + \pi r_2 \ell_2\right].$

17. $S.A._1 = 6(2 \text{ ft})^2 = 24 \text{ ft}^2$

$S.A._2 = 6(4 \text{ ft})^2 = 96 \text{ ft}^2$.

$24 \text{ ft}^2 : 96 \text{ ft}^2 = \mathbf{1 : 4}$.

19. The cross section is shown below:

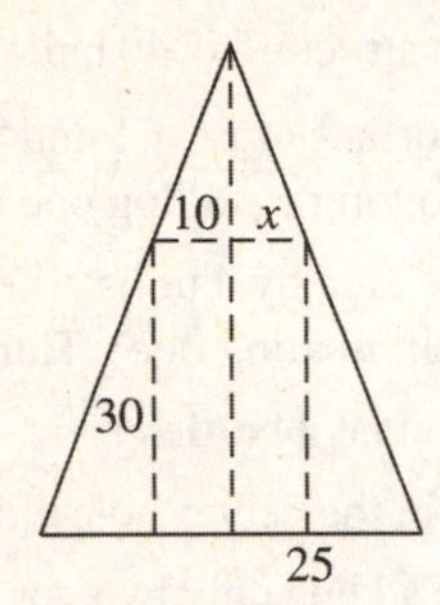

Using similar triangles: $\frac{10}{x} = \frac{40}{25} \Rightarrow x = 6.25$ cm (i.e., the radius of the cylinder).

Lateral surface area $= 2\pi rh = 2\pi(6.25)(30) =$ $\mathbf{375\pi \text{ cm}^2}$.

Review Problems

13. $d = \sqrt{10^2 + 20^2} = \sqrt{500} = \mathbf{10\sqrt{5}}$ **cm.**

15. **(a)** Change 0.6m to 60 cm. Hypotenuse = $\sqrt{60^2 + 80^2} = 100cm$.

(i) Perimeter = 60 + 80 + 100 = **240 cm.**

(ii) Area = $\frac{1}{2}(80)(6) = \mathbf{2400 \text{ cm}^2}$.

(b) Drawing an altitude to the end point of the top base formals a $45° - 45° - 90°$ isosceles triangle. The bottom base is 5 cm longer on each side than the top. Thus one leg of the isosceles triangle = 5cm. The height is also 5 cm. Use the Pythagorean Theorem to find the length of each leg of the trapezoid to be $\sqrt{5^2 + 5^2} = 2\sqrt{2}cm$.

(i) Perimeter = 20 + 10 + $2(5\sqrt{2}) = \left(\mathbf{30 + 10\sqrt{2}}\right)$**cm.**

(ii) Area = $\frac{1}{2}(10 + 20)(5) = \mathbf{75 cm^2}$.

17. The diagonal of the square at the top of the cube is $\sqrt{1^2 + 1^2} = \sqrt{2}$ units. The diagonal of the cube is the hypotenuse of the right triangle created by the diagonal of any one of the faces and the edge of two adjacent faces. Thus, the diagonal of the cube has length $\sqrt{1^2 + \sqrt{2}^2} = \sqrt{3}$ **units.**

Assessment 14-5A: Volume, Mass and Temperature

1. **(a)** $\frac{8 \text{ m}^3}{1} \cdot \frac{(10 \text{ dm})^3}{1 \text{ m}^3} = \mathbf{8000 \text{ m}^3}$.

(b) $\frac{675000 \text{ m}^3}{1} \cdot \frac{(0.001 \text{ km})^3}{1 \text{ m}^3} = \mathbf{0.000675 \text{ km}^3}$.

(c) $\frac{7000 \text{ mm}^3}{1} \cdot \frac{(0.1 \text{ cm})^3}{1 \text{ mm}^3} = \mathbf{7 \text{ mm}^3}$.

(d) $\frac{400 \text{ in.}^3}{1} \cdot \frac{1 \text{ yd}^3}{(36 \text{ in.})^3} \approx \mathbf{0.00857 \text{ yd}^3}$.

(e) $\frac{0.2 \text{ ft}^3}{1} \cdot \frac{(12 \text{ in.})^3}{1 \text{ ft}^3} = \mathbf{345.6 \text{ in.}^3}$.

3. $V_{Great\ Pyramid} = \frac{1}{3}(771^2)(486) = 96{,}299{,}442 \text{ ft}^3$.

$V_{each\ apartment} = (35)(20)(8) = 5600 \text{ ft}^3$.

$\frac{96{,}299{,}442 \text{ ft}^3}{5600 \text{ ft}^3} \approx 17{,}196.36$, or the equivalent volume of about **17,197 apartments.**

5. $\text{cm}^3 \times 0.001 = \text{dm}^3; \text{dm}^3 = \text{L}; \text{L} \times 1000 = \text{mL}$ (i.e., $\text{cm}^3 = \text{mL}$ and $\text{dm}^3 = \text{L}$).

	(a)	(b)	(c)	(d)	(e)	(f)
cm^3	**2000**	500	**1500**	**5000**	750	4800
dm^3	2	**0.5**	**1.5**	**5**	**0.750**	**4.8**
L	**2**	**0.5**	1.5	**5**	**0.750**	**4.8**
mL	**2000**	**500**	**1500**	5000	**750**	**4800**

7. $V_t = 4^3 = 64; V_2 = 6^3 = 216$. $V_1 : V_2 =$ $64 : 216 = \mathbf{8 : 27}$. (If the side lengths of the two cubes have the ratio $m : n$, their volumes will have the ratio $m^3 : n^3$.)

9. For each right rectangular prism, $V = \ell wh \Rightarrow$ $h = \frac{V}{\ell w}$. Note that in (b) and (d), units must be matched:

	(a)	(b)	(c)	(d)
Length	20 cm	10 cm	2 dm	15 cm
Width	10 cm	2 dm	1 dm	2 dm
Height	10 cm	3 dm	**2 dm**	**2.5 dm** or **25 cm**
Volume (cm^3)	**2000**	**6000**	**4000**	**7500**
Volume (dm^3)	**2**	**6**	**4**	7.5
Volume (L)	**2**	**6**	4	**7.5**

11. $V = \ell wh = (50)(25)(2) = 2500\text{ m}^3 =$ **2,500,000 L**. ($1\text{ m}^3 = 1000$ L.)

13. **(a)** Volume is **multiplied by $2^3 = 8$.**

(b) Volume would be **multiplied by $3^3 = 27$.**

(c) Volume will be **multiplied by n^3.**

15. A cross-section of the cone filled to half its height shows two similar 30° - 60° - 90° triangles. The radius at a height of 4 cm is 2 cm. Each dimension of the cup is halved, thus $\frac{1}{2}\cdot\frac{1}{2}\cdot\frac{1}{2} = \frac{1}{8}$ **the volume** of the full cup when it is filled to half its height.

17. $V_{can} = \pi r^2 h = \pi(3.5^2)(2 \cdot 3.5 \cdot 3) = 257.25\pi\text{ cm}^3$.

$V_{tennis\ balls} = 3\left(\frac{4}{3}\pi r^3\right) = 3\left(\frac{4}{3}\right)\pi(3.5^3) = 171.5\pi\text{ cm}^3$.

$V_{air} = (257.25 - 171.5)\pi = 85.75\pi\text{ cm}^3$.

$\frac{85.75\pi\text{ cm}^3}{257.25\text{ cm}^3} = \frac{1}{3} =$ **$33\frac{1}{3}$% air.**

19. $V_{prism} = AB \cdot BC \cdot AP$. $V_{pyramid} = \frac{1}{3}(AB \cdot BC \cdot AX) = \frac{1}{3}(AB \cdot BC \cdot 3AP) = AB \cdot BC \cdot AP$, or **equal volume.**

21. **(a)** **Kilograms or metric tons**. A car weighs in the thousands of pounds or tons.

(b) **Kilograms**. An adult human weighs from about 100 to 250 pounds.

(c) **Grams**. Orange juice concentrate is normally weighed in ounces.

(d) **Metric tons**. An adult African elephant can weigh as much as $7\frac{1}{2}$ tons.

23. In each metric case below, when converting from smaller to larger units move the decimal point to the left. When converting from larger to smaller units move the decimal point to the right.

(a) 15,000 g = **15** kg.

(b) 0.036 kg = **36 g.**

(c) 4320 mg = **4.320** g.

(d) 0.03 t = **30** kg.

(e) $\frac{25\text{ oz}}{1} \cdot \frac{1\text{ lb}}{16\text{ oz}} = \mathbf{1\frac{9}{16}}$ lb = **1.5625** lb.

25. $V = \ell wh = (40)(20)(20) = 16{,}000\text{ cm}^3$. 1 cm^3 of water weights 1 g. $16{,}000\text{ cm}^3 = 16{,}000\text{ g} =$ **16 kg.**

27. When converting from C to F: $F = \frac{9}{5}C + 32$.

(a) **Probably not**. $F = \frac{9}{5}(20) + 32 = 68°F$.

(b) **Yes**. $F = \frac{9}{5}(39) + 32 = 102.2°F$.

(c) **Yes**. $F = \frac{9}{5}(35) + 32 = 95°F$.

(d) **Hot**. $F = \frac{9}{5}(30) + 32 = 86°F$.

Assessment 14-5B

1. **(a)** $\frac{500\text{ cm}^3}{1} \cdot \frac{1\text{ m}^3}{(100\text{ cm})^3} =$ **0.0005 m³.**

(b) $\frac{3\text{ m}^3}{1} \cdot \frac{(100\text{ cm})^3}{1\text{ m}^3} =$ **3,000,000 cm³.**

(c) $\frac{0.002\text{ m}^3}{1} \cdot \frac{(100\text{ cm})^3}{1\text{ m}^3} =$ **2000 cm³.**

(d) $\frac{25\text{ yd}^3}{1} \cdot \frac{(3\text{ ft})^3}{1\text{ yd}^3} =$ **675 ft³.**

(e) $\frac{1200\text{ in}^3}{1} \cdot \frac{1\text{ ft}^3}{(12\text{ in})^3} = \mathbf{0.69\overline{4}\text{ ft}^3}$.

3. Volume to be heated/cooled equals the volume of the 1 ft increase in ceiling height. (2000 ft²)(1 foot increase) = **2000 ft³.**

5. $\text{cm}^3 \times 0.001 = \text{dm}^3$; $\text{dm}^3 = \text{L}$; $\text{L} \times 1000 = \text{mL}$ (i.e., $\text{cm}^3 = \text{mL}$ and $\text{dm}^3 = \text{L}$).

	(a)	(b)	(c)	(d)	(e)	(f)
cm^3	**6000**	200	**1200**	**3000**	202	6500
dm^3	6	**0.2**	**1.2**	**3**	**0.202**	**6.5**
L	**6**	**0.2**	**1.2**	3	**0.202**	**6.5**

mL	**6000**	**200**	1200	**3000**	**202**	**6500**

7. (a) Volume is proportional to s^3, where s is the length of a side. Thus the ratio of volumes would be $\left(\frac{2}{5}\right)^3$, or **8:125**.

   (b) The cones are similar, so the radii also share the ratio $a:b$. The ratio of the volumes is $\frac{\frac{1}{3}\pi r^2 \cdot \text{base}}{\frac{1}{3}\pi\left(\frac{b}{a}r\right)^2 \cdot \left(\frac{b}{a} \cdot \text{base}\right)}$, or $\boldsymbol{a^3:b^3}$.

9. For each right rectangular prism, $V = \ell wh \Rightarrow h = \frac{V}{\ell w}$. Note that in (b) and (d), units must be matched:

	(a)	(b)	(c)	(d)
Length	5 cm	8 cm	2 dm	15 cm
Width	10 cm	6 dm	1dm	2 dm
Height	20 cm	4 dm	**50 cm**	**40 cm**
Volume (cm^3)	**1000**	**19,200**	**10,000**	**12,000**
Volume (dm^3)	**1**	**19.2**	10	12
Volume (L)	**1**	**19.2**	10	**12**

11. Partition the pool into two shapes, where the volume of each is B (area of base) $\cdot$ h (height): A right rectangular prism measuring 25 m by 10 m by 2 m and a right triangular prism with base legs 2 m by 10 m and height 10 m.

$V_{rectangular\ prisim} = (25)(10)(92) = 500 \text{ m}^3;$

$V_{triangular\ prism} = \frac{1}{2}(2)(10)(10) = 100 \text{ m}^3.$

$V_{pool} = V_{rectangular\ prism} + V_{triangular\ prism} =$

$500 + 100 = \mathbf{600\ m^3}.$

13. Convert all measurement to mm. Radius of the inner circle = 20 mm and height = 20 mm.

$V_{outer\ cylinder} = \pi r^2 h = \pi(22^2)(20) = 9680\pi \text{ mm}^3.$

$V_{inner\ cylinder} = \pi(20^2)(20) = 8000\pi \text{ mm}^3.$

$V_{ring} = (9680 - 8000)\pi = \mathbf{1680\pi\ mm^3}.$

15. $1 \text{ L} = 1000 \text{ cm}^3$ and $V = \pi r^2 h$, thus $1000 = \pi(12^2)\, h \Rightarrow h = \frac{1000}{144\pi}$, or **about 2.2 cm**.

17. $V_{sphere} = \frac{4}{3}\pi(10^3) = \frac{4000}{3}\pi \text{ cm}^3$. Volume in a right circular cylinder is $\frac{4000}{3}\pi = \pi(10^2)h$, so $h = \frac{4000}{300} = \frac{40}{3}$ cm.

Then water height $= \left(20 + \frac{40}{3}\right) = \mathbf{33\frac{1}{3}\ cm}$.

19. Volume is proportional to the cube of the radius, thus radius is proportional to the cube root of volume. If volume is halved, radius is decreased by a factor of $\sqrt[3]{0.5} \approx 0.794$, or **about 0.8 times the original**.

21..

(a) **Grams**. Mustard is normally weighed in ounces.

(b) **Grams**. Peanuts are normally weighed in ounces.

(c) **Metric tons**. An Army main battle tank weighs about 60 tons.

(d) **Kilograms**. Most cats weigh about 5-10 pounds.

23. (a) 8000 kg = **8** t.

(b) 72 g = **0.072** kg.

(c) 5 kg 750 g = 5.750 kg = **5750** g.

(d) $\frac{2.6 \text{ lb}}{1} \cdot \frac{16 \text{ oz}}{1 \text{ lb}} = \mathbf{41.6}$ oz.

(e) $\frac{3.8 \text{ lb}}{1} \cdot \frac{16 \text{ oz}}{1 \text{ lb}} = \mathbf{60.8}$ oz.

25. (a) $\frac{1 \text{ ha}}{1} \cdot \frac{10{,}000 \text{ m}^2}{1 \text{ ha}} \cdot \frac{10{,}000 \text{ cm}^2}{1 \text{ m}^2} \cdot \frac{2 \text{ cm}}{1} \cdot \frac{1 \text{ L}}{1000 \text{ cm}^3} =$ **200,000 L** of rainfall.

(b) 1 L of water weighs 1 kg. 200,000 L = 200,000 kg = **200 t** of water.

27. When converting from C to F: $F = \frac{9}{5}C + 32$.

(a) **No**. $F = \frac{9}{5}(26) + 32 \approx 79° F$.

(b) **No**. $F = \frac{9}{5}(40) + 32 = 104° F$.

(c) **Chilly**. $F = \frac{9}{5}(16) + 32 \approx 61° F$.

Review Problems

15. (a) (*i*) Perimeter $= \frac{1}{2}(2\pi \cdot 6) + 2\left(\sqrt{6^2 + 8^2}\right)$ $= \mathbf{(6\pi + 20)\ cm}$.

(*ii*) Area $= \frac{1}{2}(\pi \cdot 6^2) + \frac{1}{2}(12)(8) =$ $\mathbf{(18\pi + 48)\ cm^2}$.

(b) (*i*) Perimeter $= \frac{1}{2}(2\pi \cdot 20) + 2\left[\frac{1}{2}(2\pi \cdot 10)\right]$ $= \mathbf{40\pi\ cm}$.

(*ii*) Area $= \frac{1}{2}(\pi \cdot 20^2) - 2\left[\frac{1}{2}(\pi \cdot 10^2)\right] =$ $\mathbf{100\pi\ cm^2}$.

17. **(a)** **Yes.** $1^2 + (\sqrt{2})^2 = (\sqrt{3})^2$.

(b) **No.** Sides are: $\sqrt{4^2 + 1^2} = \sqrt{17}$; $\sqrt{3^2 + 1^1} = \sqrt{10}$; and $\sqrt{3^2 + 4^2} = 5$. $(\sqrt{17})^2 + (\sqrt{10})^2 \neq 5^2$.

Chapter 14 Review

1. From Pick's theorem (Assessment 14-2B, problem 7), $A = I + \frac{1}{2}B - 1$, where $I =$ the number of dots inside the polygon and $B =$ the number of dots on the polygon boundary. Alternatively, area could be determined through the rectangle method.

(a) $A = 7 + \frac{1}{2}(5) - 1 = \mathbf{8\frac{1}{2}\ cm^2}$.

(b) $A = 5 + \frac{1}{2}(5) - 1 = \mathbf{6\frac{1}{2}\ cm^2}$.

(c) $A = 2 + \frac{1}{2}(12) - 1 = \mathbf{7\ cm^2}$.

3. Area of $\Delta ABC < \Delta ABD = \Delta ABE < \Delta ABF$. All triangles have the same base, so area is proportional only to height. Ordering by height only, ΔABD and ΔABE are equal in area.

5. **(a)** $A = \pi(4^2) - \pi(2^2) = \mathbf{12\pi\ cm^2}$.

(b) $A_{semicircle} = \frac{1}{2}\pi(3^2) = 4.5\pi\ \text{cm}^2$.

$A_{triangle} = \frac{1}{2}(6)(4) = 12\ \text{cm}^2$.

$A_{shaded\ region} = \mathbf{(4.5\pi + 12)\ cm^2}$.

(c) $A = (6)(4) = \mathbf{24\ cm^2}$.

(d) $A = \frac{40°}{360°}\pi(6^2) = \mathbf{4\pi\ cm^2}$.

(e) $A = \frac{1}{2}(2)(3) + \frac{1}{2}(3)(3) + (3)(15) + (3)(4) =$ $3 + 4.5 + 45 + 12 = \mathbf{64.5\ cm^2}$.

(f) $A = (8)(18) + \frac{1}{2}(18 + 5)(3) = \mathbf{178.5\ cm^2}$.

7. $AC = \sqrt{1^2 + 2^2} = \sqrt{5}$. $AD = \sqrt{1^2 + (\sqrt{5})^2} = \sqrt{6}$. Each succeeding segment thus adds 1 to the number under the radical; $AG = \sqrt{9} = \mathbf{3}$.

9. There are as many lateral faces as sides. Thus, there are **eight** lateral faces.

11. **No,** the base must be a simple closed carve but does not have to be a circle.

13. Answers may vary.

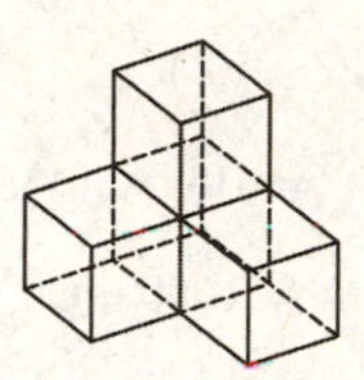

15. $A_{top/bottom} = 2\left[40^2 - 4(5^2)\right] = 3000\ \text{cm}^2$.

$A_{sides} = 4(40)(15) = 2400\ \text{cm}^2$.

$A_{total} = 3000 + 2400 = \mathbf{5400\ cm^2}$.

17. Lateral surface area $= \pi r\ell$, where $\ell = \sqrt{12^2 + 5^2} = 13$. Area $= \pi(5)(13) = \mathbf{65\pi\ m^2}$.

19. Sum the areas of the four triangles.
$A = 2\left(\frac{1}{2} \cdot 5 \cdot 12\right) + 2\left(\frac{1}{2} \cdot 12 \cdot \sqrt{20^2 - 12^2}\right) =$ $60 + 192 = \mathbf{252\ cm^2}$.

21. $\mathbf{2\sqrt{2}\ m^2}\left[h = \sqrt{3^2 - 1^2} = 2\sqrt{2} \Rightarrow A = \frac{1}{2}bh = \frac{1}{2}(2)(2\sqrt{2}) = 2\sqrt{2}\ \text{m}^2\right]$.

23. $V_{cylinder} = \pi r^2 h = 10\pi r^2$ and $V_{cone} = \frac{1}{3}\pi r^2 h$. $10\pi r^2 = \frac{1}{3}\pi r^2 h \Rightarrow 10 = \frac{1}{3}h$. $h = \mathbf{30\ cm}$.

25. Diagonal $= \sqrt{\left(8\frac{1}{2}\right)^2 + 11^2} = \frac{\sqrt{773}}{2} \approx$ **13.9 inches**.

27. **(a)** **Metric tons**, which corresponds to thousands of pounds, or tons.

(b) $1 \text{ cm} \cdot 1 \text{ cm} \cdot 1 \text{ cm} = \mathbf{1\ cm^3}$.

(c) 1 cm^3 of water weighs **1 gram**.

(d) $\mathbf{1\ L = 1\ dm^3}$, so the two have the same volume.

(e) $\frac{1 \text{ L gas}}{12 \text{ km}} = \frac{x \text{ L gas}}{300 \text{ km}} \Rightarrow 12x = 300.\ x = \mathbf{25\ L}$.

(f) **2000** a (1 ha = 100 a).

(g) **51,800** cm^3 ($1 \text{ L} = 1000 \text{ cm}^3$)

(h) **10,000,000** m^2 ($1 \text{ km}^2 = 1{,}000{,}000 \text{ m}^2$).

(i) **50,000** mL (1 L = 1000 mL).

(j) **5.830** L (1000 mL = 1 L).

(k) **25,000** dm^3 ($1 \text{ m}^3 = 1000 \text{ dm}^3$).

(l) **75,000** mL ($1 \text{ dm}^3 = 1000 \text{ mL}$).

(m) **52.813** kg (1000 g = 1 kg).

(n) **4.8** t (1000 kg = 1 t).

29. **(a)** **80 L**, or about 21 gallons.

(b) **82 kg**, or about 181 pounds.

(c) **978 g** = 0.978 kg, or about 2 pounds.

(d) **5 g**, or about 0.2 ounces.

(e) **4 kg**, or about 8.8 pounds.

(f) **1.5 metric tons** = 1500 kg, or about 3300 pounds.

(g) **180 mL** = 0.180 L.

31. **(a)** **2000 g** ($1 \text{ dm}^3 = 1000 \text{ cm}^3$, or 1000 g of water).

(b) **1000 g** ($1 \text{ L} = 1000 \text{ cm}^3$, or 1000 g of water).

(c) **3 g** ($1 \text{ cm}^3 = 1\ g$).

(d) **0.0042 kg** (1 mL of water = 1 g = 0.001 kg).

(e) $\mathbf{0.0002\ m^3}$ ($1 \text{ L} = 1000 \text{ cm}^3 = 0.001 \text{ m}^3$).